Models of Cognition
A Review of Cognitive Science

Volume One

Models of Cognition
A Review of Cognitive Science

Volume One

Noel E. Sharkey

University of Essex, United Kingdom

ABLEX PUBLISHING CORPORATION
NORWOOD, NEW JERSEY

Printed in the United States of America

Library of Congress Cataloging-in-Publication Data

Models of cognition: a review of cognitive science / Noel E. Sharkey.
p. cm.
Bibliography: p.
Includes index.
ISBN 0-89391-528-9
1. Cognition—Philosophy. 2. Psycholinguistics—Philosophy.
3. Cognition—Computer simulation. 4. Psycholinguistics—Data processing. 5. Neural computers. I. Sharkey, N. E. (Noel E.)
BF311.M566 1988
153—dc19 88-7473
CIP

Ablex Publishing Corporation
355 Chestnut Street
Norwood, New Jersey 07648

To Amanda Sharkey for Her Tireless Support

Contents

Preface

This is the first in a series of new Cognitive Science books by Ablex to deal with models of cognition. Nowadays this is a very broad subject area covering computer simulations, psychological approaches, and plain old pencil and paper formalisms. The eleven chapters presented here discuss eleven different models ranging, in their scope, from image analysis through word recognition to communication failure in dialogue. Although all of this may seem very diverse for a single volume, it is my opinion that diversity is the very essence of cognitive science.

This is a book for those who are in the modelling business—either as a builder, a user, or a spectator (not as a poser). The selection herein covers the spectrum of techniques from the purely symbolic to the purely connectionist. There is also a smattering of the *in-betweens* or hybrids which indicates either a transition or a reconciliation between the frameworks.

I would like to acknowledge the support I have received from my graduate students at Essex, in particular Richard Sutcliffe, Nice Gonzales, Nick Reeves, Paul Treffner, Bob Marsden, Wayne Wobcke, Richard Doust, Paul Hearn, Paul Freeman and especially Wendy Fryer my secretary for her patience and help. I thank *The Freaking Herberts* (our rock band) for support through the irritations (often caused by them) by keeping me playing: Pam, Charlie and Nougel Tweed, Nick Thompson, the Smiffs, Antoinette Tucker, and Phippsy. I also thank the Saffron Waldonites, John, Mick, and George for the weekend breaks and Hugie Hutchinson for nothing at all.

Contributors

Afzal Ballim, Rio Grande Research Corridor, Computing Research Laboratory, New Mexico State University, Las Cruces, New Mexico 88003.

Garrison W. Cottrell, Computer Science and Engineering, University of California, San Diego, La Jolla, California 92037.

Jerome A. Feldman, Computer Science Department, University of Rochester, Rochester, New York 14637.

Alan Garnham, Centre for Research on Perception and Cognition, Laboratory of Experimental Psychology, University of Sussex, Falmer, Brighton, UK.

Steven Greenspan, AT&T Bell Laboratories, Murray Hill, New Jersey 07974.

James A. Hendler, Department of Computer Science and Systems Research Center, University of Maryland, Baltimore, Maryland, 21201.

Daniel Morrow, Department of Psychiatry and Behavioral Sciences, Stanford University School of Medicine, Stanford, California, 94305.

Paul Munro, Department of Computer Science, University of Pittsburgh, Pittsburgh, Pennsylvania, 15261.

Jane Oakhill, MRC Perceptual and Cognitive Performance Unit, Laboratory of Experimental Psychology, University of Sussex, Falmer, Brighton, UK.

Andrew Ortony, University of Illinois at Urbana-Champaign, 51 Gerty Drive, Champaign, Illinois, 61820.

Dean I. Radin, Princeton University, Princeton, New Jersey, 08450.

Ronan Reilly, Educational Research Centre, St. Patrick's College, Dublin, Ireland.

Geoffrey Sampson, Department of Linguistics and Phonetics, University of Leeds, Leeds, UK.

Anthony J. Sanford, Department of Psychology, University of Glasgow, Scotland.

Noel E. Sharkey, Department of Computer Science, University of Exeter, Exeter, UK.

Yorick Wilks, Rio Grande Research Corridor, Computing Research Laboratory, New Mexico State University, Las Cruces, New Mexico, 88003.

David Zipser, Institute for Cognitive Science, University of California, San Diego, La Jolla, California 92037.

Introduction

Although cognitive science is still in its infancy, we have already begun to see a number of splits. Initially these splits were between what Bob Abelson has called the *neat* and the *scruffy* approach to modelling cognition. But these divisions were really a matter of cognitive style, so to speak. The neats were the formalists who wanted their theories to be very tidy and complete while the scruffies believed that it was still too early to develop rigorous formalisms for mind. They preferred instead to look at the seemingly *ad hoc* nature of human cognition and develop theories which, while not neat, were very general. This was, in some ways, really an AI versus linguistics debate with scruffy Roger Schank in the left corner and neat Noam Chomsky in the right. Psychology thought that it was the referee, but the others didn't believe that it knew the rules.

Nowadays, there is a new distinction which, on the surface, is more clear cut although it cuts across the neat/scruffy boundaries. This is the distinction between those who wish to model cognition using symbolic information processing and those who wish to use connectionist or neural network techniques. The symbolic research has dominated the study of cognition since the 1950s. In this approach, mental representations, in the form of combinations of symbols, are operated on or manipulated by explicit rules. However, since the early 1980s, the new connectionism has had a very sudden rise in popularity and is making more and more new converts every day. This approach represents an attempt to model cognition in a style similar to the nervous system, i.e., with very many simple computing elements connected in a parallel processing network. Rather than passing symbolic information to one another, these units pass only numerical activation values. The connectionist approach has also been around since the 1950s, but it got off to a false start after great initial optimism. This was mainly because of certain limitations of the early learning algorithms. As an example, the early systems could not learn Exclusive OR. Newer algorithms, such as those used in Chapters 10 and 11 have overcome many of the earlier limitations.

Some believe that the new connectionism represents a paradigm shift within cognitive science, while others believe that it is just a passing phase. Whichever it is, both the old and the new work tackle very similar issues within the same areas such as natural language understanding, parsing, human memory, visual perception, speech recognition and so on. Both groups are even informed by the same data and the same theories.

In this book we have representatives from both approaches, and some in the middle. The first four chapters are seated firmly in the symbolic tradition. Chapters 1 and 2 both contain artificial intelligence models based on classical von Neumann architectures. In the first chapter, Wilkes and Ballim present a computational approach to the ascription of beliefs in their ViewGen program. In the second chapter, Reilly presents an augmented transition network model of dialogue which is extended in an examination of communication failure. Chapters 3, 4, and 5 all present psychological evidence directed towards particular models. Morrow and Greenspan, in Chapter 3, examine the use of situation models in language understanding, whereas in Chapter 4, Garnham and Oakhill put forward evidence from anaphora research which suggests that extensions are needed to the mental models approach to language understanding. In Chapter 5, Sanford also uses evidence from anaphora research. But he suggests that a connectionist approach may be essential to natural language research. This is the first move in the book away from the standard symbolic line.

The next three chapters are all partly connectionist to a greater or lesser extent. In Chapter 6, Hendler presents a planning model which is part symbolic and part connectionist. Chapter 7, by Ortony and Radin, introduces a model of frame selection, which is an outgrowh of the spreading activation formalism in psychology, geared towards connectionism. The next chapter, by Sampson, takes syntactic parsing partly into the realms of connectionist research with simulated annealing.

The final three chapters are firmly within the connectist tradition. In Chapter 9, Cottrell, Munro, and Zipser present an extensional programming approach to image analysis. Chapter 10, by Zipser, introduces a learning technique for the extraction of spatial information from simple images. Feldman finishes the book with a discussion of a central connectionist debate on whether to use distributed localist representations, i.e., what should be represented on a unit. We turn now to review the various chapters in more detail.

A REVIEW OF THE CHAPTERS

Chapter 1: *The heuristic ascription of beliefs.* Wilks and Ballim are interested in developing a computational system which performs belief ascription in a plausible manner. They describe the ViewGen program, which embodies methods for heuristically generating nested beliefs, i.e., what some agent believes that another agent believes about a topic. These nested beliefs, they argue, are essential to many processes such as those used for discourse understanding and reasoning about other agents' reasoning. The authors focus particularly on a class of beliefs which they refer to as *atypical*, and contrast these with *typical* beliefs. For Wilks and Ballim, the ascription of beliefs depends crucially upon their typicality and upon the agent that the beliefs are being ascribed to. ViewGen generates a set of viewpoints from a set of beliefs held by the system.

Chapter 2: *An ATN-based grammer for the structural analysis of dialogue.* Reilly attempts to provide a structural description of natural language dialogue in the form of an augmented transition network (ATN). His idea is to demonstate that the grammar is sensitive to the dynamic structural variations that might be expected to occur during the time course of a dialogue. Furthermore, Reilly is concerned to show that aspects of the structural description are sensitive to characteristics of dialogue that indicate or precipitate miscommunication. After generating a number of empirical hypotheses from the model and testing them, he concludes by suggesting two main characteristics of dialogue that give rise to communication failure. These are a high rate of information transmission and a low rate of information checking.

Chapter 3: *Situation models and information accessibility.* Morrow and Greenspan examine how a psychological notion of information accessibility (in working memory) could be incorporated into a theory of discourse processing. They take the view that the goal of communication is to share a model of a particular set of situations. This view underlies their argument that writers and readers attend more to information about the described world conveyed by the text than to information about the text. Evidence is reviewed which indicates that writers use a variety of grammatical devices to express the three critical aspects of situations that combine to help specify relevance: the topic of the situation model, the perspective from which the model is constructed, and which parts of the situations are focused from the perspective. These aspects of situations, Morrow and Greenspan argue, are essential for guiding comprehension.

Chapter 4: *The everyday use of anaphoric expressions: Implications for the mental models theory of text comprehension.* Garnham and Oakhill discuss some of the problems that certain uses of anaphoric expressions cause for theories of comprehension. After a review of the linguistics literature, they outline the mental models theory of comprehension and point out some of its strengths. However, Garnham and Oakhill present evidence which appears inconsistent with the theory. This evidence is derived from two main sources: first, from a detailed examination of some examples from a corpus of anaphoric expressions which have, in the main, appeared in published texts; secondly, evidence is presented in the form of two experiments which demonstrate the violation of the anaphoric island constraint. The authors suggest ways in which the mental models theory could be extended to accommodate their findings.

Chapter 5: *Component processes of reference resolution in discourse.* Sanford reviews some of the recent work on the time-course of anaphor resolution. The evidence, he suggests, indicates that there are a number of autonomous processes which seem to be triggered by aspects of the text, and which are initiated immediately. Sanford argues that some of these processes seem to be evoked even though they lead to results which are inappropriate from a discourse viewpoint. He gives the example of the pronoun bonding effect as implicating automatic and autonomous processes rather that the computationally optimal ones in which component processes cooperate by message-passing. Sanford appeals to the evidence to argue that language understanding is a parallel process that would best be modelled in a connectionist system.

Chapter 6: *Extending marker-passing to microfeatures.* Hendler proposes a hybridization of connectionism with symbolic operations. He takes an already existing system, his SCRAPS planner, and introduces a connectionist component. What he calls symbolic marker-passing is combined with analog activation in a microfeature network. Some problems are introduced within his planning work, some of which, he argues can best be solved by one approach and some by the other. Hendler's idea is to take strengths from both symbolic and connectionist research and combine them to form an integrated AI system. The work reported here represents a demonstration that this can be done gracefully.

Chapter 7: *Simulated annealing as a parsing technique.* Sampson describes a pilot experiment designed to assess the feasibility of syntactic parsing using the computational technique of simulated annealing. This technique has recently been used in a large number of studies to find optimal solutions to combinatorial problems for which the space of possible solutions is too large to be systematically examined case by case (it is probably most famous for its use in the Boltzmann machine). Unlike standard connectionist approaches, Sampson does not use a network of interconnected units. Rather, the output takes the form of syntactic parse-trees. The system is based on statistics derived from the manual parsing of sentences totalling 40,000 words taken from the LOB corpus.

Chapter 8: *SAPIENS: Spreading activation for information encoded in network structures.* Ortony and Radin present the SAPIENS system, which utilizes a parallel spreading activation mechanism for the thorny AI problem of frame selection. One interesting difference between this and other AI research is that, in SAPIENS, facilitation of frame selection is by processes that operate only at the lexical level. To ensure that their simulations scale well, the authors use an unusually large vocabulary of 16,000 words. Each word is encoded as a single unit. These units are connected only by unlabelled connections that reflect the empirical values of their associative strengths. A second difference between this and other AI systems is the concern to model psychological data. To this end, Ortony and Radin examine the behavior of the system as it might be used in a number of context sensitive tasks such as lexical disambiguation, sentence verification, and cued recall.

Chapter 9: *Image compression by back propagation: An example of extensional programming.* Cottrell, Munro, and Zipser present a powerful new computational technique which they call *extensional programming.* They use this name because the programs learn from examples (a variant of back propagation is used). Extensional programming is compared to a traditional approach to the well-understood problem of bandwidth compression for image analysis. This is an example of the encoder problem in which a network is trained to perform an identity map through a narrow channel. Since the authors use the input to the network as the teaching signal, we can regard their program as a self-organizing system that finds an efficient representation of its environment. The results indicate that a very simple learning algorithm (back propagation) can be programmed to do image compression about as well as more traditional methods.

Chapter 10: *Programming neural nets to do spatial computations.* The main focus of Zipser's chapter is on programming networks to extract spatial information from simple images and to use this information to compute spatial relations. Zipser is concerned with the problem of the teacher in real neural networks. He obviates the need for a teacher here by using the input stimulus itself, or a simple transformation of it, as the target response. Using this procedure, combined with back propagation, he demonstrates how hidden units can be programmed to represent either the location of objects in the image or the relationships between objects, depending on the training details. When the functions being taught are continuous, the networks learn to generalize from finite learning sets to make good approximations to the function for all stimulus patterns. Zipser claims that the networks he has generated have a neurophysiological flavor, and that the individual units have properties similar to those found in the visual system.

Chapter 11: *The neural representation of conceptual knowledge.* Feldman discusses the neural representation of conceptual knowledge and examines two ways of encoding memories: *localist* and *distributed.* Localist or punctate encoding refers to the encoding of a concept or a proposition on a single network unit. This contrasts with distributed encoding, which means that many units are implicated in the encoding of a single concept. Feldman balances arguments for and against each position and concludes that neither extreme is consistent with what is known about brain and cognition. He argues instead for the middle ground of *compact* encoding (really derived from the punctate model), which is particularly interesting from the point of view of structured learning. Finally, he considers distributed models to be of great importance when reinterpreted as studying the system properties of neural networks.

CHAPTER 1

The Heuristic Ascription of Belief

YORICK WILKS AND AFZAL BALLIM
Rio Grande Research Corridor
Computing Research Laboratory
New Mexico State University

1. INTRODUCTION

We describe a computational model of beliefs for natural language understanding, planning, and reasoning. In particular we are interested in generating complex belief structures known as *points of view* (what an agent believes about some topic). The model is based upon prior work reported in Wilks and Bien (1979, 1983).

A general algorithm for generating points of view, based upon the notion of default reasoning, is presented and the complexities of ascribing belief to agents is discussed. The algorithm presented is embodied within a program for generating points of view known as *ViewGen*, and an example is given of viewpoint generation by *ViewGen*.

ViewGen is a program which models the beliefs of an agent known as the system. It is capable of generating nested points of view about a topic upon demand. *ViewGen* is written in Prolog.

The layout is as follows: Section 2 presents background material for this chapter; Section 3 describes the process of constructing a point of view; Section 4 discusses cases where a default should be overridden; Section 5 presents a worked example in *ViewGen*; finally, section 6 summarizes the chapter.

2. BACKGROUND

A *point of view* is a set of beliefs held about some topic by an agent. A *nested point of view* is a set of beliefs about some topic that one agent believes another agent believes . . . another agent holds. For example, agent A may

Figure 1. A's view of B's view of C's view of a Topic.

believe that agent B believes that agent C has some set of beliefs about world hunger. We use a shorthand (as in Figure 1) to represent a nested point of view.

2.1. Relationship to Other Work

In general our work has been, since that of Wilks and Bien (1979), to construct a formalism and programs that capture the heuristic belief ascriptions that individuals actually perform in the process of understanding and participating in dialogue: that is to say, contentful, concrete beliefs and not merely metabeliefs about the reasoning architecture of others, activities we suspect are rarely, if ever, undertaken in fact. Thus concern has been less with the powers of particular notations and proofs of their adequacy (as has been the central feature of the work of Creary, 1979; Moore & Hendrix, 1979; Konolige, 1982, 1983, 1984; and Attardi & Simi, 1984), than with the *content* of belief ascription.

In that sense, our work has been closer in spirit to that of Perrault and his colleagues (e.g., Perrault & Allen, 1980, Cohen & Levesque, 1980), though without their (then) commitment to the language of speech act theory and, most importantly, without their key assumption that the partitions within nested beliefs are all present at the beginning of the belief ascription procedures. Our work makes no such assumption: Nested beliefs are not merely accessed but constructed and maintained in real time, a position we find both computationally and psychologically more plausible. The Gedanken Experiment here is to ask yourself if you already know what Mr. Reagan believes the Ayatollah believes about Col. Gaddafi. Of course you can work it out, but how plausible is it that you have already precomputed such nested belief spaces?

The work of Maida (1983, 1986) clearly shares many of the concerns of the current work: his diagramatic representation of nested beliefs are isomorphic to those of Wilks and Bien (1979) and Shadbolt (1983). He shares, for example, our concern with intensional rather than explicitly extensional representations, as does the SNePS group (Shapiro & Rapaport, 1986). However, his concerns are still those of Martins and Shapiro (1983; see below), namely, the problem of shared reasoning strategies between believers and how, for example, you could establish that a dialogue partner also used *modus ponens*. We argue, on the contrary, that this phenomenon is best handled by general default assumptions, as are the concrete contents of belief.

It seems to us much more natural to assume by default that the other's general strategies are like ours, unless we have real information to the contrary. Indeed, no finite set of dialogue observations ever could establish conclusively that another believer was using *modus ponens*. That being so, concentration on such issues that are not susceptible of proof seems to us only to delay the central issue which is how to infer heuristically the actual contentful beliefs of other believers. However, Maida (1983) is concerned with the very important, but we believe quite separable, issue of a heuristic rule for identifying intensional individuals under different descriptions.

Martins and Shapiro (1983) is, despite its title, not really concerned with recursive nesting, nor with the identification of spaces with individuals (or classes), but with the construction of union-of-belief-sets spaces so that agents can mutually plan tasks. It is not clear that this is a belief system at all, since there is never any need to consider the use of a BELIEVE or KNOW predicate; and the union-of-belief spaces do not, in any case, correspond to individual believers.

Konolige's (1983) work has strong similarities to that just noted: He considers what he calls *views*, e.g., v=John,Sue,Kim, which are exactly what we shall call the nesting of beliefs, i.e., John's view of what Sue believes Kim believes. Konolige has no construction for that set, but is concerned with deduction within such sets and with problems with the assumption that each individual has the same deductive system. This is better covered, we suggest, as with Maida above, by some general rationality postulate.

Konolige's main concern is to give a resource-limited account of a deduction schema, so as to avoid the assumption (present in the possible world models of Moore et al.) that one must believe all the consequences of one's beliefs. This is undoubtedly important as a process account of deductive systems, but not relevant to the issue of the construction of the content of individual views.

3. HEURISTICS FOR CONSTRUCTING POINTS OF VIEW

Two questions need to be answered about nested beliefs. Firstly, when should a nesting of beliefs be constructed? Secondly, what decides the particular nesting (who is in it and what it is about)? In order to answer these questions, first assume that the nested beliefs are being used in a dialogue system which is capable of performing the role of a participant in the dialogue.

Wilks (Wilks & Bien, 1979, 1983) suggests a number of strategies for deciding when to construct a nesting and what nesting to construct, using two strategies: namely the *presentation strategy* and the *insertional strategy*. The important question for a presentation strategy is: Given incoming information about an individual, how many levels of nesting should the system construct?

Two forms of presentation strategy are distinguished: *a minimal strategy*, which corresponds to the shallowest nesting with no level which corresponds to the speaker (such a strategy would be used where one is not interpreting the input in terms of the speaker, for example where the speaker is a person delivering a lecture and the hearer is taking notes verbatim); and a *standard presentation strategy*, which is to construct a nesting corresponding to the speaker, and then to construct levels corresponding to mentioned individuals,[1] appropriately. As pointed out in Wilks and Bien (1983), this standard strategy allows a hearer to either disbelieve a speaker or co-operate with him or her, as he or she chooses, in much the same way as that proposed by Taylor and Whitehill (1981).

The insertional strategy is designed to handle the question of where to store beliefs asserted by another agent. This strategy is discussed elsewhere (Wilks and Bien, 1983, Ballim, 1986).

3.1 A Contextual Focus of Beliefs

A point of view is a collection of beliefs about some topic. The question arises of how to select this collection of beliefs.

The set of beliefs with which we begin the process of generating nested points of view should be restricted in some manner to only those beliefs which are, or may be, useful at that point to the system which is using the beliefs. In our case we are considering a language comprehension system as using beliefs to explicate a dialogue. Therefore we must have a mechanism which is capable of deciding the relevance of beliefs held, within the system, to the dialogue in question.

Our solution to this problem is essentially simple. The system's belief space is divided into a number of topic-specific subspaces. These subspaces may be thought of as a less permanent version of *frames* (Minsky, 1975, Charniak, 1978) or more suitably in terms of Wilks (1977) as pseudotexts (henceforth, PTs). In effect, a PT is a set of unsorted, unrefined items of knowledge.

These Pts are general items and are not only stored for individual human

---

[1] It should be noted that a nesting need not correspond solely to individuals. We have no qualms about constructing an environment to represent the beliefs of groups of individuals, such as the view of retired generals about world terrorism.

beings, but also for groups of humans, objects, and abstract ideas. Their hierarchial and inheritance relations are discussed in Wilks (1977).

Our method of restricting the initial set of beliefs is to form this set from a very small number PTs; for example if the system is a medical diagnostician, being used to advise a patient on thalassemia, then the initial set of beliefs may be just those concerning thalassemia. We term the initial belief set the *contextual focus*. We now turn to the process of constructing a point of view from a given contextual focus.

3.2 Generation of Points of View

Points of view are represented by structures know as *environments*. An environment is a tuple $E = (A, \Psi)$ where A is a sequence of agents and Ψ is a set of propositions. If $A = \langle_{o}, i_1, \ldots, i_n \rangle$ and there are sentences $P \in \Psi$ then the environment is equivalent to a set of sentences of the form $B_{io}(B_{i_1}(\ldots (B_{i_n P}) \ldots))$, where B_{inP} can be interpreted as agent i_η believes ρ. This may be phrased as i_o believes that i_1 believes that . . . i_n believes p.

The process of generating an environment may be regarded as a decision mechanism that ascribes beliefs from one environment to another environment.

Assume that our contextual focus is as shown if Figure 2. This represents the beliefs held by the system that an agent known as the User dislikes another agent knows as Frank; that the system likes Frank; that the Earth is round; and that the system believes that Frank believes the Earth is flat. Given this set of beliefs, how do we generate what the system believes Frank believes about the contextual focus?

The proposal is that a nesting should be generated by a form of default reasoning (see Wilks and Bien, 1979, 1983) using a default rule for ascription of beliefs. The default ascriptional rule is to assume that one's view of another person's view is the same as one's own *except where there is explicit*

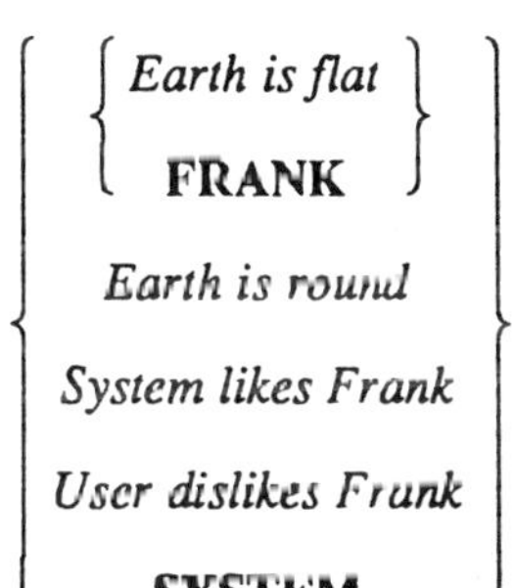

Figure 2. A Contextual Focus of Belief.

Earth is flat

System likes Frank

User dislikes Frank

FRANK

Earth is round

System likes Frank

User dislikes Frank

SYSTEM

Figure 3. The Result of Applying the Default Rule to Figure 2.

evidence to the contrary. Applying this rule to Figure 2 we get the system's view of Frank's view, shown in Figure 3.

Let us examine this example closely. The belief of the system in Figure 2 that the system likes Frank has been ascribed into the PT for Frank; i.e., it is now a belief of the system that Frank believes that the system likes him. The same is true of the system's belief that the user dislikes Frank. However, the system's belief that the Earth is round has been superceded by the already existent belief that the system believes Frank holds, that the Earth is flat. The result in Figure 3, and we may now reason with the environment of the system's views of Frank's views,[2] (which we shall call $System_{Frank}$).

We are thus able to generate points of view for our minimal presentation strategy. The problem which now remains is how to generate nested points of view in accordance with our standard presentation strategy? We use a method called *pushing down environments*.[3] Pushing down of one environment inside another means resetting values in the environment being pushed down. The transitory object achieved by this method we shall interpret as begin our outer environment holder's view of the inner environment. Suppose we want to construct the system's view of the user's view of Frank; this is done in two stages as follows: first, by constructing the system's view of Frank, secondly by constructing the system's view of the user and then pushing the former down into the latter.

Suppose that we wish to construct the system's view of the user's view of Frank's view about some context (we shall write this as $System_{User_{Frank}}$ (*'context'*) as shorthand). To construct this, we firstly construct the system's view of the user's view. We have available the system's view of Frank's view,

[2] In Figure 3, the environment in question is that portion which is inside the section marked *FRANK*.

[3] Two distinct forms of pushing down environments are identified in Ballim (1986), only one of which is discussed here.

Figure 4. A Contextual Focus about the Shape of Earth.

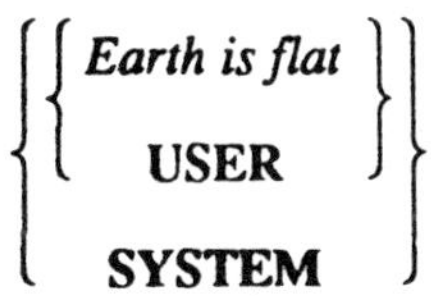

Figure 5. The System's View of the User's View of the Shape of Earth.

Figure 6. The System's View of Frank's View of the Shape of Earth.

Figure 7. The System's View of the User's View of Frank's View of the Shape of the Earth.

which is then pushed down into the former view. For example presume that we have the contextual focus shown in Figure 4.

We construct the system's view of the user's view according to our default rule of ascription to get Figure 5.

We have available the system's view of Frank's view which is shown in Figure 6. Now, pushing the environment of Figure 5 down into the environment of Figure 6 results in the environment shown in Figure 7.

To construct a deeper nesting, we continually apply this method. So, to construct $System_{UserFrank_{System(\text{'}context\text{'})}}$ we push the system's view down into the user, push the result of that down into Frank, and finally push the result of that operation down into the system. The mechanism is described in more detail in Ballim (1986).

4. THE NATURE OF COUNTER EVIDENCE TO DEFAULT ASCRIPTION OF BELIEF

Thus far we have shown how the default rule can ascribe belief to an agent if there is no further information (as in ascribing the belief that the System dislikes Frank, to Frank, in Figures 2 and 3) and how the rule can be overridden by an a priori belief which contradicts the belief which we are attempting to ascribe (as in "Earth is flat" overridding "Earth is round" in Figures 4 and 5). We now consider more complex cases.

4.1 Blocking Beliefs that an Agent is Unaware of

Ascribing belief should be blocked in cases where a belief exists that an agent does not know something, or does not have a belief on some topic. Consider the scope of negation on the *know* operator. Given the following:

Smith NOT–know p (1)

Smith know NOT–p (2)

It is parsing/translation issue to decide which form an input should take, but, representationally, (1) implies lack of knowledge while (2) implies knowledge of a negative proposition, hence (1) should *not* be ascribed to Smith, while (2) should be (cf. Wilks, 1986).

We feel that this also holds true for *beliefs*. So for the following:

Smith NOT–believe p (3)

Smith believe NOT–p (4)

(3) should not be ascribed, while (4) should be. In other words, NOT-know and NOT-believe are explicit counterevidence that we should not ascribe a certain belief to a certain agent.

Furthermore, example (1) is more complex than it appears. If the system believes that Smith does not know proposition *p*, then the implication is that the system believes[4] that it does know p. Not only must (1) not be promoted, but it must also cancel the promotion of the system belief that *it* knows p.

4.2 Atypical Beliefs

We now turn our attention to a class of beliefs called *atypical beliefs*, which need a rule which is the opposite of the default rule for ascribing belief.

An atypical belief is a belief which is held by an agent but would not generally be held by other agents. The class of atypical beliefs covers such areas as self knowledge, secrets, expertise and knowledge of uncommon domains (such as the believer's hobbies, skills, personal medical history, etc.).

So, for example, the belief that the Earth is flat is atypical. An important point must be made here. In terms of a specific agent, the foregoing definition is insufficient. For a belief to be considered atypical with respect to an agent, the agent must believe the belief to be atypical; i.e., I my believe you to have an atypical belief, however, I may also believe that you think it is a typically held belief. So, while I believe it to be atypical, you believe it to be typical.

Expert belief, because uncommonly held, is a type of atypical belief that presents special problems. Consider the case of medical knowledge. It is possible to reason about a treatment for an illness without knowing the details of the treatment. We may discuss a cure for tuberculosis without knowing the cure. This example shows that simply preventing beliefs on expertise from being ascribed to an agent is often undesirable, even though the agent is not believed to be expert on the topic of beliefs.

Another type of atypical belief is belief which cover *uncommon domains*. An uncommon domain is a topic about which a particular agent has a very detailed set of beliefs that most other agents are not aware of, for example, beliefs about a particular agent's childhood. Beliefs on uncommon domains are similar to beliefs on expert domains because both types are uncommon. A belief about some agent's phone number can be treated as a belief on expertise. The experts are those people who know what the agent's phone number is, but nonexperts are still capable of reasoning about the agent's phone number.

For the class of atypical beliefs the rule should be *not* to ascribe unless one has explicit evidence to justify ascribing the belief. The problem is one of representing and handling a wide range of types of atypical beliefs. The introduction of metabeliefs in one possible solution, because metabeliefs enable explicit representation of atypical beliefs (i.e., for atypical belief p, have

[4] This implication is valid where p is a proposition of the form *knowing that*, but would not necessarily be valid for p is p is of a *knowing value* or a *knowing skill* form.

the metabelief *atypical (p))*. Due to the wide range of atypical beliefs, and to the problems that they pose, we use a special representation to handle them.

4.3 Intensional Descriptions and Atypical Belief

McCarthy and others have suggested that lambda expressions be used to represent knowledge of values. Wilks (1986) proposed that expertise may be expressed within a system by use of lambda expressions with restrictions on the capable evaluators of each such lambda expression. Knowing or having the belief represented by the lambda expression means that the agent is capable of evaluating the expression. So the representation for a cure for tuberculosis is:

$$(CURE\text{-}FOR\ Tuberculosis)\ BE\ (\lambda(x)\ (CURE\text{-}FOR\ x)\ Tuberculosis) < MDs \quad (5)$$

where the only capable evaluators are those known to be MDs (medical doctors). These lambda expressions can be viewed as intensional expressions, i.e. expressions which return the value of some intensional description (cf. Maida, 1983, Haas, 1986). Beliefs involving such intensional expressions are generally atypical, as shown by the problems of iterated propositional attitudes which involve intensional descriptions (Creary, 1979, Maida, 1983).

We feel that the problems of expressing beliefs about intensional descriptions are so closely aligned with those of atypical beliefs in general that a representation for differing beliefs about the referent of intensional descriptions can be used for atypical beliefs.

4.3.1. A Taxonomy of Metabelief's about Values

The form of representation suggested by Wilks (1986) allows expressions which can only be evaluated by specific agents or classes of agents. This is a first step towards representing atypical belief; however, it is insufficient.

Consider the major factor that makes a belief atypical; some agent believes that a belief held by another agent is not commonly held. A belief about another belief is known as a metabelief. It is the metabeliefs about a belief that mark a belief as atypical. An atypical belief can have a large number of metabeliefs associated with it.

These metabeliefs can often be classified according to the relation that they define between an agent and the belief with which they are concerned. We propose a taxonomy of metabeliefs (beliefs about what agents believe is the value of some expression), as shown in Figure 8.

Consider a belief about an agent's phone number (call this belief P), i.e., the belief is that the agent's phone number is some value. An agent *(A)* may have a number of metabeliefs concerning this belief P. In Figure 8 we see

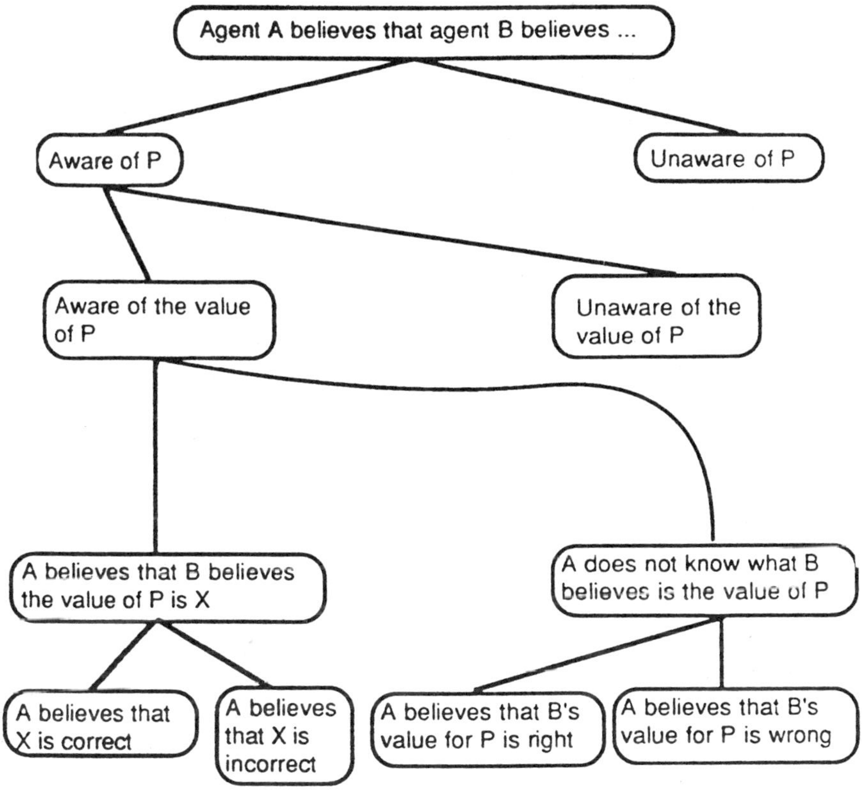

Figure 8. A Taxonomy of Metabeliefs about Values.

that agent *A* may believe another agent *(B)*[5] to be either aware or else unaware of the belief (P). Somebody who does not know John can hardly have beliefs about John's phone number, and so that person should be unaware of any beliefs about John's phone number.

Further types of metabelief are possible if A believes B is aware of P. A may believe that B is unaware of the value of P. For example, I may believe that you know that John has a phone number, but that you do not know what his phone number is.

If agent A believes that agent B does know the value of P, then *A* may either know what *B* believes is the value, or else *A* may now know what *B* believes is the value of P. I may know that you believe John's phone number

[5] The other agent may be *A* himself or herself. In that case the metabelief is one of introspection; however, it would be absurd for agent *A* to have a belief that he or she is unaware of belief P, so this situation is now allowed. Introspection is discussed in detail in Konolige (1985) and in Maida (1986).

to be 505-526-5444, or I may simply now know what you believe is his or her phone number.

If *A* knows what *B* believes is the value of P, then *A* may decide whether *B* is correct or incorrect in his or her belief, according to what *A* believes is the correct value; i.e., if I believe that John's phone number is 505-525-5444, and I believe that you believe it's 505-526-5444, then I believe that you are incorrect in your belief about John's phone number.

If *A* does not know what *B* believes is the value of P, it is possible that *A* can use other information to determine whether *B* is correct or incorrect in his or her belief about the value of P. For example, if you have not seen John in years, and he has moved house since you last saw him (but you do not know that he has moved house) although you believe that you know his or her phone number, I believe that you have his or her phone number wrong. I can believe this even if I do not know what his or her previous phone number was.

The ability to represent metabeliefs, of the type described above, is crucial to the process of ascribing belief. *Lambda formulas* are an extension of lambda expressions which allow the metabeliefs of a belief to be represented with the associated belief.

4.4 Lambda Formulas

Lambda expressions are finite partial functions. They are computable and map into a finite range of values. We extend the lambda expression representation to a Lambda Formula so we can represent opposing views on the value of such an expression.[6] For example,

(PHONE—NO—OF John) Be (λ(x)(PHONE–NO–OF x) John)<John> (6)

represents an expression which evaluates to John's phone number for John, but is unevaluable to any other agent. The imposition of John as a capable evaluator of the lambda expression (henceforth λ-expression) is equivalent to adding an extra parameter to the expression, the extra parameter being the agent performing the evaluation.

The list of capable evaluators is not necessarily a constant, because our beliefs may vary regarding who are the capable evaluators of an expression.

---

[6] It may appear that talking about the value of a proposition is only valid for knowledge of values, such as phone number, etc., but not for propositions of the form *John is here*; however, these propositions may be treated in one of two ways: (a) they may often be rewritten to reflect a value (e.g., location of John is *x*), or (b) the truth or falsity of the proposition may be treated as its value. Coupling the latter case with *unknown* values (see section 4.5) provides us with the power of a Kleene three-valued logic, or a model logic with a *not-know-whether* operator.

$$(\mathit{CURE{-}FOR\ Tuberculosis})\ \mathit{BE}\ \ (\lambda(x)\ (\mathit{CURE{-}FOR}\ x)\ \mathit{Tuberculosis})$$

$$\begin{bmatrix} <\{\mathit{MDs, John}\},\ v_0>, \\ <\{\mathit{Dan, Tibetan\text{-}Priest}\},\ v_1>, \\ \\ <\{\mathit{Susan, Paul}\},\ v_n>, \end{bmatrix}$$

Figure 9. Augmented Lambda Expression for the Cure for Tuberculosis

Also different evaluators may evaluate the expression to different values, which is also subject to each individual's beliefs.

We want to express different agents as being capable of evaluating such an expression but to different values. We propose a more complex form of the capable evaluators list proposed by Wilks (1986). A typical expression in the augmented representation is shown in Figure 9.

Figure 9 contains *n* sets of capable evaluators and the values to which they evaluate the expression. The basic set of capable evaluators has been replaced by a set of pairs, consisting of a set of capable evaluators and the value that they return. A typical entry, for example, $<\{\text{Susan, Paul}\}, v_n>$ is called a Capable_ Evaluators-Value pair (abbreviated to CEV pair). It is stipulated that each value v_i must be of the same structure and that the λ-expression returns this structure.

The set of CEV pairs is a set of metabeliefs about the belief represented by the λ-expression. The CEV pairs in Figure 9 represent the metabeliefs that the system believes the agents in those CEVs to be aware of there being a cure for tuberculosis, that they know a cure for tuberculosis, and that the system knows what they each believe is the cure for tuberculosis. Only one of the values v_o, . . . , v_n is believed by the system to be the correct value. The other values are considered incorrect. This covers the section *A knows what B believes is the value of P* in Figure 8.

The set of CEV pairs is also a function table representation of the λ-expression. We refer to the λ-expression and its function table as a λ-formula.

More formally, a λ-formula is a tuple $\Lambda = (\lambda, \Gamma)$, where λ is a λ-expression and Γ is a function table. A function table Γ is defined to be a set of tuples of the form $\Xi = (\Delta, \upsilon)$, where Δ is set a set of agents, known as capable evaluators, and υ is a value returned by the λ-expresion.

4.5. Unknown Values

In the original representation, values which are unknown to an agent are represented by the agent not being a capable evaluator. In other words, awareness of the value of a λ-expression is represented by being able to evaluate the λ-expression. In Figure 8 this would be equivalent to replacing the section below *aware of the value of P* by *the correct value*. The original representation does not handle problems of awareness of P with sufficient power.

Being capable of evaluating the expression is still the criterion for knowing what the expression represents. Now, however, by the introduction of what we call *uncertain values* and *uncertain-but-incorrect values*, we are able to express more with the representation.

An uncertain value is a value (of a CEV pair) which is unknown to an agent, i.e., the agent is aware that the value exists but not what the value is. This uncertain value may or may not be equal to a known value or some other uncertain value. Thus, the agent can represent the value that another agent believes is the value of a λ-expression without knowing what the value is.

Uncertain-but-incorrect values are known by the agent to be different from the value that the agent believes to be correct. In other words the value, while unknown to the agent, is believed by the agent to be wrong (the metabeliefs that A believes that what B believes is the value of P is wrong, even though A does not know what B believes is the value of P).

An example is in order here. We annotate our representation by using Ψ to represent an uncertain value and $\bar{\Psi}$ to represent an uncertain-but-incorrect value (Figure 10).

How do we interpret this structure? If this is a belief of the system, then it may be interpreted as follows: The system believes that Mary and itself know the correct value[7] for Mary's phone number, namely v_o. Dan is believed to know Mary's phone number to be v_1; i.e., he knows the incorrect value, a value known to the system, but does not know that value to be incorrect. Susan and Paul both know the same phone number for Mary which may or may not be correct as the system does not know what they believe Mary's phone number to be. Finally, Fred and John know Mary's phone number to be a value that the system does not know; however, the system believes Fred and John to have the wrong phone number.

One problem of λ-formulas is that given some λ-expression in a nesting such that the agent at the bottom of the nesting can evaluate it, but an agent higher up in the nesting is unaware of the value of the expression, how then can the evaluation of the expression within the nesting be prevented?

For example, the λ-expression $(\lambda(x).P(x))[<\{Dan, System\}, v_o>]$, represents some function on x which Dan and the system are capable of evaluating, and resides in an environment for what the system believes Susan be-

---

[7] Whichever set, if any, that the system lies within defines the *correct* value from the system's point of view.

$$(PHONE\text{–}NO\text{–}OF\ Mary)\ Be\ (\lambda(x)\ (PHONE\text{–}NO\text{–}OF\ x)\ Mary)\begin{bmatrix} <\{Mary, System\}, \nu_0>, \\ <\{Dan\}, \nu_1>, \\ <\{Susan, Paul\}, \psi_0>, \\ <\{Fred, John\}, \bar{\psi}_0> \end{bmatrix}$$

Figure 10. Mary's Phone Number Example with Uncertain Values

lieves Dan believes. Susan is not a capable evaluator of the expression and so the expression should not be evaluable, even though it lies within an environment which has Dan as the innermost agent.

Should evaluation be prevented? It seems obvious that the evaluation to a *specific value* should indeed be prevented. The representation allows for unknown values by means of uncertain values. What do the uncertain values represent? They are intended to allow the system to represent values that it doesn't know, but whose existence it is aware of. Thus the system is able to reason about these values without knowing what they are. In this case we wish to represent Susan being unaware of the value of P. A value called an *unknown* value is used to represent this.

Unknown values are values which are unknown with respect to the environment in which they appear. Hence, if $\lambda(x).P(x))[<\{Dan, System\},\mu>]$ lies within an environment which represents what the system believes Susan believes Dan believes, the expression evaluates to an unknown value (represented by μ).

We have demonstrated representations of all cases where agent B (Figure 8) is believed to be aware of P. Next, we consider the case where B is unaware of P.

4.6 Awareness, and Explicit and Implicit Mention of Agents

A question is *how to represent an agent being aware of the proposition represented by a λ-formula*? Our solution is this: An agent is said to be aware of the proposition represented by a λ-formula if and only if that agent is either explicitly or implicitly mentioned in the set of all Δs for the λ-formula; otherwise, the agent is said to be unaware of the proposition.

In Section 4.1, we say that a λ-formula consists of a λ-expression and a function table, and that the function table is a set of tuples $\Xi = (\Delta, \upsilon)$. The union of all such Δs for a given λ-formula, $(\bar{\Delta} = \bigcup_{all\ \Delta s\ in\ \lambda} \Delta)$, is the set of all capable evaluators for the given λ-formula. An agent is explicitly mentioned if the agent is a member of $\bar{\Delta}$ and is implicitly mentioned if the agent is a member of a class of agents[8] that is explicitly mentioned in $\bar{\Delta}$. This may be seen in Figure 9 where MDs are capable evaluators. Hence, any agent who is an MD is capable of evaluating Figure 9. MDs are explicitly mentioned in this example, someone who is an MD is implicitly mentioned.

An agent who is neither explicitly mentioned is deemed to be unaware of the proposition that the formula represents. In the case where an agent is implicitly mentioned in more than one class of agent, the most specific class is chosen to determine that agent's view.

It may appear that this representation requires the enumeration of countless number of agents. This is avoided by the use of classes of agents. If the majority of agents believe a particular λ-formulas to return the value v_i, then this may be represented by having a Ξ with average-man as a member of its Δ, and with its υ equal to ν_i. This is the least specific class of agents possible.

4.7. Function Transformations

The use of λ-formulas poses a question: What affect does ascribing a λ- formula to an agent have upon the λ-formula?

Ascribing a λ-formula often requires altering the formula. For example, ascribing a formula to an agent who does not know the value it returns for any agent will require changing the formula to reflect this situation.

Ascribing a λ-formula to an agent thus involves changing the function table for the formula. This is equivalent to saying that ascribing a λ-formula to an agent involves changing the function that the formula represents.

A function table Γ is a set of tuples $\Xi = (\Delta, \upsilon)$. The transformation function will typically involve merging some of these Ξs and introducing new ones. We are currently investigating the role that the semanatics of the λ-expression plays in determining the transformation. That the semantics of the λ-expression *does* play a role may readily be seen by considering two examples. Consider two expressions, one of which pertains to Mary's phone number, and the other to a facet of Mary's health. The heuristic assumptions that the function makes is different in these two cases by virtue of the nature of the information that they contain. We can assume that Mary knows her own phone number but not that she accurately knows the state of her health.

Function Transformations on λ-formulas are discussed in more detail by Ballim (1987).

4.8. λ-expressions and Metabeliefs

We can see how λ-formulas solve the problems which were discussed in Section 4. Section 4.1 describes beliefs which block the ascription of other be-

---

[8] Meaning postulates are used to determine if an agent is a member of a particular class of agents.

liefs. The role of this class of beliefs is totally subsumed by λ-formulas as this class describes lack of awareness of some propositions; lack of awareness is relized within λ-formulas by an agent being neither an explicit nor implicit agent of the formula's capable evaluators set ($\bar{\Delta}$). Atypical beliefs are characterized by their associated metabeliefs. λ-formulas are very effective at representing metabeliefs of atypical beliefs.

While it may appear at first glance that the introduction of λ-formulas has created more problems than it has solved, this is not true. The problem of representing atypical and expert belief is a major problem in comparison to the problems presented by λ-formulas. Furthermore, the problems presented by λ-formulas have provided important insights into belief systems and the nature of metabeliefs about atypical beliefs.

5. A DESCRIPTION OF VIEWGEN AND A MEDICAL EXAMPLE

ViewGen is a program which generates nested points of view. The program has a set of beliefs which are considered held by an agent known as the system. These beliefs are partitioned into topics (held in labelled PTs). An example call to the program is shown in Figure 11.

The point of view represented in Figure 11 is the system's view about thalassemia. This view consists of one belief, namely that it is a genetic disorder.

ViewGen operates by forming a contextual focus from the PTs of the terms

```
| ?- viewgen([system],[thalassemia],_).

Viewpoint of [system]
 is

[
  thalassemia  is_type_of  lambda(type_of(thalassemia),genetic_disorder)
]

yes
| ?-
```

Figure 11. The System's View about Thalassemia[9]

[9] In the examples that we use, beliefs are simply propositions which reside within a point of view. This proposition are skolemised so that existentially quantified variables are replaced by skolem constants, and all other variables are universally quantified within a viewpoint. Variables are terms which begin with a capital letter.

```
lambda(type_of(thalassemia), D,
                [
                 [ [system, medically_informed_person],
                                  genetic_disorder ],
                 [ [avg_man], disease ]
                ])
```

Figure 12. Lambda Formula for the type of thing that Thalassemia is

in the list that forms the second argument to the program. The first argument is a list of agents. This list represents the nesting that is required. The third argument is a variable which is instantiated to the structure that the program returns.

In the following example assume that the system is a medical expert conducting a dialogue with a married couple who are seeking advice about thalassemia. The system has a lambda formula for the intensional description *type_.of(thalassemia)* shown if Figure 12.

With the information that thalassemia is a type of genetic disorder the system is able to generate what it believes about them both. This point of view is shown in Figure 13.

The point of view in Figure 13 contains the belief that thalassemia is a genetic disorder. In addition it contains a complex belief the child of two people who both have a genetic disorder will suffer from the same genetic disorder.[10]

Suppose that the couple in question are called Paul and Susan, and that the system has the beliefs about them shown in Figure 14.

The important beliefs here are that the system believes that Susan is a medically informed person, and that they both suffer from thalassemia. We presume that we have reasoning mechanisms which allow us to reason with the viewpoints in Figures 13 and 14 to hypothesise that any offspring of the pair will also suffer from thalassemia (something which they need to be warned about). If we presume that the system follows Gricean maxims, then we need to generate what the system believes each of them believes about thalassemia. These points of view are shown in Figures 15 and 16.

In Figure 15 the belief that thalassemia is a type of genetic disorder has been ascribed to Susan. This is because the system believes that Susan is a medically informed person, and the system believes that medically informed people know that thalassemia is a genetic disorder (see Figure 12). Paul, however, is not believed to be a medically informed person. He is simply an average man.[11] The average man view of thalassemia is that it is a disease.

[10] This is an extremely simplistic belief, but suffices for the purpose of this example. A real medical expert using *ViewGen* would contain far more complex beliefs than this.

[11] *ViewGen* has a lattice of classes and a set of meaning postulates which allow Paul to be seen as an average man.

```
| ?- viewgen([system],[thalassemia,genetic_disorder],_).

Viewpoint of [system] about [thalassemia,genetic_disorder]
 is

[
  ( ( D is_type_of genetic_disorder and
     ( A1 suffers_from D and
       ( A2 suffers_from D and
         ( A1 married_to A2 and A3 child_of  [A1,A2] ))))
   implies A3 suffers_from D )

  thalassemia is_type_of lambda(type_of(thalassemia),genetic_disorder)
]

yes
| ?-
```

Figure 13. The System's view of Thalassemia and Genetic_disorder(s)

```
| ?- viewgen([system],[susan,paul],_).

Viewpoint of [system] about [susan,paul]
 is

[
  susan is_type_of medically_informed_person
  susan married_to paul
  paul suffers_from thalassemia
  susan suffers_from thalassemia
]

yes
| ?-
```

Figure 14. The System's view of Susan and Paul

```
| ?- viewgen([system,susan],[thalassemia],_).


Viewpoint of [system,susan] about [thalassemia]
 is


[
  thalassemia is_type_of lambda(type_of(thalassemia),genetic_disorder)
]


yes
| ?-
```

Figure 15. The System's View of Susan's View of Thalassemia

This is just the view that the system believes Paul has about thalassemia (see Figure 16).

To further complicate matters, in a dialogue between three agents it can be necessary for one agent to model what the second believes the third believes, or what the third believes the second believes about the topic. So we generate what the system believes Susan believes Paul believes about thalassemia, and what the system believes Paul believes Susan believes about thalassemia. These point of view are shown in Figures 17 and 18, respectively.

```
| ?- viewgen([system,paul],[thalassemia],_).


Viewpoint of [system,paul] about [thalassemia]
 is


[
  thalassemia is_type_of lambda(type_of(thalassemia),disease)
]


yes
| ?-
```

Figure 16. The System's View of Paul's View of Thalassemia

```
| ?- viewgen([system,susan,paul],[thalassemia],_).

Viewpoint of [system,susan,paul] about [thalassemia]
 is

[
  thalassemia is_type_of lambda(type_of(thalassemia),disease)
]

yes
| ?-
```

Figure 17. The System's View of Susan's View of Paul's View of Thalassemia

In Figure 17, using the beliefs about Susan and Paul, the system generates that Susan believes that Paul believes that thalassemia is a disease. In Figure 18, however, the system has generated that Paul believes that Susan believes thalassemia is some kind of complex thing that he doesn't know about, represented by [u,1], which is uncertain value. In other words, according to the default rule, Paul believes Susan is a medically informed person (since there is no evidence to contradict this), and, hence Paul believes that Susan has a more complex belief about thalassemia than he does.

This example demonstrates the importance of nested beliefs in reasoning and dialogue, and to show the utility of our program in a system to perform reasoning about other agents, or to explicate dialogue.

```
| ?- viewgen([system,paul,susan],[thalassemia],_).

Viewpoint of [system,paul,susan] about [thalassemia]
 is

[
  thalassemia is_type_of lambda(type_of(thalassemia),[u,1])
]

yes
| ?-
```

Figure 18. The System's View of Paul's View of Susan's View of Thalassemia

6. CONCLUSION

This work claims only to be a first implementation of a "belief engine" (see Maida, 1986) that contains plausible heuristics or the default ascription of concrete, contentful beliefs. Such a process will be needed by any project that proposes (e.g., Pollack, 1986) to model the interaction of agents planning on the basis of differing beliefs and plans. We believe no other system has yet tackled this problem in a practical way. Many extensions will be required to the current work, particularly in the treatment of the identification of intensionally distinct by extensionally identical individuals and classes.

REFERENCES

Attardi, G., & Simi, M. (1984). Metalanguage and reasoning across viewpoints. *Proceedings of ECAI-84* (pp. 315–324).

Ballim, A. (1986). *Generating points of view* (Memoranda in Computer and Cognitive Sciences, MCCS-86–68. Las Cruces, NM: Computing Research Laboratory, New Mexico State University.

Ballim, A. (1987). The subjective ascription of belief to agents. In J. Hallam & C. Mellish (Eds.), *Advances in artificial intelligence* (pp. 267–278). Chichester, England: John Wiley.

Charniak, E. (1978). On the use of knowledge in language comprehension. *Artificial Intelligence, 11,* pp. 225–265.

Cohen, P., & Levesque, H. (1980). Speech acts and recognition of shared plans. *Proceedings of the third Biennial Conference, Canadian Society for Computational Studies in Intelligence* (pp. 263–271).

Creary. L. (1979). Propositional attitudes: Fregean representation and simulative reasoning. *Proceedings of IJCAI-79* (pp. 176–181). Tokyo, Japan: Morgan Kaufmann.

Haas, A. (1986). A syntatic theory of belief and action. *Artificial Intelligence, 28,* 245–292.

Konolige, K. (1982). Circumscriptive ignorance. *Proceedings of AAAI82* (pp. 202–204).

Konolige, K. (1983). A deductive model of belief. *Proceedings of IJCAI-83* (pp. 377–381). Karlsruhe, West Germany.

Konolige, K. (1984). *Belief and incompleteness* (SRI Report No. 319). Stanford, CA: Stanford Research Institute.

Konolige, K. (1985). A computational theory of belief introspection. *Proceedings of IJCAI-85* (pp. 502–508). Los Angeles, CA.

Maida, A. (1983). Knowing intensional individuals. *Proceedings of IJCAI-83* (pp. 179–183). Karlsruhe, West Germany.

Maida, A. S. (1986). Introspection and reasoning about the beliefs of other agents. *Proceedings of Cognitive Science Society* (pp. 187–195).

Martins, J., & Shapiro, S. (1983). Reasoning in multiple belief spaces. *Proceedings of IJCAI-83* (pp. 370–373). Karlsruhe, West Germany.

Minsky, M. (1975). A framework for representing knowledge. In P. H. Winston (Ed.), *The psychology of computer vision.* New York: McGraw-Hill.

Moore, R., & Hendrix, G. (1979). *Computational models of belief and the semantics of belief sentences* (SRI Technical Note No. 187). Stanford, CA: Stanford Research Institute.

Perrault, R., & Allen J. (1980). A plan-based analysis of indirect speech acts. *American Journal of Computational Linguistics, 6,* 167–182.

Pollack, M.E. (1986). A model of plan inference that distinguishes between the beliefs of actors and observers. *Proceedings of the Association for Computational Linguistics* (pp. 207–214).

Shadbolt, N. (1983). Processing reference. *Journal of Semantics, 2* (1), 63–98.

Shapiro, S.C., & Rapaport, W.J. (1986). SNePS considered as a fully intensional propositional semantic network. *Proceedings of AAAI86,* (pp. 278–283).

Taylor, G.B., & Whitehill, S.B. (1981). A belief representation for understanding deception. *Proceedings of IJCAI-81* (pp. 388–393). Vancouver, Canada.

Wilks, Y.A. (1977). Making preferences more active. *Artificial Intelligence, 8,* 75–97.

Wilks, Y.A. (1986, June). *CRL work on beliefs and computation.* Paper presented at the Workshop on the Foundations of Artificial Intelligence, University of Naples, Naples, Italy.

Wilks, Y.A., & Bien, J. (1979). Speech acts and multiple environments. *Proceedings of IJCAI-79* (pp. 451–455). Tikyo, Japan.

Wilks, Y.A., & Bien, J. (1983). Beliefs, points of view and multiple environments. *Cognitive Science, 8,* 120–146.

CHAPTER 2

An ATN-Based Grammar for the Structural Analysis of Dialogue*

RONAN REILLY
Educational Research Centre
St Patrick's College

The purpose of this chapter is to present a dialogue grammar for use in the structural analysis of natural language dialogues. It has been implemented as an augmented transition network (ATN). The system proposed here can be used to calculate a normative profile of the structure of a number of dialogues. Alternatively, the structure of dialogues gathered under different conditions can be statistically compared using the output from the ATN analysis.

The work upon which this chapter is based is the dialogue analysis system of Burton (1981), which is a modified version of the system proposed by Sinclair and Coulthard (1975) for the analysis of classroom discourse. Burton's analysis system consists of five hierarchically related levels: the interaction (the largest unit), the transaction, the exchange, the move, and the act. An interaction consists of a number of transactions, a transaction consists of a number of exchanges, and so on. The smallest *interactive* unit in this system is the move, and is equivalent to an individual turn in the dialogue. A move can consist of a number of linguistic and nonlinguistic "acts" which are realized either as utterances or physical actions.

By using an ATN, it is possible, not only to represent the structural elements of the dialogue grammar, but also to implement the network computationally and use it in the statistical analysis of act sequences in real dialogues. Furthermore, the very process of implementing an ATN provides a useful means of detecting inconsistencies in the grammar.

An ATN consists of a network of nodes representing states, a set of directed arcs connecting the states, and a set of arbitrary actions associated

* This research was supported in part by the ESPRIT programme of the Commission of the European Communities (ref. B/84 P527).

with each arc. Nodes and arcs alone would comprise a simple transition network. The association of arbitrary actions to the arcs connecting one state to the next provides the augmentation which gives the networks their name. An ATN can be used to parse strings by using elements of the input string in conjunction with conditions on the arcs to dictate the transitions from state to state. The ATN "consumes" the input string from left to right, and produces as output a structural analysis of the input. In our case the input to the ATN will be a sequence of dialogue acts and the output will be a structural analysis in terms of interactions, transactions, exchanges, and moves. A secondary form of output will be information concerning what arcs have been crossed. When accumulated over a number of dialogues, this information can be used to determine the frequency structure of typical dialogues.

THE DIALOGUE GRAMMAR

Augmented Transition Networks

For the sake of clarity the ATN grammar is represented in the form of eight subnetworks (cf. Figure 1). The first subnetwork deals with interactions, the second with transactions, the third and fourth with exchanges, and the remaining five with moves. It is possible to represent the grammar as one large ATN, but it would have been complicated to follow, and would not have preserved the hierarchical nature of the grammar.

Each state in the ATN is indexed by a subnetwork name and a number of the form <name>/n. For example, the second state in the conversational exchange subnetwork is CX/1 (the state numbers start at 0). We use JMP for *jump*, SK (not to be confused with Sk for *sketch*) for *seek* or *push*, and SND for *send* or *pop*. What follows is a more detailed explanation of the traversal conditions which can be associated with an arc in the network:

JMP: This is the null condition and indicates that the arc with which it is associated can be crossed without any conditions being met. In this network it is used to indicate the optionality of an input element.

SK <arg>: This condition indicates that the subnetwork indexed by <arg> must be successfully traversed before the arc associated with the SK condition can be crossed. When a SK arc is encountered control is passed to the indexed subnetwork, and is returned to the calling network once the subnetwork has been attempted. If the attempt is successful, the arc in the calling network is crossed; otherwise, the next arc at the current state in the calling network is tried.

SND: This condition is closely associated with the SK condition. It can only be traversed if a move, exchange, or transaction

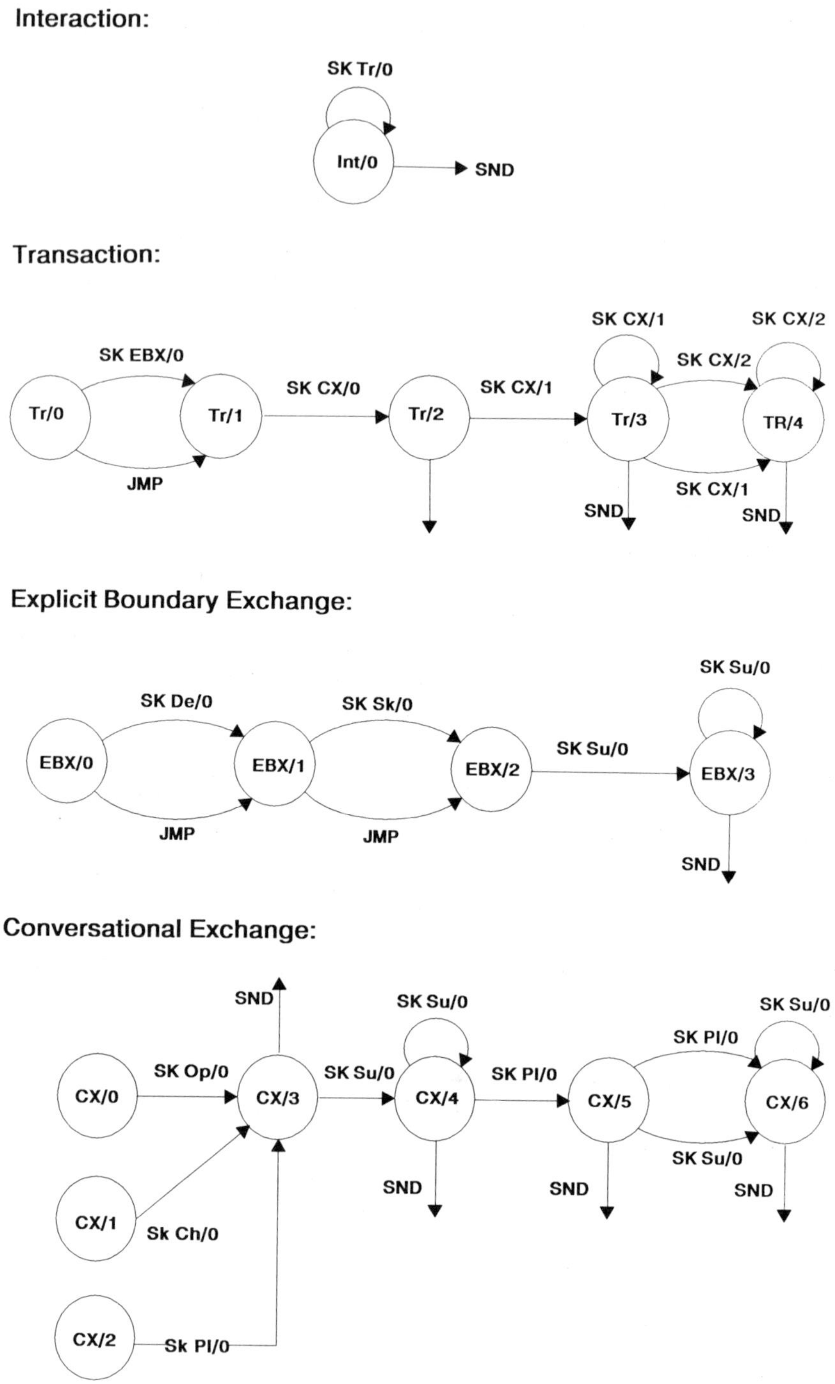

Figure 1. An Augmented Transition Network Dialogue Grammar

Delineating Move:

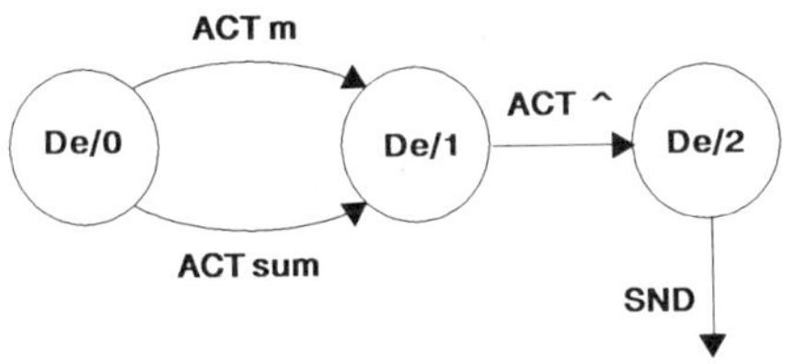

Sketching Move:

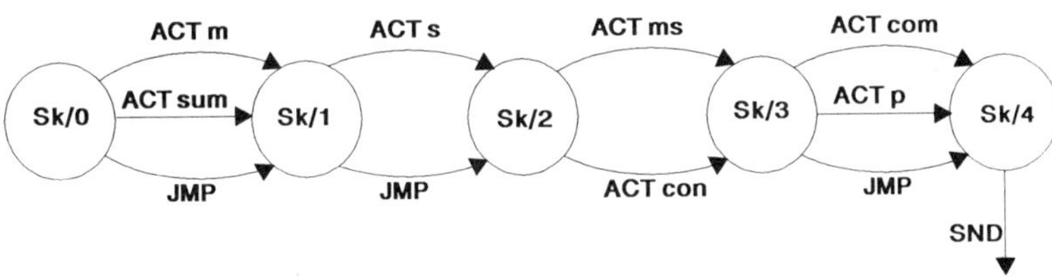

Opening Moves:

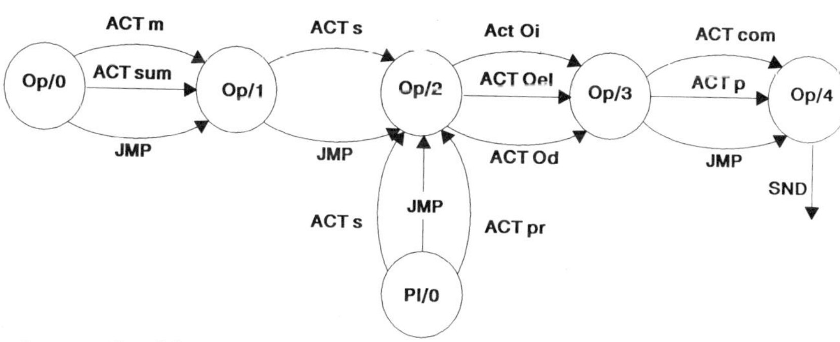

Supporting Move:

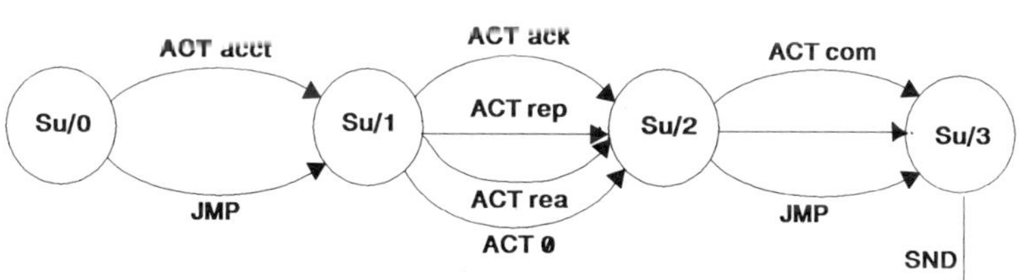

Challenging Move:

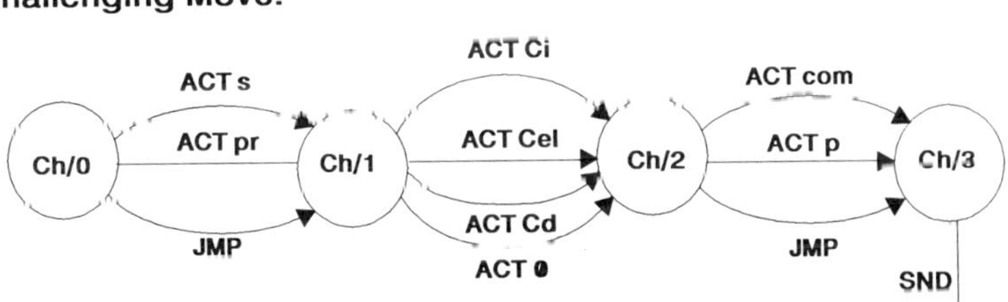

boundary has been reached, indicating that all elements in the move, exchange, or transaction have been accounted for by the subnetwork in question. An appropriate structural description of a subsequence of the act input is returned to the calling network. This partial description is integrated into the overall description being generated by the network as a whole.

ACT <arg>: This is the lowest level condition occurring in the ATN and is found in the subnetworks dealing with moves. The ACT condition indicates that the associated arc can only be traversed if the current element of the input is an act of type <arg>.

A detailed description of the various elements of the ATN is provided in the following sections.

Acts

In Burton's scheme, an *act* is a way of classifying utterances which occur in a conversational move. The act is a level above that of the utterance, and a move can consist of a number of acts. For example, the following is an opening move by a teacher in a classroom dialogue consisting of two acts (in brackets):

A:	those letters have a special name	(Starter)
	do you know what it is	(Elicitation)

Most acts are realized by a wide range of utterance types. For example, the starter act can be realized by either a statement, question, command, or a moodless item.

Burton uses some of the 22 acts proposed by Sinclair and Coulthard (1975). However, in order to extend the application of their scheme to conversational situations outside the classroom, a number of Sinclair and Coulthard's acts were either altered or deleted by Burton. We in turn will not be using Burton's entire set of acts, but will delete those which we think are irrelevant to the types of dialogue we will be considering.

What follows is a list of the 15 acts and their definitions which will be used in or classification system (after Burton, 1981, pp. 76–78):

Marker

A marker is realized by a closed class of items, e.g., *well, OK, now, good, all right*, and various expressive particles. Its function is to mark boundaries in the discourse and to indicate that the speaker has a topic to introduce. This act is represented as *m* in Figure 1.

Summons

A summons is realized by a closed class of verbal and nonverbal items, e.g., the use of the name of another participant, or mechanical devices like door bells and telephone bells. Its function is to mark a boundary in the discourse, and to indicate that the producer of the item has a topic to introduce once he has gained the attention of the hearer. This act is represented as *sum*.

Silent Stress

A silent stress is realized by a pause, indicated in text by an exclamation mark, or following a marker. It functions to highlight a marker or summons when they act as the head of a boundary exchange. It is represented as ^.

Starter

A starter is realized by a statement, question, command, or moodless item. Its function is to provide information about, direct attention to, or thought towards an area, in order to make a correct response to the coming initiation more likely. It is represented as *s*.

Metastatements

A metastatement is realized by a statement, question, or command which refers to a future event in the ongoing talk, or a request for speaker's rights. Its function is to make clear the structure of the immediately following discourse, and to indicate the speaker's wish for an extended turn. It is represented as *ms*.

Conclusion

A conclusion is realized by an anaphoric statement, which can be seen as the complement to metastatements, in that its function is to make clear the structure of the immediately preceding discourse. It is represented as *con*.

Informatives

Informatives are realized by statements whose sole function is to provide information. The appropriate response is the giving of attention and indication of understanding. There are two main classes of informative, challenging informative (*Ci*) and opening informatives (*Oi*). This distinction will be elaborated on in the discussion of dialogue moves.

Elicitations

Elicitations are realized by questions. Their function is to request a linguistic response, Occasionally they may be realized by a command requesting a linguistic response. As with informatives, there are two classes of elicitation, challenging (*Cel*) and opening (*Oel*).

Directives

Directives are realized by commands, and function to request a nonlinguistic response. There are two types of directives, challenging (*Cd*) and opening (*Od*).

Comment

A comment is realized by a statement, question or command, or moodless item, and functions to expand, justify, or provide additional information to a preceding informative or comment. It is represented by *com.*

Accept

An accept act is realized by a closed class of items, e.g., *yes*, *OK*, *uhah*, *I will*, *no* (where the preceding utterance was negative). Its function is to indicate that the speaker has heard and understood the previous utterance and is compliant. It is represented as *acct*.

Reply

A reply is realized by a statement, question, moodless item, or nonverbal surrogates such as nods. Its function is to provide a linguistic response appropriate to a preceding elicitation. It is represented by *rep*.

React

A react is realized by a nonlinguistic action. Its function is to provide an appropriate nonlinguistic response to a preceding directive. It is represented by *rea*.

Acknowledge

An acknowledge act is realized by a closed class of items such as *yes*, *OK*, *uhuh*, and other expressive particles. Its function is to show that an informative has been understood, and its significance appreciated. It is represented by *ack*.

Preface

A preface is realized by combinations of placement markers, self-referential metaterms, and metareference to preceding parts of the dialogue. It function is to show that a diverted topic is being reintroduced. It is represented as *pr*.

Prompt

A prompt is realized by a closed class of items such as *go on, what are you waiting for, hurry up*. Its function is to reinforce a preceding directive or elicitation. It is represented as *p*.

Null

The null act is is used to account for moves which are negatively realized by utterances, such as a long periods of silence or inaction. It is represented by the symbol Ø.

Moves

Moves are the basic units of a dyadic exchange and can consist of a number of acts. The following is a description of the moves in our dialogue grammar.

Delineating and Sketching Moves

Delineating and sketching moves occur at the boundary of exchanges. They serve to get the attention of the listener, to direct attention to an idea, and to either sketch out roughly the structure of the subsequent discourse or to wind up a previous exchange by summarizing the structure of the preceding discourse. Delineating moves in particular involve attention getting acts such as summonses and markers. Sketching moves can involve metastatements on the discourse structure to come, or comments on the discourse structure that has gone.

The dialogue D1 illustrates a sketching move in a database query context. Note that the elicitation act is not part of the move.

		Utterance	Act	Move
D1	A:	OK	m	) sket.
		student numbers	s	)
		I want to ask a few	) ms	)
		questions about them.	)	
		How many girls took	) Oel	open.
		Academic Irish in 3rd year '83?	)	

A point to note in the above example is how the classification framework can cope with utterances which would be considered ill-formed if treated at a syntactic level.

Opening Moves

Opening moves can, in the absence of delineating or sketching moves, occur as the first move in a transaction. They are topic–carrying items and do not refer to the immediately preceding utterance.

Supporting Moves

Supporting moves can occur after all the other moves listed above. Delineating, sketching, opening, and challenging moves can all be followed by a supporting move. However, there is a dependency between acts which in turn determines permissable patterns of moves. Certain initiatory moves can be followed only by specific supporting moves: The following dependencies exist:

marker	→	acknowledge
summons	→	accept
metastatement	→	accept
informative	→	acknowledge
elicitation	→	reply
directive	→	react

When this dependency structure is violated it marks the termination of an exchange, and the start of new one.

The following fragment of dialogue, a continuation of the database query dialogue in D1, illustrates the production of a supporting move in response to an opening move which contains an elicitation act. The supporting move takes the form of an optional accept followed by a reply.

		Utterance	Act	Move
D2	A:	How many girls took academic Irish in 3rd year '83	) Oel)	opening
	B:	Processing . . . 0	acct rep	) supp.)

Challenging Moves

Whereas supporting moves facilitate the maintenance of topic continuity in a dialogue, challenging moves serve the purpose of suspending the topic and withholding support. In Burton's description, challenging moves can occur after any move except a supporting move. However, it is possible to imagine a dialogue where this exception does not hold:

		Utterance	Act	Move
D3	A:	All man-made flying	) Oi	opening
		machines are heavier	)	
		than air	)	
	B:	OK	acct	supp.
		But wait a minute	s	) chall.
		What about hot air balloons	Cel	)

In D3 we have B responding initially with a supporting move, and then making a challenging move. Such a pattern would be common in dialogues where complex information is being conveyed and where participants might have to make a supporting move before fully processing an informative so as not to interrupt the flow of discourse. It could be argued that this is a discourse performance phenomenon, and that the classification system should only be concerned with discourse competence. However, as Coulthard and Brazil (1981) have pointed out, "the concept of competence, as it has been understood since Chomsky set it in sharp contrast to performance, may ultimately be unhelpful in our field [of discourse analysis]" (pp. 82–83). One of the main reasons why this distinction is not valid for discourse data is because, in contrast to sentence data, speakers seem to have weaker intuitions about when discourse is ill-formed. Intuitions concerning ill-formedness are a crucial source of data for the construction of a competence theory.

The following dialogue, a continuation of D2, illustrates a challenging move:

		Utterance	Act	Move
D4	A:	How many girls took	) Oel	opening
		academic Irish in 3rd year '83	)	
	B:	Processing . . .	acct	) supp.
		0	rep	)
	A:	That doesn't seem right	Ci	chall.

Bound-opening and Re-opening Moves

Bound-opening moves are continuations of opening moves, and occur after an initial opening move has been supported. They serve to maintain the topic of the previous opening move. Re-opening moves reinstate a topic after it has been either diverted or delayed by a challenging move. In the ATN of Figure 1, re-opening and bound-opening moves have been subsumed within a new category: postinitial opening, which has an entry in the opening subnetwork at PI/0. This is to distinguish them from an initial opening move, which have an entry point at OP/0.

What distinguishes bound-opening from re-opening moves is the context in which they occur: whether the opening move follows a supporting move or a challenging move.

The following piece of dialogue illustrates both types of move:

		Utterance	Act	Move
D5	A:	How many girls took academic Irish in 3rd year '83	) Oel)	open.
	B:	Processing . . . 0	acct rep	) supp.)
	A:	That doesn't seem right	Ci	chall.
	A:	How many boys took Academic Irish in 3rd year	) Oel)	re-op.
	B:	Processing . . . 0	acct rep	) supp.)
	A:	Of course! Academic Irish isn't taken in 3rd Year I really want to know how many girls took Professional Irish in 3rd Year '83	) pr))) Oel))	) bo-op.)))

Exchanges

An exchange is composed of moves. Burton distinguishes between two types of exchange: explicit boundary exchanges (EBX subnetwork) and conversational exchanges (CX subnetwork). Explicit boundary exchanges occur, as their name suggests, at the boundary of transactions. They can include delineating and/or sketching moves which must be supported by another speaker. Conversational exchanges consist of chains of opening, challenging, and/or supporting moves.

The dialogue fragment in D5 consists of three conversational exchanges, as follows:

Exchange	1:	opening → supporting
	2:	challenging
	3:	re-opening → supporting → bound-opening → . . .

The first exchange consists of an opening and supporting move. The second consists solely of a challenging move. The third exchange starts with a re-opening move, and is followed by a supporting and bound-opening move. Whether the second exchange is really and exchange at all is questionable. Intuitively, an exchange should consist of more than one move, or otherwise be indicative of uncooperative dialogue. In D5, a cooperative listener would have responded to the speaker when challenging move was made. Indeed, the very lack of response was in some way informative. There is possibly some justification for proposing that the null move occurred here, thus creating an exchange with two moves.

Transactions and Interactions

Transactions and interactions are the final two levels of the classification hierarchy. Transactions consist of patterns of exchanges, and interactions consist of unordered strings of transactions.

A Sample Application

The dialogue extract shown in D5 consists of the following sequence of arts:

(Oel acct rep Ci Oel acct rep pr Ocl)

Note that it is not necessary to indicate the turns of the speakers, because in most cases a well-formed move represents a speaker's turn. The well-formedness constraints on what acts can make up a move automatically distinguished between turns. In other words, moves are equivalent to turns.

The ATN implementation provides three types of output: a parse structure of the act sequence, a trace of the parse through the network, a table of transition counts and probabilities, and a few simple statistics on the frequency of certain elements in the parse.

The following is a bracketed parse of the act sequence:

```
(int
    (tr
        (cx
            (op (head (act oel)))
            (su (prehead (act acct)) (head (act rep))))
        (cx (ch (head (act ci))))
        (cx
            (pi (head (act oel)))
            (su (prehead (act acct)) (head (act rep)))
            (pi (prehead (act pr)) (head (act oel)))) ))
```

As can be seen, the dialogue in D5 consists of one transaction, which itself consists of three conversational exchanges. The terms *pre-head* and *head* refer to the optional and compulsory parts of the move respectively. A move can also have an optional posthead, which is not shown in the above example.

A table of transition probabilities would not be very informative at this stage because one needs to parse a substantial dialogue to get a representative picture. This is our next task.

DIALOGUE ANALYSES

Normative Analysis

The Corpus

The dialogue used in this analysis is one of a number collected in an experiment to study the nature of dialogue between an expert and nonexpert user of a database system. Nonexpert users were given a problem to solve which involved accessing a student records database. He or she could enlist the help of an individual who was expert in dealing with such databases.

The particular dialogue chosen consisted of 192 utterances. It was chosen because the subject took the task to completion. The dialogue had a natural pattern of initiation, development, and completion. This was not always the case with the rest of the corpus. Often the task proved too difficult for subjects, and their dialogue had to be broken off abruptly.

Applying the Dialogue Grammar

Using the dialogue act categories described above, the selected dialogue was analysed into a sequence of acts. Some concern has already been expressed about the subjective nature of the categorization process (Toolan, 1985) and about its inherent circularity (Levinson, 1983). In general, little difficulty was encountered in determining the act-to-utterance correspondence. Where most of the problems lay was in determining what moves were taking place. However, since the identity of some moves determines the identity of the compulsory, or head, act (e.g., a Cel can only occur in a challenging move), the process of act and move assignment is not completely independent.

One of the problems encountered in applying the grammar to natural dialogue was the frequent occurrence of self-interruption, and indeed, interruption by the other dialogue participant. Take for example the following extract:

D6 Because it hasn't—I don't know—because it hasn't given a breakdown

In D6 do we treat the whole utterance as a reply (rep), or do we ignore the interrupted phrase "Because it hasn't—" and assume that the reply is "I don't know"? How then do we deal with the last part of the utterance? Is it part of the reply, and therefore a rep, or is it in addition to the reply and therefore a comment (com)? The approach adopted here was to try to ascertain the underlying intention of the utterance, and to use our knowledge of that to determine the category of act to which the utterance belonged. This solution echoes Levinson's complaint that we need to take account of the move to determine the category of the act. Similarly, by taking account of the speaker's intentions, we are having to refer to a wider context to disambiguate the act.

The utterances in D7 illustrate some other problems in applying the grammar. First, there is the problem of dealing with interruption. In a number of cases, an interruption takes place before a move is completed (a slash indicates an interruption).

D7 A1: So
B2: /So which one do you want? The whole file or
A3: /The whole file first.
B4: Fair enough To do that you are going to have to tell it to expect em . . . 100 cases.

For example, utterance A1 is interrupted before completion. From the context of the dialogue it looks like it might turn out to be an opening elicitation (Oel), but we can only code it as a supporting move of the form (acct null), where null is an empty act which can form part of a supporting move, but no other. Although we would have liked to code A1 as a marker followed by an elicitation (m Oel) in order to preserve the intentionality of the phrase, we have to draw the line at inferring too much beyond the linguistic data We can code B2 as (Oel com) and A3 as (rep). Neither of these utterances present any great problems. However, B4 is problematic. This brings us up against the problem of move/turn equivalence. The phrase *Fair enough* can be thought of as an acknowledgement (ack) of the preceding reply (rep). However, if the turn represented by B4 is to be equivalent to just one move, it can only be followed by a comment (com). But the rest of the utterance is obviously not a comment, it is an informative, which constitutes a new move by the same speaker. Alternatively, we could classify *fair enough* as a marker (m), the rest of the utterance as an opening informative (Oi), and then we would have just one move (a PI opening). Thus, preserving the move/turn equivalence. However, this solution ignores the acknowledgement function of the first part of the utterance. In coding the transcripts, we went for the solution which preserved the move/turn equivalence.

Quantifying Structural Properties of the Dialogue

If the dialogue grammar proposed here is to be of any use, it must be sensitive in some way to the variations in dialogue structure, both as it develops over time and while the participants are dealing with communication breakdown. There is evidence (for example, Anderson & Garrod, 1987) that the structure of a task-oriented dialogue changes as the two participants become familiar with each other's style of communication. An important aspect of task-oriented dialogue is that initially there is a complex interaction while the participants negotiate a common vocabulary with which to describe objects and operations involved in the task. As the dialogue proceeds the exchanges become more compact.

Given that we can obtain measures of dialogue structure, we are in a position to suggest a number of hypotheses which might be tested using various structural indices. For example, with regard to the structural feature of *compactness*, which has been introduced at an intuitive level only and which we would link with other features such as *density*, *information-load*, and so on, it would be important to find out whether the particular feature we have in mind could be modeled by simple measures such as the number of exchanges per transaction, or perhaps the number of moves per exchange. To explore the construct validity of such measures, we computed them for beginning, middle, and end sections of the dialogue, arguing that dialogue should be more diffuse towards the beginning and gradually become more compact as it proceeds. Early on in a dialogue we would expect a relatively large number of requests for clarification and information, while there would be a decrease in these as the dialogue proceeded. Such requests for clarification entail shifts in topic, and therefore would be manifested as increased numbers of exchanges per transaction early on in the dialogue. As the dialogue preceeded, the number of exchanges per transaction would decrease.

Similarly, and index based on the number of moves per exchange, or the length of an exchange expressed in some other terms, such as duration, could be expected to capture important aspects of compactness, thus providing a useful instrument for shadowing changes in compactness in the course of dialogue. (We would anticipate that certain kinds of go-ahead signals would be linked to satisfactory levels of compactness, and its reciprocal, redundancy.) The viability of these measures also can be tested by computing them for beginning, middle, and end sections of dialogue, since compactness, insofar as it can be understood at an intuitive level, is certainly a property which should increase as the dialogue proceeds. There is large amount of information being imparted to the nonexpert by the expert at the beginning of the dialogue, increasing the number of moves per topic, which in turn is closely related to exchange units. This would simply be because, at each stage in the expert's elucidation, the nonexpert would be providing go-ahead signals to the expert. The amount of information being imparted as the dialogue proceeded would decrease. There would be a comensurate decrease in the number of moves per exchange.

Measures of compactness based on the frequency of challenging move can also be considered. We expect to find large number of challenging moves at the beginning of the dialogue, since they often have the function of reducing the density of information, and that this number should decrease as the dialogue progressed. Conversely, measures based on the frequency of supporting moves might be expected to show an inverse relationship with density.

If, as we are suggesting, dialogue shows a systematic change in its density or compactness as it proceeds, this is a phenomenon which might also be captured in the frequency of occurrence of particular acts. We should expect to find a large number of elicitations, for example, at the beginning of the dialogue, with the specific function of preventing a high level of compactness, which then decrease as the dialogue proceeds. For the same reason, informatives, and in particular challenging informatives, should also be more frequent at the beginning of the dialogue than elsewhere.

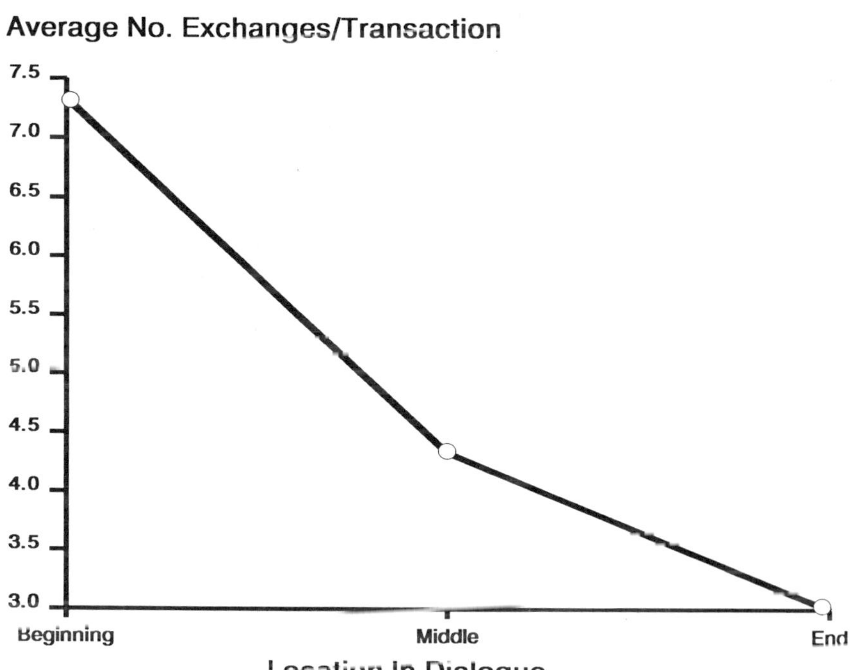

Figure 2. Average Number of Exchanges Per Transaction in Sample Dialogue Broken Down by Location

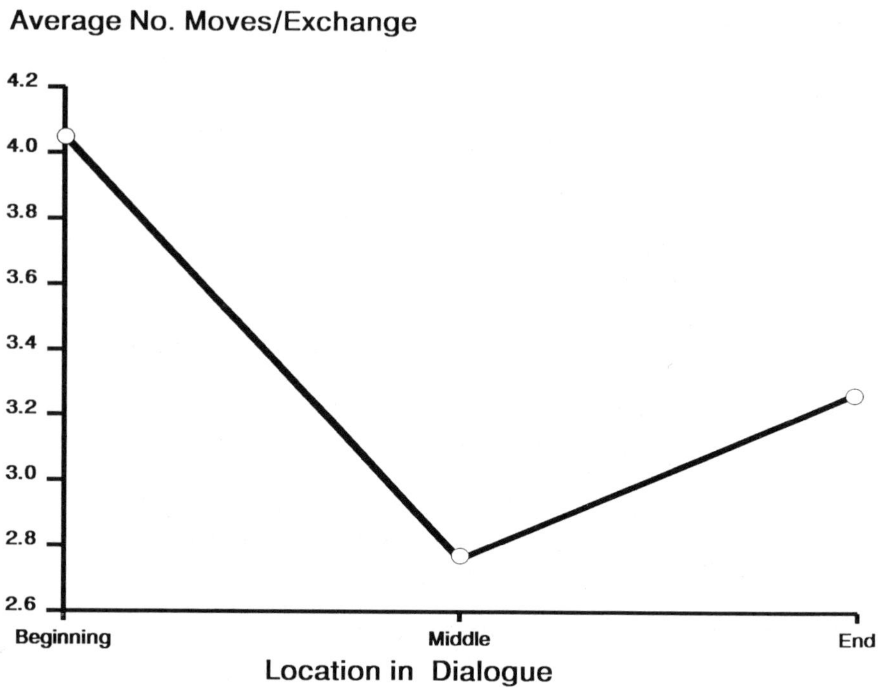

Figure 3. Average of Number of Moves Per Exchange in Sample Dialogue Averaged Over Transaction and Broken Down by Dialogue Location

Some Exploratory Analyses of the Data

the figures presented in Tables 1 and 2 support the initial plausibility of such measures as indicators of one aspect or another of the structural property of dialogue compactness. In order to allow a comparison between different stages in the dialogue, the act sequence derived from the analysis of the transcript was split into thirds. Altogether, there were 312 acts. The beginning, middle, and end act sequences consisted of 105, 106, and 101 acts, respectively. We then computed the measures mentioned above, using the output from the ATN-based parser. It can be seen from the figures in Tables 1 and 2, and their graphic representations in Figures 2 and 3, that there is

Table 1. Average Number of Exchanges per Transaction from Analysis of Sample Dialogue

Location	No. of Trans	Mean Exs	SD
Beginning	3	7.33	4.04
Middle	6	4.33	3.88
End	7	3.00	2.71

Table 2. Average of the Average Number of Moves per Exchange from Analysis of Sample Transcript

Location	No. of Trans	Mean Moves	SD
Beginning	3	4.05	1.40
Middle	6	2.76	0.97
End	7	3.27	2.51

moderate support for the proposed measures of compactness. Generally they show the kind of change from beginning to end sections of dialogue which we would be expected of such measures. Unfortunately, neither of the trends were statistically significant.

As well as a set of simple descriptive statistics, the ATN provides a table detailing the number of transitions made over each arc and the probability of transition for each arc given a specific node. Indices based on the occurrence of specific acts can be readily computed from the table.

As can be seen from Figure 4, there appears to be no variation in the number of either opening, supporting or challenging moves across the three stages of the dialogue. The most frequently occurring move at all stages in the dialogue is the supporting move, followed by the opening and challenging moves, respectively. The frequencies for each move and act at the different dialogue stages are given in Table 3.

Figures 5 to 7 look in more detail at the act structure of the moves, and at their variation through the three stages of the dialogue. Note that in these figures we only show the compulsory acts (heads) which had nonzero frequencies at all stages of the dialogue. The graphed data are given in Table 3. Within the supporting moves at the transition from beginning to middle stages there is an apparent shift in the dominant form of support from replies

Table 3. Frequencies of Main Moves and Act Heads by Dialogue Location

		Dialogue Location		
Move	Act Head	Beginning	Middle	End
Opening	Oi	5	11	10
	Oel	13	14	11
	Total	21	25	22
Supporting	ack	12	21	21
	rep	22	15	15
	Total	41	39	39
Challenging	Ci	4	9	3
	Cel	11	5	5
	Total	15	15	8

Note: The the act totals are not always equal to the sum of the act counts because of low frequency acts which are not shown in the table.

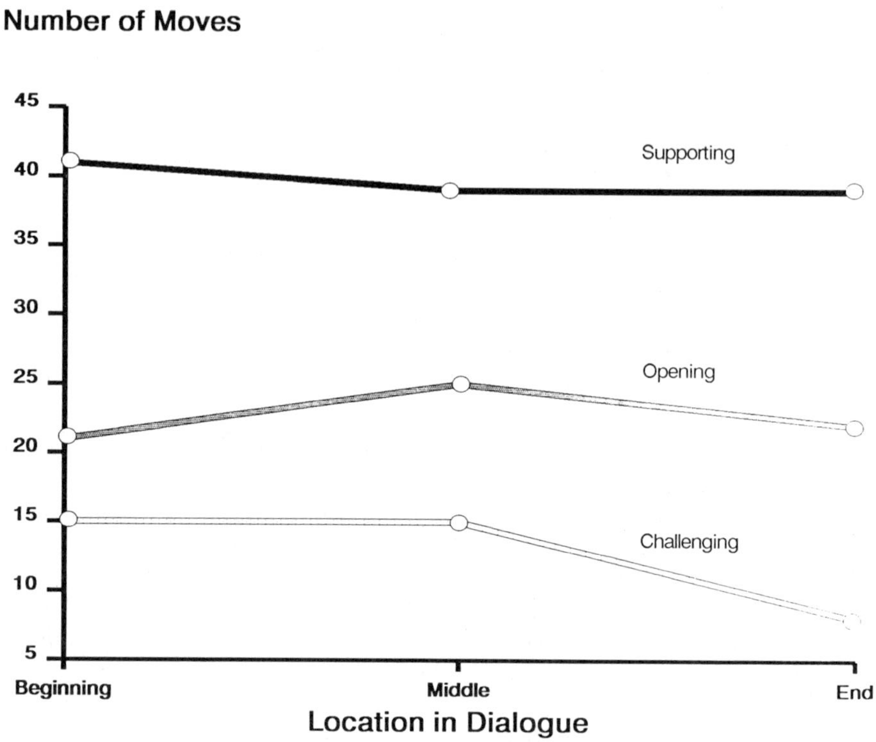

Figure 4. Breakdown of Move Type by Dialogue Location

(rep) to acknowledgements (ack) as the dialogue proceeds. This is perhaps not too surprising. Early on in the dialogue, as the expert is responding to the nonexpert's requests for information, he provides extensive amounts of information. As the nonexpert becomes more proficient in dealing with the database, the amount of feedback information required, and supporting moves, degenerate to simple go-ahead signals realized by a small class of linguistic response (e.g., *OK*, *yes*, *alright*). The change in pattern is marginally statistically significant (Chi-square = 4.91; df = 4; $.10 > p > .05$).

The two most frequent acts within the opening moves (the data from PI and ordinary openings are combined in Figure 6) are the informative (Oi) and the elicitation (Oel). Elicitations are the commonest form of opening across all stages of the dialogue. However, informative openings appear to be less frequent at the beginning of the dialogue, than at latter stages. This might be due to the nonexpert being less in control of the direction of the dialogue in its early stages, and taking more control as it proceeds, and as she becomes more focused on solving the problem she was set. However, this change in act frequency is not statistically significant.

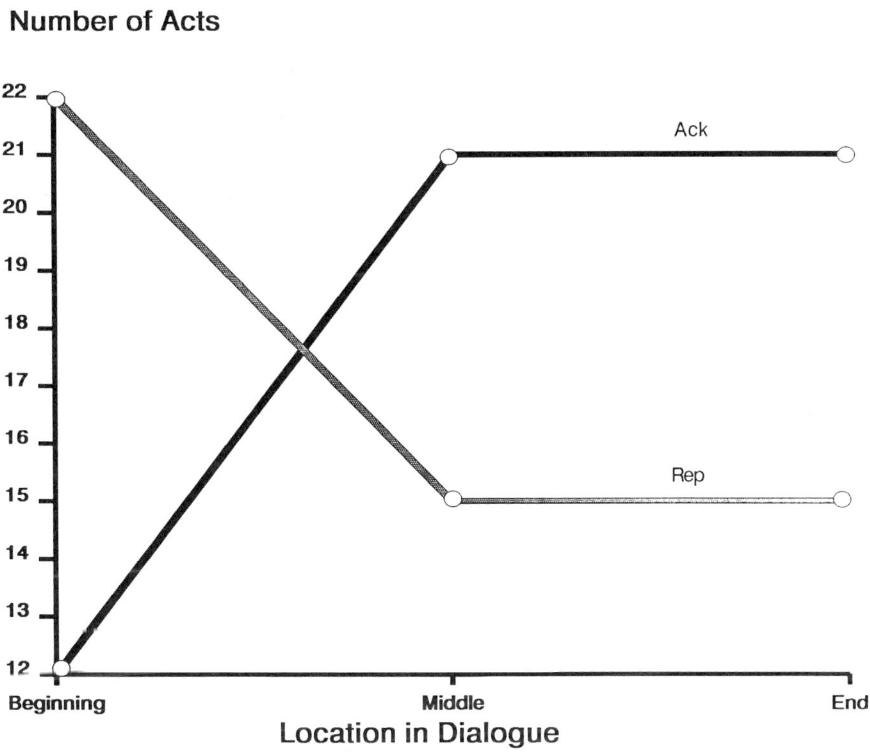

Figure 5. Breakdown of Main Acts in Supporting Moves of Sample Dialogue by Dialogue Location

The pattern of the act structure within challenging moves (Figure 7) varies considerably from stage to stage of the dialogue. The two most frequent challenge acts are, as with the opening move, elicitation (Cel) and informative (Ci). At the beginning of the dialogue the most common act is the elicitation, at the middle stage it is the informative, and at the end of the dialogue the pattern reverts once more to roughly what it was at the beginning, although there is an overall decrease in the number of challenging moves. This pattern is marginally significant (Chi-square = 4.31; df = 2; $.10 > p > .05$). At the first two stages the pattern mirrors that found for the supporting moves. This reflects a reciprocal relationship between elicitations and replies, and between informatives and acknowledgements. In hindsight, the trailing off of challenging moves as a whole is perhaps to be expected in a dialogue in which there is an increasing degree of agreement between both participants over the goals of the dialogue and how best to achieve them.

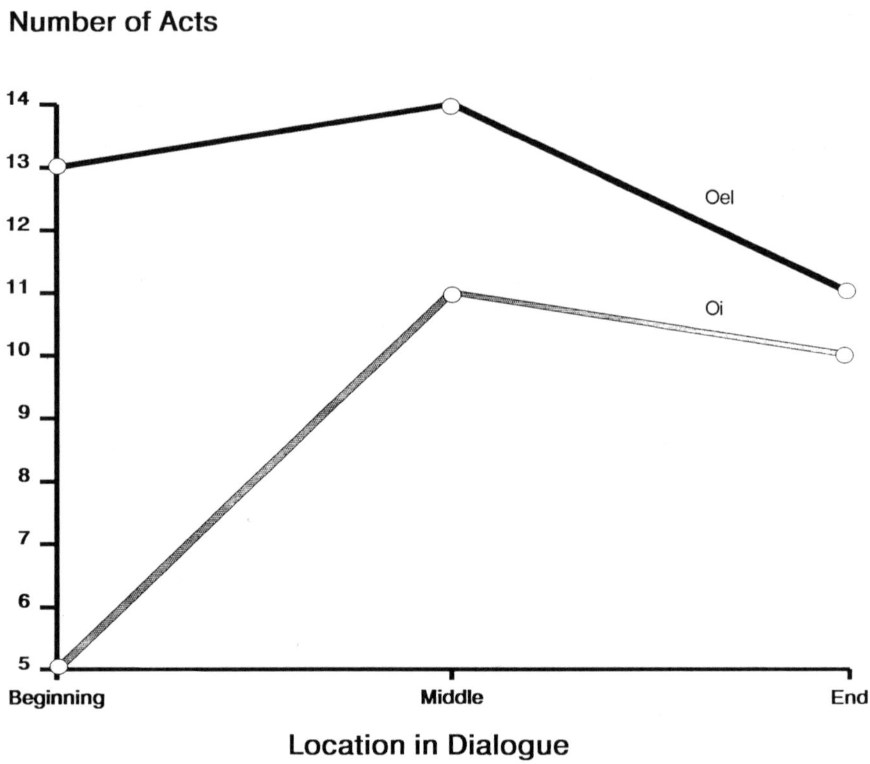

Figure 6. Breakdown of Main Acts in Opening Moves of Sample Dialogue by Dialogue Location

Normative Conclusions

A number of our initial hypotheses were not borne out by our analysis, but in interesting ways. For example, we expected to find a general decrease in the number of elicitations as the dialogue progressed. However, this was only true of challenging elicitations, and not opening ones. The reason for this might be because challenging elicitations represent the interruption of a topic because something has been misunderstood, whereas opening elicitations represent a smoother transition in topic. One expects, therefore that the number of misunderstandings should decrease as the dialogue proceeds, but that the number of normal topic shifts should remain fairly constant. The prediction that informatives should be more frequent at the beginning of the dialogue than at the end is not borne out. In fact, both opening and challenging informatives reach a peak at the middle of the dialogue. This increase probably represents, as already mentioned, the contribution of the nonexpert to the control of the dialogue, as she focuses on solving the problem.

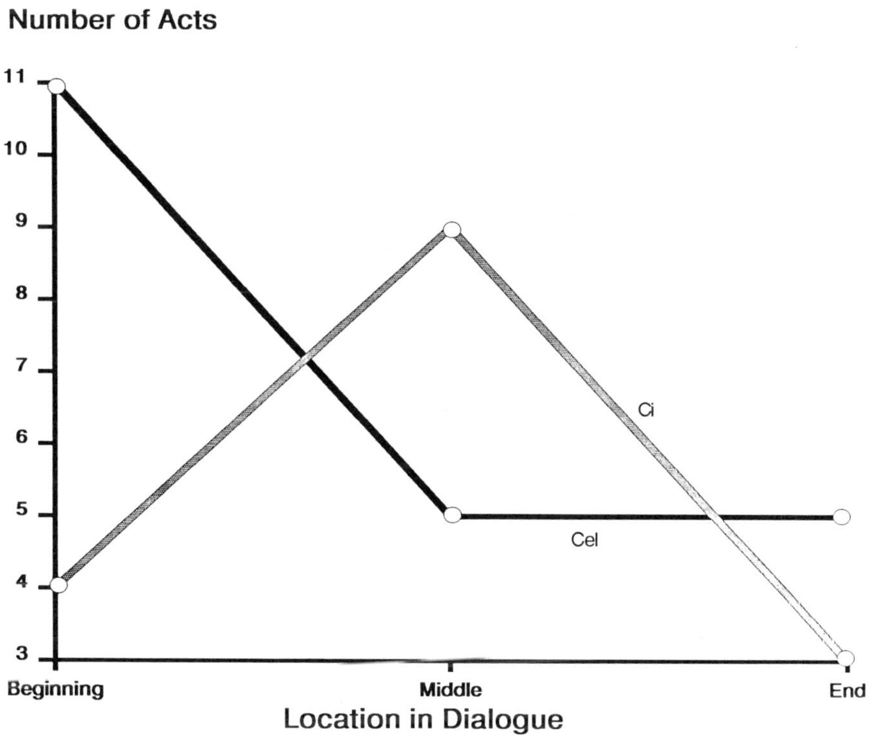

Figure 7. Breakdown of Main Acts in Challenging Moves of Sample Dialogue by Dialogue Location

This suggests that global indices based on act-frequencies may not reveal any importance features of dialogue. Their importance, rather, resides in their relationship with the moves which they implement, and very likely it is only those statistics which acknowledge this interdependence which will gain access to the casual structure of dialogue. We are thinking in particular of contingency statistics, reporting, for example, the frequency of a particular act, given that the move is a challenge, a support, or whatever.

Miscommunication Analysis

The Corpus

In addition to a straightforward normative analysis, we were also interested in finding out whether the various categories and statistics produced by the dialogue analysis might show differences between normative passages of dialogue, and passages of dialogue in the process of failure. We therefore applied the dialogue grammar to passages of failing dialogue to see how best the phenomenon of failure might be conceptualized and quantified within the terms of the grammar.

A number of short extracts from other dialogues in the database experiment were examined for occurrences of miscommunications. The criterion for inclusion as an example of a dialogue miscommunication was an acknowledgement by one or other participant that a misunderstanding had taken place. As a result, the set of examples may be unrepresentative of miscommunications as a whole. In particular we leave out miscommunications which minimally disrupt the flow of dialogue, such as those which are corrected by a request for clarification of a previous utterance. However, failures that admit of nondisruptive correction are generally input level failures (Ringle & Bruce, 1982) and, as such, are more tractable. Model failures are more problematic, and, when detected, more disruptive. They are, therefore, the primary focus of this analysis.

In each example, the section of dialogue extracted started at the point where the failure-inducing topic was first introduced, and ended where the failure was resolved. Eight dialogue examples of this sort were collected and analyzed.

Quantitative Aspects of Miscommunication

Several summary dialogue structure statistics were presented in the previous section, where we concluded that they appeared to capture at least some structural properties of discourse, such as the transition from beginning to end sections. In this section we compute the same statistics and apply alternatively to normative and failing passages of dialogue. As in the earlier section, our intention is not to test specific hypotheses about the discourse structure of failed dialogue, but rather to look for indications that the measures proposed are workable.

We decided on the normal dialogue transcript as the source of a control sample of dialogues. More accurately, we used as a source the act analysis of the transcript. The eight sample miscommunications were analysed. These produced act sequences of varying length. Eight samples were then taken randomly from the normal act sequence and were matched for their act length with the miscommunication examples. The control and miscommunication examples were then input to the ATN program. Because of the short length of the miscommunication dialogues, the statistics on numbers of exchanges per transaction and average number of moves per exchange were not considered reliable. This analysis will concentrate, therefore, on the tabular output of arc transition frequencies. Arc counts were accumulated over all eight examples and all eight controls, yielding two tables.

Move structure. Table 4 gives the comparative distribution of the three main move types—opening, supporting, and challenging. No strong pattern

Table 4. Comparison Between Miscommunication Examples and Controls in Terms of Counts of the Main Moves

Move	Examples	Controls
Opening	50	47
Supporting	65	63
Challenging	4	14

is evident, apart from the fact that miscommunication appears to be characterized by a reduction in challenging moves. The drop in challenging moves produces a marginally significant Chi-square (Chi-square = 5.58; df = 2; $.10 \rangle p \rangle .05$).

One possible explanation for this apparent relationship between miscommunication and a decrease in challenging moves could be that challenging moves serve an important role in preventing misunderstanding occurring. There is an increased likelihood of miscommunication when there is a reduction in challenges, for whatever reason.

Act structure. The three main move-types in both the example and control dialogues were opening, supporting and challenging moves. Each of these moves can be made up of a series of acts, some of which are optional, and one of which is compulsory. Acts which are compulsory are called the head of the move. This analysis of the act structure will concentrate on the two most frequent act heads for each move. A table of the data with which we will be concerned is given in Table 5. The data are graphed in Figures 8 to 10.

If we take the opening move first, the graph in Figure 8 suggests that there is a decrease in opening elicitations and an increase in opening informatives in the miscommunication examples. If we assume that elicitations come primarily from the nonexpert, and informatives from the expert, it could be that miscommunication occurs when there is a high rate of information transmission from the expert to the nonexpert, and fewer checks being

Table 5. Comparison Between Miscommunication Examples and Controls in Terms of Most Frequent Act Heads in Main Moves

Move	Act Head	Examples	Control
Opening	Oi	21	15
	Oel	23	27
Supporting	ack	29	25
	rep	22	31
Challenging	Ci	2	4
	Cel	2	9

made on the information by the nonexpert. Unfortunately, this trend is not statistically significant.

The idea that high information flow and a consequent reduction in information checking precipitates miscommunication is also suggested by the frequency pattern of the commonest act heads within the supporting move. Figure 9 shows that acknowledgements are more frequent than replies in the examples, whereas the reverse is the case in the control dialogues. Acknowledgements consist of a closed class of items, and they follow an informative. Replies follow an elicitation, from which it follows that informatives may be more common in the lead up to dialogue failure.

Finally, there is a suggestion that the miscommunication sample contains a smaller number of challenges than normal dialogue. As can be seen from Figure 10, there is an overall drop in the number of challenging moves in the examples. However, the biggest drop occurs for the challenging elicitation category. Since the challenging elicitation is more likely to be a move of the nonexpert than the expert, this finding is suggestive of a falling off in information checking prior to dialogue failure.

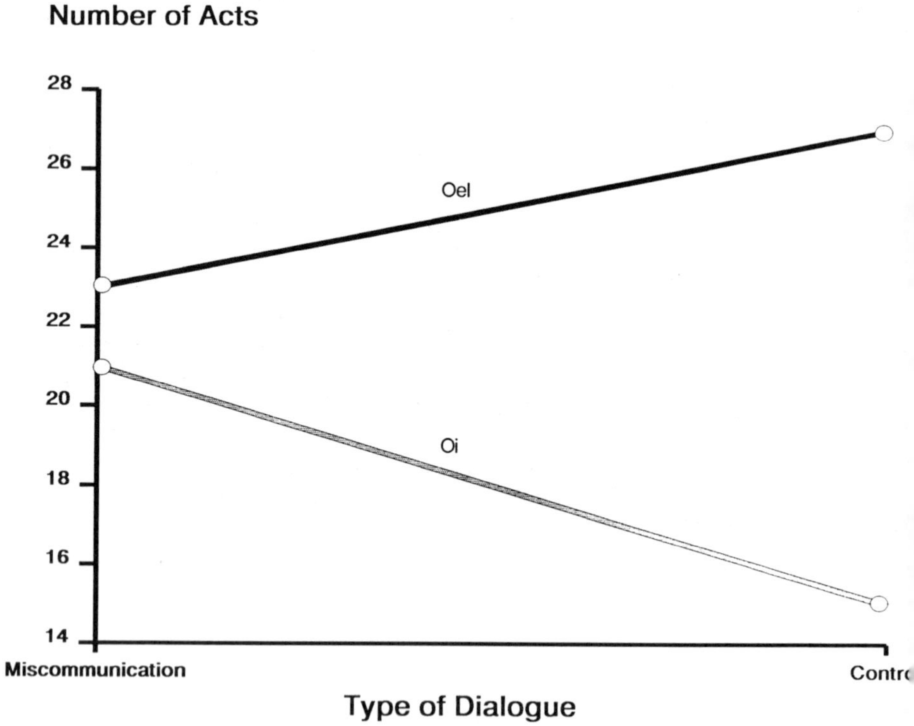

Figure 8. Comparison Between Miscommunication and Control Dialogues in Main Acts of Opening Move

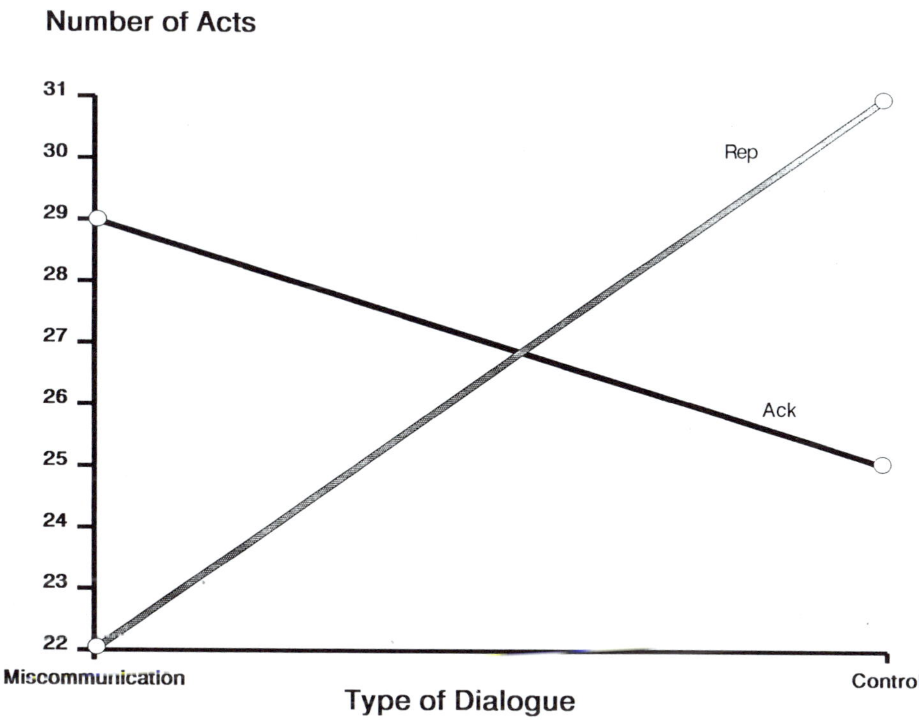

Figure 9. Comparison Between Miscommunication and Control Dialogues in Main Acts of Supporting Move

Miscommunication Conclusions

In conclusion it seems that the structural analysis of dialogue, as implemented in the ATN grammar, is sensitive to the structural changes that precipitate communication failure. The characteristics of a dialogue that are likely to give rise to communication failure are a high rate of information transmission and a low rate of information checking.

GENERAL SUMMARY

Three main aims motivated this chapter. First, it was intended to provide a reasonably complete structural description of natural language dialogue within an ATN framework. It was then hoped to show that the grammar was sensitive to the dynamic structural variations that occur during the time course of a dialogue. Finally, it was intended to show that aspects of the structural description were sensitive to those characteristics of dialogue that indicated or precipitated miscommunication. A number of empirical hypoth-

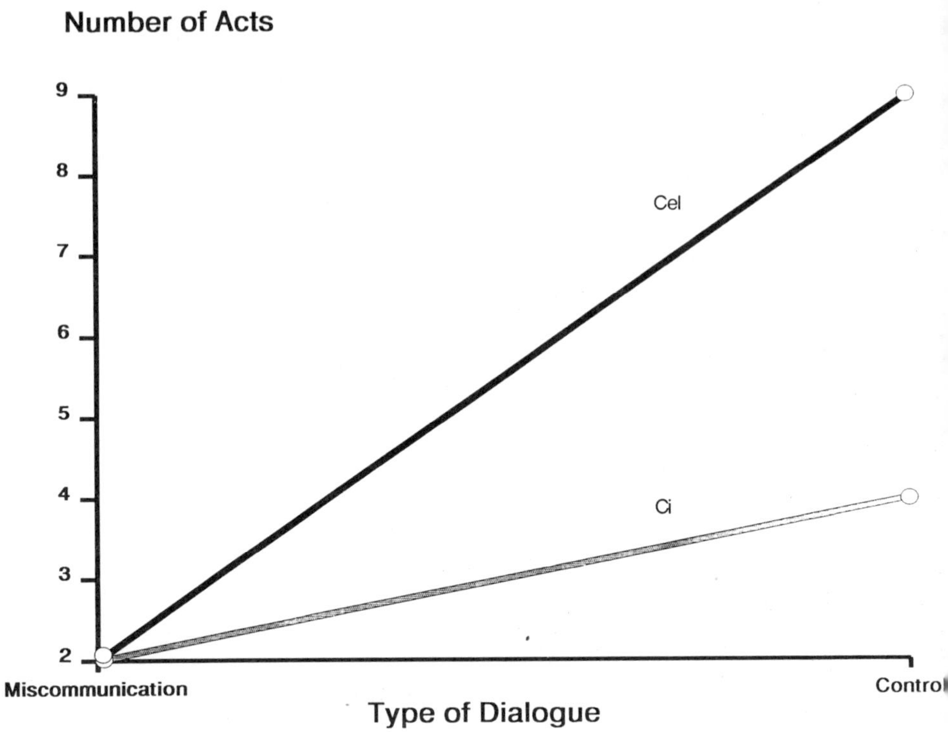

Figure 10. Comparison Between Miscommunication and Control Dialogues in Main Acts of Challenging Move

eses were generated to test the properties of the grammar and to ascertain the degree to which the aims outlined above were achieved. The results are summarized below.

The trend in the number of exchanges per transaction supported the hypothesis that there should be a decrease in transaction complexity as the dialogue progressed. However, none of the differences between the means reached statistical significance, mainly because the number of transactions involved in the averaging was too small to yield a reliable set of differences. The pattern of results from the analysis of the number of moves per exchange showed no discernible or significant trend.

Regarding the move- and act-based hypotheses outlined above, none appeared to have been borne out. However, analysis of the act structure of the moves did reveal some unexpected findings. We expected that there would be a general decrease in the number of elicitations as the dialogue progressed. However, this was only true of challenging elicitations, and not opening ones. The reason for this was probably that challenging elicitations represent the interruption of a topic when something has been misunderstood, whereas opening elicitations represent a smoother transition in topic. One would expect, therefore, that the number of misunderstandings should decrease as the dialogue proceeds, but that the number of normal topic shifts should remain fairly constant. The prediction that informatives should be more frequent at the beginning of the dialogue than at the end, is not supported. In fact, both opening and challenging informatives reach a peak at the middle of the dialogue. This increase probably represents the contribution of the querant to the control of the dialogue as she focuses on solving the problem.

In order to test the discourse model for sensitivity to variations in dialogue structure that occur during communication breakdown, we need to be able to compare the miscommunication examples with some kind of control sample of dialogue. An analysis of the act sequence from the eight sample miscommunications was compared with eight sample sequences taken randomly from the act sequence representation of the main dialogue sample. The sample was matched for act length with the miscommunication examples. ATN arc counts were accumulated over all eight examples and all eight controls, yielding two tables.

Taking opening moves first, there was a decrease in opening elicitations and an increase in opening informatives in the miscommunication examples. If we assume that elicitations come primarily from the querant (Q), and informatives from the oracle (O), it could be that miscommunication occurs when there is a high rate of information transmission from O to Q, and when fewer checks are being made on the information by Q.

The idea that high information flow and a consequent reduction in information checking precipitates miscommunication gets further confirmation when we examine the frequency pattern of the commonest act heads within the supporting move. Acknowledgements are more frequent than replies in the examples, whereas the reverse is the case in the control dialogues. Acknowledgements consist of a closed class of items, and they follow an informative. Replies follow an elicitation. Therefore, informatives must be more common close to, and during, miscommunication.

Finally, the most common act heads within the challenging move show a distinct pattern of change during miscommunication. There is an overall drop in the number of challenging moves in the examples. However, the biggest drop occurs for the challenging elicitation category. Since the challenging elicitation is more likely to be a move of Q rather than O, this finding is consistent with a dialogue in which there is a reduction in information checking.

In summary, the ATN discourse model appears to be sensitive to the structural changes in the dialogue that precipitate communication failure. The characteristics of a dialogue that are likely to give rise to communication failure are a high rate of information transmission and a low rate of information checking.

REFERENCES

Anderson, A., & Garrod, S.C. (1987). The dynamics of referential meaning in the spontaneous conversation: Some preliminary studies. In R. Reilly (Ed.), *Communication failure in dialogue and discourse.* (pp. 161–183). Amsterdam, Netherlands: North-Holland.

Burton, D. (1981). Analysing spoken discourse. In M. Coulthard & M. Montgomery (Eds.), *Studies in discourse analysis* (pp. 61–106). London: Routledge & Kegan Paul.

Coulthard, M., & Brazil, D. (1981). Exchange structure. In M. Coulthard & M. Montgomery (Eds.), *Studies in discourse analysis* (pp. 82–106). London, England: Routledge & Kegan Paul.

Levinson, S.C. (1983). *Pragmatics.* London: Cambridge University Press.

Ringle, M.H., & Bruce, B.C. (1982). Conversation failure. In W.G. Lehnert & M.H. Ringle (Eds.), *Strategies for natural language processing* (pp. 203–221). Hillsdale, NJ: Erlbaum.

Sinclair, J., & Coulthard, M. (1975). *Towards an analysis of discourse.* London: Oxford University Press.

Toolan, M. (1985). Analysing fictional dialogue. *Language and Communication, 5,* 193–206.

CHAPTER 3

Situation Models and Information Accessibility*

DANIEL MORROW
Stanford University

STEVEN GREENSPAN
AT&T Bell Laboratories

The same situations can be described in many ways. Particular objects, actions, and locations may be highlighted in one description but not in another. The way in which a writer describes a situation reflects the perspective from which the writer regards the situation. This perceived spatial and temporal organization of a situation is conveyed through a variety of linguistic cues against a background of shared knowledge. The reader uses semantic and grammatical information to reconstruct the representation of the situation, which includes the perspective on the situation as well as the objects and actions of the situation.

During comprehension, readers can only focus their attention on part of the information conveyed by a narrative. They only process a small segment of the narrative at any particular moment, since narratives are read in a linear fashion. Furthermore, they do not have all of the information that is important for interpreting the narrative immediately available in memory, since only a small part of the preceding narrative remains readily available in active or working memory (Kintsch & van Dijk, 1978; Sanford & Garrod,

*While writing the present chapter, Daniel Morrow was supported by National Institute of Mental Health postdoctoral training grant #MH 15157–08, and by NIMH grant #13950–16 to Gordan H. Bower at Stanford University. Steven Greenspan was supported by NIMH postdoctoral training grant #T32 NS–07134–06, and by grant NIH–NS–12179 to David Pisoni at Indiana University. Steven Greenspan is currently at ATT Bell Laboratories, Napierville–Wheaton Rd., Napierville, Illinois, 60566.

1981). The problem of availability or accessibility of information is compounded by the fact that a narrative often implies more information than it explicitly states. Thus, understanding must be highly inferential, since readers use large amounts of knowledge about described situations to understand a narrative. If it is the case that, at any point in comprehension, some information is more accessible and prominent in processing than other information, it is also often the case that the writer only intends part of the information to be most relevant. From a particular perspective, some objects, actions, and locations are more important or relevant to the described situation. The reader must attend to the relevant parts of the situations so that they become readily accessible from memory during comprehension. If the writer indicates the relative importance of objects and actions in situations from a particular perspective, the reader's task of reconstructing the writer's representation will be greatly simplified.

The relationship between information accessibility and relevance is illustrated by the following passage.

1. The fireman rushed down the street, past the bank and into the warehouse. He worked swiftly and then as the smoke cleared, he noticed an unconscious child lying on the floor in front of him. He knelt beside him and felt his pulse.

Comprehending this passage requires inferring that the fireman put out a fire in the warehouse, probably with a hose, which in turn requires accessing information about fires and firemen. Therefore, a large amount of knowledge may be accessible, because it is potentially relevant to interpreting the passage. Thus, readers must identify which information is not intended by the writer as relevant to interpreting the discourse. For example, in the situation described by the first sentence of the passage, the bank and street are not as relevant as other parts of the situation, since they do not require elaboration via inferencing (substituting the noun phrase *the empty lot* for *the bank* has little effect on the interpretation of the narrative). These decisions depend on adopting a perspective on the situation. Moreover, interpreting expressions such as *he* and *in front of him* requires understanding the situation from a perspective. The reader must be aware that, during the described action, the fireman is located inside the warehouse, that the actions taking place within the warehouse cannot be seen from outside of the warehouse, and, therefore, that the situation is described from a perspective that is at, or near, the location of the fireman. In short, the passage must be understood from a particular perspective, and this perspective in turn helps indicate which part of the situation is most relevant and should be focused on to elaborate via inferencing.

In the present chapter, we examine how accessibility or prominence of information during comprehension is determined by properties of described situations. These properties specify which information is most relevant to the narrative. We examine how this relevance is linguistically signalled in order to guide comprehension. Our approach combines a psychological account of discourse processing and representation with constructs from linguistics, rhetoric, and artificial intelligence. Unfortunately, terms relating to information accessibility, such as *focus* or *foreground*, have been used in various ways by researchers in different traditions. Therefore, it is important to first provide a brief outline of these other approaches and to distinguish our use of the terms from theirs.

Accessibility in Other Theories of Discourse Processing

Artificial Intelligence.

Many artificial intelligence theories of discourse understanding have recognized that understanding requires background knowledge about the objects and actions mentioned in the discourse (Charniak,, 1972; Minsky, 1975; Schank & Abelson, 1977). These theories have helped explain how implicit information in a discourse is inferred in order to construct a coherent interpretation of the discourse. However, if the ability to integrate information is limited by the capacity of working memory, then the amount of background information and inferencing used to interpret the discourse must also be limited. Example 1 shows that not all of the information in a narrative is equally relevant (i.e., knowing about banks does not contribute to understanding the actions of the fireman). Ideally, background knowledge about objects and actions that are not very relevant to the discourse should have a negligible role in inference making and other processes. AI approaches made the problem of information accessibility more manageable by assuming that background knowledge is structured into units organized around objects and actions. However, in these representation theories, there was little discussion of how readers identify and keep track of relevant information during comprehension.

Theories of referent assignment faced the problem most directly. They sought to explain how readers choose referents for definite noun phrases and pronouns when the referent is not uniquely determined by syntactic and lexical constraints (e.g., Hirst, 1981). Consider example 2.

2. John went with Harry to Joe's house. He felt happy.

Who is referred to in the second sentence? Syntactic and lexical factors do not determine whether John, Harry, or Joe is the referent. Several AI approaches proposed that at any point in the discourse, a single candidate, or small set of candidates, are in *focus*. Focus, in these theories, referred to the entities mentioned in a discourse that are most accessible from the discourse representation at a particular point in the discourse. Thus the theories equated focus with accessibility, that is, a state of processing. The construct was developed in order to minimize the computational cost of referent assignment by reducing the number of candidates to be searched and the amount of inferencing needed to check the appropriateness of each one (Grosz, 1981; Sidner, 1984).

Earlier AI models relied on recency for choosing referents (Winograd, 1972), thereby assuming that focus is determined by temporal properties of the text: Recently mentioned information is more active in memory and thus more accessible. Later theories emphasized that understanders actively direct their attention to parts of the world conveyed by the text in order to build a coherent representation one piece at a time. In this view, focus is determined by the structure of the situations described by discourse. However, focus was identified with only one aspect of the described world: the most recent topic of discourse. *Topic* has been variously defined by linguists and psychologists, but it is usually equated with what is being talked about in a sentence or discourse. For sentences, the topic is often the first mentioned entity in a sentence (e.g., Reinhart, 1982). For example, John is the topic in sentence 3a, and Joe's house is the topic in sentence 3b.

3a. John went with Harry to Joe's house.
3b. Joe's house was where John went with Harry.

For discourse as a whole, the topic is often a character or a whole situation, such as John's terrible day at work (van Dijk, 1979).

Analyses of discourse in AI often assume that at any point in the discourse, entities serving as the topic are most focal. Grosz (1981) examined dialogues in which an expert told a novice how to assemble a water pump, and argued that the focus was the current step in the task, or the current topic as defined by the task structure. In analyzing conversations, Reichman defined focus as the current topic of conversation (Reichman, 1978). In analyzing apartment descriptions and task-oriented conversations, Linde assumed the topic was the last mentioned major discourse unit, either major rooms in the apartment descriptions or goals and plans in conversations (Linde, 1979; Linde & Goguen, 1978). Sidner (1984) distinguished *discourse focus* and *agent focus*. Discourse focus is the topic of the whole discourse (or the topic of a section of discourse) and tends to occur as the object of the verb. The agent focus tends to be the actor of the sentence, usually mentioned as subject.

By identifying focus with topic, these theories assume that information accessibility is primarily determined by a single aspect of the described world, and this assumption leads to an oversimplified theory of how focus is signalled. Reichman (1978) points to the importance of cue words such as *by the way* and *meanwhile* for signalling a switch away from, or back to, the topic (the information in focus). Sidner argues that semantic and syntactic properties of sentences signal focus. It is signalled by syntactic constructions such as *there*-insertion sentences and by grammatical roles associated with verbs, the agent focus usually being associated with sentence subject and the discourse topic with the verb object.

Equating topic and focus also oversimplifies the role of accessibility in discourse processing and representation. In order to integrate focus into a theory of discourse representation and the procedures for constructing the representation, the AI theories have construed focus as the active part of a partitioned semantic network (Grosz, 1981; Robinson, 1981), a moving point of attention associated with the current node being added to a hierarchically organized network (Linde & Goguen, 1978), or the position of a concept in a push-down stack (Sidner, 1984). In all these cases, focus seems an ad hoc addition to representational theories developed for other purposes.

Psychological Theories of Discourse Processing.

Psychological theories of discourse comprehension borrowed the notion of focus from artificial intelligence and linguistics to help explain how a discourse representation is constructed during understanding. They assumed that, at any moment in processing a text, part of the previously constructed representation is most active in working memory and thus is more accessible and influential in processing the text than other parts of the representation. The highly active parts of the representation are considered to be focal (Carpenter & Just, 1977; Kintsch & van Dijk, 1978; Sanford & Garrod, 1981; Engelkamp & Zimmer; 1983). Thus it is very similar to the notion of *foregrounding* as used by Chafe (1972, 1974). Focus has been represented either as the contents of working memory (Sanford & Garrod, 1981; Kintsch & van Dijk, 1978), or as the currently activated region of an associative network (Sharkey & Mitchell, 1985; Sharkey & Sharkey, 1987).

Engelkamp and Zimmer (1983) emphasize that activation is not an all-or-none mechanism, that portions of a representation can be activated to varying degrees. The degree of activation should be related to its relevance at a particular point in the text. Moreover, they argue that concepts can be unactivated, activated to various degrees, and that one or more of the concepts can be the focus of attention, which is the most activated part of a representation. They suggest that the theme (Halliday, 1967), or topic (Chafe, 1976; Reinhart, 1982) of a sentence constitute a particular type of focus. Although nontopicalized concepts can be focal (in addition to the topic), the topic is the prototypic focus of attention.

In summary, both AI and psychological theories of discourse processing equate focus with information accessibility. Moreover, they often identify focus with topic (but see Engelkamp & Zimmer, 1983), with AI theories stressing global topic and psychological theories stressing sentence topic. They both stress the role of recency of mention in the text for establishing an entity as focal. Thus they assume focus is strongly influenced by order of mention (e.g., Clark & Sengul, 1979; Lesgold, Roth, & Curtis, 1979), or by both order of mention and topic status (Kintsch & van Dijk, 1978; Sanford & Garrod, 1981). The reliance on recency of mention suggests that readers focus on the temporally ordered text itself more than the described situations.

These theories correctly point to the critical role played by the dynamics of information accessibility in constructing a discourse representation. However, they fail to explain the dynamic nature of discourse comprehension because they only associate focus with accessibility of information about the text rather than about the situations conveyed by the text (e.g., Clark & Sengul, 1979), or they identify focus with a single aspect of the described world: the topic or theme (e.g., Grosz, 1981; Sanford & Garrod, 1981). By itself, discourse topic only informs the reader at a general level about the situations described by a discourse. Therefore, the notion of topic is too undifferentiated to adequately explain the dynamics of information accessibility during comprehension. For example, the topic of the passage presented in the introduction might be *A fireman fights a fire*, but this does not explain the difference in relevance between the bank and the inside of the warehouse. Because accessibility is identified with topic in AI and psychology, it has been readily assimilated to a number of diverse representational schemes, rather than helping to constrain a theory of discourse representation. Similarly, the theories do not adequately explain how writers and speakers linguistically indicate which parts of the situation that they are describing are most relevant to the discourse, and therefore how readers know which information should be most prominent and accessible as the readers construct the discourse representation. Many of these drawbacks can be eliminated by analyzing accessibility within a broader framework that distinguishes different aspects of discourse that are critical for determining relevance.

A Framework for Investigating Comprehension: Communication and Situation

Levels of Discourse.

It is useful to distinguish three broad levels of discourse production (Chatman, 1978; Morgan & Sellner, 1980; Brewer & Lichtenstein, 1981). The first level concerns the *content* of the message speakers or writers want to communicate. This content is a representation of the situations they want to describe. Because writers think about the situations from a particular vantage point (in terms of attitude, spatial location, or both), perspective is part of the content of a message.

The second level of discourse involves the *presentation plan* by which a situation representation is mapped into a discourse. Writers must decide which aspects of the situations to describe, and in how much detail. Furthermore, they must decide the order in which to describe situations since complex situations must be described in a linear order. They can describe a sequence of events in the same order in which the events occurred in the world, or in some other order. For example, if a story culminates in a murder, the murder can be described at the end of the story, or it can be mentioned at the beginning of the story, followed by a flashback to the events leading up to the murder. The possible mappings from the writer's situation representation to the discourse allow for different narrative styles. Finally, just as perspective helps define the situation that the speaker wants to describe, it also helps organize the way in which the speaker communicates the situation, since the speaker can change his or her perspective on the same content, depending on his or her communicative intent. Therefore, perspective is part of the presentation plan as well as part of the speaker's representation of the situations.

Once the situations are organized into a discourse, they must be expressed by a linguistic medium. Thus, the third level of discourse organization concerns the linguistic devices used to express the discourse. Content is most often expressed by lexical and grammatical units, although perspective can be conveyed by order of mention, stress, and other nonlinguistic aspects of the text. Recent work has stressed the role of grammatical devices for signalling perspective and other aspects of situations that determine what should be relevant during comprehension (e.g., Hopper, 1979; Talmy, 1978, 1983; MacWhinney, 1984; Morrow, 1986a). The presentation plan is signalled by textual devices like order of mention (whether it matches order of occurrence in the world) as well as lexical and grammatical units such as conjunctions. We discuss relationships between grammar and accessibility in more detail in the section entitled *Situation models and accessibility*.

Situation Models.

Recent developments in psychological theories of discourse representation provide a means of incorporating the three levels of discourse organization into a theory of discourse processing. Discourse comprehension often culminates in a mental model of the situations described by discourse, that is, a situation model. This model is constructed by integrating linguistic knowledge, general background knowledge about actions and objects, and knowledge of the specific referent situation, including perspective on the situations (Johnson-Laird, 1983; Sanford & Garrod, 1981; van Dijk & Kintsch, 1983). The model represents the described situations and their role in the presentation plan. At the same time, a model of the text itself is often constructed. This text model contains lexical and syntactic information as well as information about the order of mention of words and other text units. It is usually retained just long enough to construct the situation model, or to get back to the text in order to change the situation model if it is incorrect (Jarvella, 1971; Sachs, 1967; Johnson-Laird, 1983). It can be retained over a longer period if the situation model cannot be constructed, or if the text itself is an important part of the message or situation, as in jokes or poems (e.g., Keenan, MacWhinney, & Mayhew 1977; Mani & Johnson-Laird, 1982).

The theory of mental models provides a framework that helps clarify issues concerning information accessibility during discourse processing: how different aspects of discourse interact to determine accessibility, and how writers signal what information should be most accessible in the reader's situation model.

Writers often intend to communicate a set of situations, and readers understand the resulting discourse by constructing a situation model. Therefore, writers and readers usually coordinate by attending to the objects and actions comprising situations, rather than to the text conveying the situations. Writers intend readers to take a perspective and focus on some situations and situation participants more than on others. Thus, perspective and focus determine what is relevant in the described world and what should be accessible from the reader's situation model during comprehension. However, what is (or should be) relevant in the described world is not always what is most accessible at a particular point in the text. Therefore, it is important to distinguish information that is accessible because it is relevant to the described situations, and information that is accessible during the processing of the text. For example, a situation that is not highly relevant to the narrative (i.e., background information) may be accessible as the reader processes the background information. In the following example,

4. Jeff flirted with Hannah as they watched some people ice-skating.

the skaters are backgrounded in the described situation, but information about them will be accessible to some extent while the reader processes the final phrase of the sentence. Thus, as a part of the text processing, accessibility is often determined by the order in which information occurs in the text. The writer must signal which parts of the described world are most relevant at each point in processing, so that the reader can keep track of what they should be focusing on as they build the situation model. When accessibility reflects relevance to the described situations, it is determined by the con-

Table 1. Determinants of Accessibility

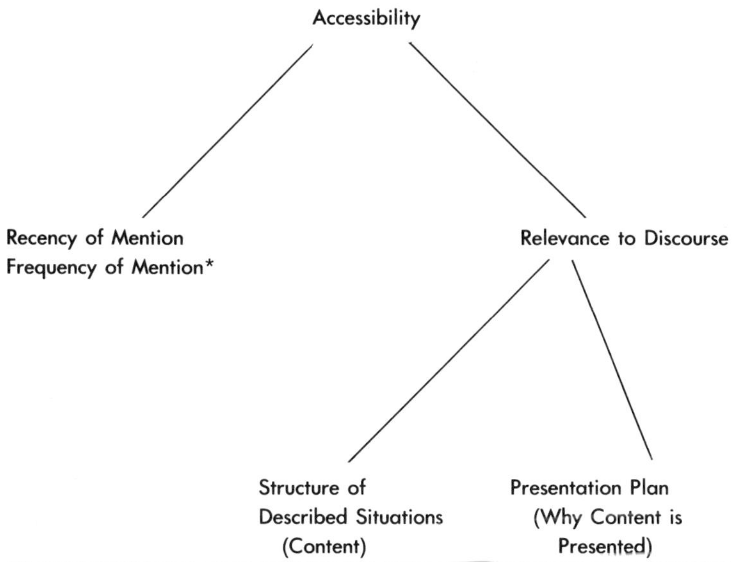

*Recency and frequency of mention are properties of the linear text. Although they sometimes serve as a cue to relevance, more often they compete with cues to relevance as another possible determinant of accessibility. For example, two characters may be repeatedly mentioned in a discourse, but, if one is referred to by a full name or noun phrase while the other is pronominalized, the former character will be less accessible, because the referring expressions presuppose that it is not relevant, even though it is frequently mentioned (e.g., Reichman, 1978). Thus, recency and frequency of mention are often not sufficient to determine accessibility.

tent of the situations and by the role these situations play in the writer's presentation plan (See Table 1 for a summary of determinants of accessibility). As mentioned earlier, many previous theories of referent assignment tended to conflate textual and situational determinants of accessibility by developing reference resolution mechanisms based solely on text-based concepts like recency (e.g., Clark & Sengul, 1979). In our view, situational and textual determinants can be disassociated. For example, Morrow (1985a) demonstrated that pronoun referent assignment is affected more by situational factors than by recency (see below).

Relevance in Situations.

Relevance in the described world arises from the interplay between perspective, topic, and focus at both global and local levels of discourse (see Table 2). Writers describe situations from particular perspectives or points of view, and they expect readers to adopt the same perspective. They take a

Table 2. Situational Determinants of Relevance

Dimension of Situation	Level of Discourse	
	Global	Local
Perspective	Relation of writer to discourse, or vantage point from which discourse is described	Vantage point from which situation is described. Typically identified with protagonist's perspective
Topic	What the discourse is about, summary of main point. Often indicated by title. Example—protagonist's goals and main actions; object of description	What the sentence is about, or what the writer takes as the central entity of the situation at a particular moment in the narrative
Focus	Part of discourse that is salient within frame of reference set up by topic and perspective, or main point of discourse. Example—narrative plot	Part of situation that is salient from perspective of topic. Example—parts of situation salient from *Here/Now* point of narrative

global perspective on the entire world, or set of situations, and a local perspective on the individual situations comprising the world. The global perspective is the relationship between the writer and the discourse topic, what the discourse is about. For example, a narrative might be about Bob, or Bob's trip to the beach. The writer can tell the story from Bob's perspective, from the perspective of another character in the story (e.g., Bob's wife or boss), or from the perspective of an omniscient narrator.[1] The global topic, together with the global perspective, serves as a frame of reference from which other aspects of the world are understood. As such, they constrain what can be in focus, since writers and readers focus on situations or parts of situations relative to the global topic and perspective. In narratives, the sequence of events involving the global topic (most often, the narrative protagonist) will often be focal. These events form the plot or foreground of the narrative. Thus, the global topic is what the narrative is about, and the global focus is what happens during the narrative. Foreground events move the narrative forward in time, while events that serve as the setting for, or comment upon, these events are backgrounded, or nonfocal (Hopper, 1979; Labov, 1972). The global focus can be more restrictively defined as the causal chain of events in the narrative (Trabasso, Secco, & van den Broek,

[1]Note that *perspective* denotes a conceptual reference point for interpreting situations, and that visual point of view is only one kind of perspective. In addition, for the purposes of the present chapter, we do not distinguish narrative voice, or the narrator's point of view in telling the story, from the character's perspective or point of view, which is part of the story itself rather than the way in which the story is presented (see Chatman, 1978, for a discussion of this distinction).

1983; Omanson, 1982), or as the chain of events through which the protagonist reaches his or her goal (Glowalla & Colonius, 1982). In descriptions, the global topic is generally the object of description and the global focus is the features of the object that the writer describes. In expository text, the topic is generally the point of debate and the focus is the main line of argument. Thus, the global focus develops the global topic, or point, of the discourse.

Writers also take a perspective on each situation in a narrative, and this is usually associated with the topic of the sentence(s) describing the situation. If the global topic is the writer's center of interest for the whole discourse, the local topic reflects the writer's interest at a particular point in the discourse. The local topic is the part of the situation around which the description of the situation is organized, and from which parts of the situation are focused (MacWhinney, 1977; Zubin, 1979). Thus, the sentences describing the local situation are often taken to be "about" the local topic. In narratives, the local topic is often a character, usually the protagonist. The local focus is defined as what is salient within the perspective of the topical character.

This definition of local topic and focus is related to, but contrasts with, linguistic definitions of topic and focus. Some linguists and psychologists identify focus with the new information in a sentence (Akmajian, 1979; Clark & Clark, 1977; Yekovich, Walker & Blackman, 1979). In their use of the term, *focus* refers to a phrase in the sentence that contains the intonation center (the point of greatest acoustic-phonological stress). It carries the asserted information in a sentence and therefore contrasts with the presupposed information. Presupposed information is usually associated with the sentence topic or theme, (Halliday, 1967; but see Reinhart, 1982). Since in many sentences presupposed information precedes asserted information, the focus of the sentence often coincides with the last noun phrase in the main clause of the sentence (in active sentences, the grammatical object).

This use of the term *focus* contrasts with its use by many psychologists (Engelkamp & Zimmer, 1983; Sanford & Garrod, 1981; Johnson-Laird, 1968; Olson & Filby, 1972) who use it to refer to the accessibility of information during text processing, and usually identify it with the discourse or sentence topic. Although both accounts of focus agree that the focus of the sentence is what readers should attend to, the first account of focus identifies it with the last noun phrase in the main clause of a sentence, but the second account identifies it with the first noun phrase.

It is clear that readers must attend to more than one part of a sentence (e.g., Engelkamp & Zimmer, 1983). In comprehending a sentence, readers must identify the topic of the sentence and what is predicated of the topic in order to integrate the sentence into the situation model. Thus, both topic and focus should determine information accessibility. According to the present approach, topic and focus are properties of the situation model rather than parts of sentences, although linguistic cues may encourage readers to make an entity described by the sentence either topical or focal in the model (See Greenspan & Segal, 1984; Kates, 1981, for similar conclusions). As part of the situation, focus does not correspond to any one part of a sentence. For example, consider the following:

> 5a Peter didn't feel like doing his homework.
> He looked out the window at the street.

According to analyses that identify focus with new information, the focus of the second sentence is *looked out the window at the street*, or *window* and *street*. However, according to the present approach, street is focused relative to the perspective of the topic, Peter. Furthermore, the focus depends on the sentence predicate, which is critical for expressing situation structure. For example, in example 5b

> 5b Peter didn't feel like doing his homework.
> Instead, he looked at the window overlooking the street.

the window is more focal than the street. Similarly, example 6a focuses the bedroom, but example 6b focuses the kitchen.

> 6a Peter walked from the kitchen into the bedroom.
> 6b Peter was walking through the kitchen toward the bedroom.

These examples suggest that focus in situation models is not simply new information, nor a function of word order, but rather is determined by the topic's (in this case, Peter) relation to the described situation. The spatial and temporal location of the topic entity's perspective relative to the described situations defines the *Here/Now* point of the narrative (Almeida & Shapiro, 1983; Segal, Bruder, & Daniels, 1985). Example 6 suggests that focus depends in part on this point. Narratives consist of a sequence of situations that develops the situation model in space and time. In a murder mystery, the narrative may begin in an old mansion and end in a court room several months later. The foregrounded events develop the plot and move the location of the action from one place and time to another. However, at any point in the narrative, the situations are described from the perspective of a particular *Here/Now* point. As the narrative progresses, this point changes. In example 6a, the *Here/Now* point shifts from the kitchen at one time, to the bedroom at a later time. In many narratives, the location of the protagonist is closely related to the narrative's *Here/Now* point, and changes in the *Here/Now* point are signaled by changes in the protagonist's location. Of course, in some narratives the *Here/Now* point and the protagonist's location can be disassociated (as in narratives with an omniscient narrator not participating in the action of the story). Because this kind of narrative leads to a more complex situation model that is not well understood, we will restrict ourselves to examining situation models based on narratives in which the *Here/Now* point and protagonist location are closely related.

Thus far, we have argued that local focus in particular, and relevance in general, depends on the structure of the described situations. In addition, relevance depends on the writer's presentational plan. Entities are relevant when they play a critical role in realizing the writer's purpose. For example, a character's limp may not be very important in the world itself, but the writer may use it to illustrate an aspect of the character's personality that is important to the point of the story. Also, a situation may be foreshadowed by a discourse, and its mention prior to its occurrence in the world makes it relevant because it creates suspense. Discourse is usually designed so that entities are relevant both because of their role in the world and in the presentation plan. Thus the presentation is organized around topical entities such as protagonists. They occupy important roles in the important events of the world, and their experience is often the reason for why the narrative is told. Of course, there are exceptions, such as when a character that plays a minor role in the world serves as narrator.

Our approach to topic, perspective, and focus makes a number of predictions about discourse processing that we discuss in the next section.

Situation Models and Accessibility

In this section, we review evidence that writers use particular linguistic devices to signal which objects and actions are most relevant to described situations, and that readers attend to these objects and actions, making them prominent in processing and thus highly accessible from the unfolding situation model. We show that devices such as grammatical morphemes (function words like prepositions and conjunctions and inflections like number and tense markers) and grammatical constructions are crucial for expressing topic, perspective, and focus, and therefore function as cues to accessibility.

To construct the appropriate model, a reader must identify what the model is about (i.e., the topic), the perspective from which to construct the model, and the parts of the situations that are focal from this perspective. By doing this, the reader knows what is relevant at each moment of comprehension. This facilitates constructing the model, since relevant entities are likely to be important for the next part of the model, so updating the model with new information will proceed smoothly. Readers must also keep track of the global topic, perspective, and focus in order to construct a coherent model and to discover the point the writer is making by presenting the discourse (Kintsch & van Dijk, 1978). We will review evidence that readers integrate different kinds of cues in an efficient manner in order to determine what is relevant in the described situation and therefore what information should be accessible. They often rely on grammatical cues to identify the perspective from which to interpret described objects and actions as situations, and to identify what should be focal in the situations (Morrow, 1986a). Therefore, grammar provides important cues to relevance. When grammatical cues do not clearly indicate what is relevant, readers often turn to cues provided by the surface text, such as recency of mention of the words.

Narrative comprehension is organized around important or central characters and events. Readers must identify the narrative protagonist(s), the character serving as discourse topic (e.g., van Dijk, 1979). They keep track of protagonists as these characters participate in the foreground events comprising the plot, or global focus (Morrow, 1985a). Protagonists are grammatically marked as both global and local topic. They are marked as sentence topic by being mentioned in subject position and in main clauses (Bates & MacWhinney, 1982; Zubin, 1979), and are referred to by attentuated anaphoras such as pronouns, which presuppose that they are accessible from the situation model (Givon, 1983; Chafe, 1974, 1976). Nonprotagonists are mentioned in object position and referred to by more explicit anaphoras like names or definite descriptions. Speakers use these devices to keep protagonists and nonprotagonists distinct and to cue listeners that the protagonist is more relevant (Karmiloff-Smith, 1981; Marslen-Wilson, Levy, & Tyler, 1982; Francik, 1985). Protagonist perspective is linguistically signalled by deictic expressions (see below) as well as by presentational plan strategies, such as how information is presented about the protagonist's internal states (e.g., Chatman, 1978; Cohn, 1978).

A number of grammatical devices also indicate the foreground situations that protagonists tend to participate in. The sequence of foreground situations comprising the plot is indicated by the presentational device of matching order of mention with order of occurrence in the world (Labov, 1972; Grimes, 1975; Hopper, 1979). The distinction between foreground and background situations is also signalled by verb tense. In English, simple past verbs (e.g., *walked*) often pick out the foreground section and verbs with other tense/aspect markers (e.g., *was walking*) express background situations (Hopper, 1979; Labov, 1972; Kamp, 1979; Almeida & Shapiro, 1983). At the sentence level, conjunctions combine with verb aspect to specify foreground/background relations among situations. For example, *Harry walked into the store and Bill laughed* expresses two foreground events, but *While Harry was walking into the store, Bill laughed* expresses a different situation, since the first event is backgrounded to the second.

Protagonist location in foreground events often defines the narrative *Here/Now*, the perspective from which situations are viewed (Chatman, 1978; Segal et al., 1985). Readers construct the model from this vantage point, updating the model by advancing the *Here/Now* point as each foreground event occurs. At any point in comprehension, readers focus on parts of the situations from the perspective point.

The *Here/Now* point is signalled in part by devices that specify protagonist location. Prepositions specify roles of situation participants. For example, in *The fireman rushed down the street, past the bank, and into the warehouse—down* specifies the street as the path, *past* specifies the bank as a path landmark, and *into* specifies the warehouse as the goal. The aspect of the verb gives the temporal contour of the situation. The past tense verb in the example sentence specifies a completed situation, but if the verb was progressive, it would have specified an incomplete situation. Together, prepositions and verb aspect specify the location of the *Here/Now* relative to an explicitly mentioned place. The fireman, and presumably the *Here/Now* point, is located inside the warehouse, the goal of the completed event.

The *Here/Now* serves as a reference point for locating and orienting other objects in the situation. For example, writers and speakers use grammatical devices to indicate the way in which layout descriptions are organized around this reference point (e.g., Linde & Labov, 1975; Ehrich, 1982; Klein, 1982; Wunderlich & Reinelt, 1982). In the following passage (from Linde & Labov, 1975).

7. You walked in the front door,
   There was a narrow hallway.
   To the left, the first door you came to
   was a tiny bedroom.
   Then there was a kitchen. . .

grammar indicates that the reader or listener is the protagonist whose perspective is necessary for interpreting the description, and it also indicates what is focal from this perspective or *Here/Now* point as it moves through the described layout. For example, pronouns and demonstratives not only refer to participants and their spatial location in the layout, they indicate the appropriate perspective from which to view the scene (e.g., the use of *you* to engage the addressee's perspective). Prepositions, verbal inflections, and other morphemes express the location of parts of the apartment relative to this perspective (*To the left*. . .). They also indicate direction and other dynamic properties of the tour, which moves the perspective point (*Then there was a kitchen*. . .).

Readers use the information conveyed by particular combinations of prepositions and verb aspect markers to construct models organized around protagonist location. Morrow (1988) presented motion sentences embedded in short narratives to subjects along with a diagram of a house serving as the setting for the described situations. Subjects had to indicate on the diagram where the mover was located in the situation described by each sentence. Constructions expressing a completed situation and marking the goal as a bounded region (*walked into*) located the mover inside the goal area, while constructions expressing a progressive situation (*walking to*, *walking toward*) located the character on the path, heading toward the goal. The particular path location depended on the second preposition in the sentence. Consider the following sentences (assume there is a living room connecting the kitchen and bedroom).

> 8a. John was walking through the kitchen to the bedroom.
> 8b. John was walking out of the kitchen to the bedroom.
> 8c. John was walking from the kitchen to the bedroom.

In sentence 8a, John is inside the kitchen, while in sentence 8b he is at the kitchen exit, and in sentence 8c he is inside the living room. Simple past sentences with *to* located the character at the entrance to, but not inside, the goal room. These results show that readers determine protagonist location, and therefore the *Here/Now* point, from grammatical cues.

Protagonist location helps determine which objects and actions are most accessible from the situation model. Because protagonists often serve as the global topic and help define the *Here/Now* at a particular point of discourse, they are relevant to the narrative world and should be accessible from the unfolding situation model. Readers often adopt the protagonist's perspective, and this determines the inferences that are made to integrate the sentences into the situation model, as well as which information is best remembered after the narrative has been read (Bower, 1978; Black, Turner, & Bower, 1979; Anderson & Pichert, 1978; Morrow, 1985a). Readers choose protagonists in foreground events over nonprotagonists as referents for ambiguous pronouns occurring as sentence subject, independent of the order of mention of the situations that the characters participate in, and they complete sentences so that the sentences are about protagonists (Morrow, 1985a; Anderson, Garrod & Sanford, 1983). Moreover, when a long time has elapsed between the current and previous foreground situations, the protagonist is more accessible than a nonprotagonist from the model in working memory (Anderson et al., 1983). Also Morrow, Greenspan, and Bower (1988) show that, when the protagonist is mentioned in a foreground event and nonprotagonist in a background event occurring before the foreground event, so that the *Here/Now* point is clearly with the protagonist, information about the protagonist's location is more accessible than information about the nonprotagonist's location, even when the nonprotagonist is last mentioned.

Readers focus on other parts of situations, from the protagonist's perspective. Therefore, objects located near the protagonist should be as accessible as the protagonist. In Morrow (1985b), subjects memorized a house layout and then read narratives that took place in the house. In each narrative, there was a critical motion sentence followed by a sentence with a target anaphor. Readers tended to choose objects located near the protagonist as referent for the anaphor. Consider the following examples.

9a. John was walking past the kitchen to the bedroom.
He noticed the rug was dirty.

9b. John walked past the kitchen into the bedroom.
He noticed the rug was dirty.

In example 9a, readers thought the second sentence referred to a rug near the kitchen or elsewhere on the path, while in example 9b they thought the second sentence referred to a bedroom rug. In general, preposition/verb combinations that clearly locate the protagonist inside the goal usually focus the goal room, while sentences locating the protagonist on the path focus path rooms. However, simple past sentences with *to* focus the goal when the source room is mentioned by *from*, but the path near the goal entrance is focused when the path is mentioned by either *past* or *through*. Referent choice in these sentences reflects more than character location, since Morrow (1988) showed that all three types of sentences with *to* located the character at the entrance to the goal. Thus, readers not only consult the protagonist's location in the situation model in order to determine which parts of the situation are relevant, they also consider the role that the objects play in the presentation plan, assuming that the writer intends an explicitly mentioned path to be a relevant part of the model.

Morrow (1985b) also showed that referent decisions were not influenced by the order of mention of the prepositional phrases, suggesting that accessibility was determined by the location of the protagonist in the situation model, not by the position of the words in the text. Furthermore, the study shows that accessibility is not simply a function of grammatical roles, such as sentence object, or of pragmatic roles, such as new information, since accessibility changes with the structure of the described situation even though the grammatical or information structure of the sentence remains the same.

Grammar provides a means for referring to both focal and nonfocal objects within a situation. *This* refers to objects near the protagonist, which are focal, and *that* to far, nonfocal, objects. Once again, order of mention of the words is not important, since *that* often refers to objects located far from the protagonist in the situation, not to objects mentioned further back in the text (e.g., Casad & Langacker, 1982; Morrow, 1985b).

More direct evidence that protagonist location determines which part of the situation model is most accessible from working memory is provided by Morrow et al. (1987). Subjects memorized building diagrams and then read narratives taking place in the buildings. The narratives were periodically interrupted by the names of two objects from the buildings, and subjects had to decide whether the objects were from the same room, or from different rooms in the buildings. Readers made 'same' responses more quickly when the objects were from the room containing the protagonist at that point in the narrative, and there was some evidence that response times increased with the distance between the probed room and the protagonist location.

However, this was only the case when the character was located in the goal room of a completed motion event, which clearly advanced the *Here/Now* point. For example, in 10a

10a. Then Joe walked from the conference room into the library.

objects from the library were most accessible from the unfolding situation model. Moreover, accessibility did not depend on recency of mention, since the character location room was most accessible even when it was not mentioned last in the sentence, as in example 10b.

10b. Then Joe walked into the library from the conference room.

Therefore, at least when reading simple narratives, readers take the perspective of the protagonist and focus on the part of the situation model near the protagonist's location. People also take a perspective when solving problems, which they indicate by using deictic words such as *this* or *there* when describing their solution to the problem. They are more likely to make mistakes when trying to solve part of the problem that is far from the perspective location, suggesting that the parts of their problem representation near the perspective point are most accessible from working memory (Hutchins & Levin, 1982).

In addition to determining which objects in a situation are most focal, protagonist location determines which aspects of objects are focal. When people read *John was walking over the hill and . . .* , they chose *he thought it was very high* rather than *he thought it was very wide* as the most appropriate continuation. However, when the preposition in the first clause of the sentence is changed to *around*, the preferred continuation reverses. Similarly, *The children walked across/along the stream* focused either on the width or length of the stream (Morrow, 1982). This shows that different aspects of objects can be focal depending on their orientation with respect to protagonist perspective.

Thus far, we have provided evidence that readers often adopt the perspective of protagonists in foreground events and focus on other parts of situations from this vantage point. This strategy for determining accessibility works well as long as the protagonist participates in foreground situations that are clearly delineated from background situations, so that protagonist location clearly defines the *Here/Now* point. Accessibility is less easy to determine when protagonists participate in background situations and nonprotagonists suddenly appear in the narrative foreground, so that it is difficult to identify the *Here/Now*. For example, in the following complex sentence,

11a. While Tom was going toward the ferris wheel,
Harry went toward the exhibition hall.

Tom, the protagonist as established by the previous part of the narrative, occurs in a background situation serving to locate the foreground situation. When the foreground and background situations occur simultaneously, as in the example, both situations can be viewed from the same *Here/Now* point and therefore can be integrated into the model from the same perspective. In this case, readers maintain the protagonist's perspective even though that character has been backgrounded (Morrow, 1985a), suggesting that readers try to maintain a consistent perspective when reading narratives (also see Black et al., 1979). However, readers tend to choose the last mentioned character, whether protagonist or nonprotagonist, as referent for the subject pronoun of the following sentence. Thus, in example 11a, they choose Harry as referent (Morrow, 1985a). Therefore, they maintain the global perspective associated with Tom, the protagonist, but choose Harry as referent, both because he is last mentioned in the sentence and the local topic of the foreground situation, and so is highly accessible from the model at that point in the narrative. When the order of the situations is reversed, as in example 11b,

11b. Harry went toward the exhibition hall
while Tom was going toward the ferris wheel.

readers choose Tom as referent, presumably because he is now last mentioned as well as the global topic of the discourse. This explanation of the pattern of referent choices suggests that when grammatical cues to global and local relevance conflict, readers resort to the recency of mention of information in the text to determine accessibility in the situation model.

If the structure of the local situations clearly determines which character is most relevant, cues provided by the global topic are overridden. In this case, readers rely on the structure of the situations rather than the order of mention of information in the text to determine accessibility. For example, when the protagonist participates in a situation that is highly backgrounded relative to the foreground situation containing the nonprotagonist, readers tend to adopt the nonprotagonist's perspective and choose him or her as referent. Thus, in example 11c,

11c. After Tom had gone toward the ferris wheel,
Harry went toward the exhibition hall.

readers tend to use Harry's perspective as well as choose him as referent, regardless of the order of mention of the characters (Morrow, 1985a). In this case, the nonprotagonist is accessible, not only because he is the local topic of the foreground event, but because readers must adopt his perspective to integrate the situations into the model. The nonprotagonist's location defines the *Here/Now* point because the background event occurred prior to the Now of the narrative. Thus, readers shift from a global perspective when necessary to integrate the situations into the model. Similarly, people shift their perspective on a problem when their previous perspective precludes solving the problem (Hutchins & Levin, 1982).

In summary, readers often use linguistic, especially grammatical, cues to situation structure in order to identify which objects and actions should be accessible from the situation model they are constructing. These cues indicate the objects and actions the writer intends as relevant in the described situations, so readers make them accessible during comprehension. Readers resort to recency of mention and other aspects of the surface text when the cues based on situation structure conflict. They also use recency when the situations provide few cues to the relevance of entities in the discourse world. For example, places are intrinsically nonsalient parts of discourse, and, when serving as setting in narratives, they also play a noncentral role in the narrative. When readers must choose a place as referent for a referring expression presupposing that the place is relevant to the situation (i.e., a pronoun serving as subject of the sentence), they choose recently mentioned places in narratives (Morrow, 1985a, 1986b). However, places are more central to descriptions in which they serve as the object of the description and thus are the discourse topic or theme. In this case, thematic status rather than recency determines referent choice (Morrow, 1986b). This provides further evidence for a hierarchy of accessibility determinants: Readers try to determine what should be most accessible by consulting the structure of the situation model, which is expressed by grammatical and other linguistic cues. They only resort to recency of mention or other aspects of the surface text when they are not certain what should be relevant in the described situations.

Summary

In this chapter, we have sketched an approach that helps to more fully integrate the notion of information accessibility into a theory of discourse processing. We defined accessibility in terms of the dynamics of information availability in working memory during discourse processing. We distinguished between accessibility of information about the text and information about the world conveyed by the text, and argued that writers and readers usually attend to information about the described world since the goal of communication is to share a model of a particular set of situations. We also pointed out that accessibility can be determined by properties of the surface text, such as recency of mention, or by properties of the described situations that specify what is relevant in the described world. We distinguished three critical aspects of described situations that combine to help specify relevance: The topic of the situation model, the perspective from which the model is constructed, and which parts of the situations are focused from the perspective. We reviewed evidence that writers use a variety of grammatical devices to express these critical aspects of situations, and that readers rely on these cues to identify what should be accessible from the situation model they are constructing during comprehension. Furthermore, when these cues clearly specify what is relevant in the described situations, the cues are more important for determining accessibility than the recency of mention of words or clauses in the text. Therefore, to successfully construct a situation model, readers not only have to identify the described objects and actions, they have to identify the topic, perspective, and focus that define these objects and actions as the situations perceived by the writer. Because these aspects of situations help define what is relevant in situations, and therefore what should be accessible during comprehension, they are essential for guiding comprehension.

REFERENCES

Almeida, M.J., & Shapiro, S.C. (1983, May). Reasoning about the temporal structure of narrative texts. *Proceedings of the fifth annual conference of the Cognitive Science Society*. Hillsdale, N.J.: Erlbaum

Akmajian, A. (1979). *Aspects of the grammar of focus in English*. New York: Garland Publishing.

Anderson, A., Garrod, S.C., & Sanford, A.J. (1983). The accessibility of pronominal antecedents as a function of episode shifts in narrative text. *Quarterly Journal of Experimental Psychology, 35A*, 427–440

Anderson, R.C., & Pichert, J.W. (1978). Recall of previously unrecallable information following a shift in perspective. *Journal of Verbal Learning & Verbal Behavior, 17*, 1–12.

Bates, E., & MacWhinney, B. (1982). Functionalist approaches to grammar. In E. Wanner & L. Gleitman (Eds), *Language acquisition: State of the art* (pp. 173–218). Cambridge, England: Cambridge University Press.

Black, J.B., & Turner E., & Bower, G.H. (1979). Point of view in narrative comprehension, memory and production.. *Journal of Verbal Learning and Verbal Behavior, 18*, 187–198.

Bower, G.H. (1978). Experiments on story comprehension and recall. *Discourse Processes, 1*, 211–231.

Brewer, W.F., & Lichtenstein, E.H. (1981). Event schemas, story schemas, and story grammars. In J. Long and A. Baddeley (Eds.), *Attention & performance IX* (pp. 363–379). Hillsdale, NJ: Erlbaum.

Carpenter, P.A. & Just, M.A. (1977). Integrative processes in comprehension. In D. Laberge & S.J. Samuels (Eds.), *Basic processes in reading: perception and comprehension* (pp. 217–241). Hillsdale, NJ: Erlbaum.

Casad, E., & Langacker, R.W. (1982). *Inside and outside in Cora grammar.* Unpublished manuscript, University of California at San Diego.

Chafe, W. (1972). Discourse structure and human knowledge. In J.B. Carroll & R. Freedle (Eds.), *Language comprehension and the acquisition of knowledge* (pp. 41–69). Washington, DC: V.H. Winston & Sons.

Chafe, W. (1974). Language and consciousness. *Language, 50,* 111–133.

Chafe, W. (1976). Giveness, contrastiveness, definiteness, subjects, topics and point of view. In C.N. Li (Ed.), *Subject and topic.* New York: Academic Press.

Charniak, E. (1972). *Toward a model of children's story comprehension.* Unpublished doctoral dissertation, Massachusetts Institute of Technology.

Chatman, S. (1978). *Story and discourse: Narrative structure in fiction and film.* Ithaca, NY: Cornell University Press.

Clark, H.H., & Clark, E.V. (1977). *Psychology and language.* New York: Harcourt.

Clark, H.H., & Sengul, C.J. (1979). In search of referents for nouns and pronouns. *Memory & Cognition, 7,* 35–41.

Cohn, D. (1978). *Transparent minds: Narrative modes for presenting consciousness in fiction.* Princeton, NJ: Princeton University Press.

van Dijk, T.A. (1979). Relevance assignment in discourse comprehension. *Discourse Processes, 2,* 113–126.

van Dijk, T.A., & Kintsch, W. (1983). *Strategies of discourse comprehension.* New York: Academic Press.

Ehrich, V. (1982). The structure of living space descriptions. In R.J. Jarvella & W. Klein (Eds.), *Speech, place, and action.* (pp. 219–249) New York: Wiley.

Engelkamp, J., & Zimmer, H.D. (1983). *Dynamic aspects of language processing: Focus and presupposition.* Heidelberg, Germany: Springer-Verlag.

Francik, E.P. (1985). Referential choice and focus of attention in narratives (Doctoral dissertation, Stanford University). *Dissertation Abstracts International, 46,* 1363B.

Givon, T. (1983). Topic continuity in spoken English. in T. Givon (Ed.), *Topic continuity in discourse: Quantitative cross-language studies: Typological studies in language* (Vol. 3, pp. 347–363). Amsterdam, Netherlands: J. Benjamin's.

Glowalla, U. & Colonius, H. (1982). Toward a model of macrostructure search. In A. Flammer & W. Kintsch (Eds.), *Discourse processes,* (pp. 111–123). Amsterdam, Netherlands: North-Holland.

Greenspan S.L., & Segal, E.M. (1984). Reference and comprehension: A topic-comment analysis of sentence-picture verification. *Cognitive Psychology, 16,* 556–606.

Grimes, J. (1975). *The thread of discourse.* The Hague, Netherlands: Mouton.

Grosz, B. (1981). Focusing and description in natural language dialogues. In A. Joshi, I. Sag, & B. Nash-Webber (Eds.), *Linguistic structure and discourse setting* (pp. 84–105). Cambridge, England: Cambridge University Press.

Halliday, M.A.K. (1967). Notes on transitivity and theme in English. *Journal of linguistics, 3,* 199–244.

Hirst, G. (1981). *Anaphora in natural language understanding: A survey.* Lectures notes in computer science (Vol. 119)(G. Goos & J. Hartmanis, Eds.). Heidelberg, Germany: Springer-Verlag.

Hopper, P. (1979). Aspect and foregrounding in discourse. In T. Givon (Ed.), *Syntax and semantics. Vol. 12:Discourse and syntax* (pp. 213–241). New York: Academic Press.

Hutchins, E.L., & Levin, J.A. (1982). *Point of view in problem solving.* Technical report, Laboratory of Comparative Human Cognition, University of California, San Diego.

Jarvella, R.J. (1971). Syntactic processing of connected speech. *Journal of Verbal Learning & Verbal Behavior, 10,* 409–416.

Johnson–Laird, P.N. (1968). The interpretation of the passive voice. *Quarterly Journal of Experimental Psychology, 20,* 69–73.

Johnson–Laird, P.N. (1983). *Mental models.* Cambridge, MA: Harvard University Press.

Kamp, J.A.W. (1979). Events, instants, and temporal reference. In R. Bauerle, U. Egli, & A. von Stechow (Eds.), *Semantics from different points of view.* Berlin: Springer-Verlag.

Karmiloff-Smith, A. (1981). The grammatical marking of thematic structure in the development of language protection. In W. Deutsch (Ed.), *The child's construction of language* (pp. 121–148). London: Academic Press.

Kates, C.A. (1981). *Pragmatics and semantics: An empiricist theory,* Ithaca, NY: Cornell University Press.

Keenan, J.M. MacWhinney, B., & Mayhew, D. (1977). Pragmatics in memory: A study of natural conversation. *Journal of Verbal Learning & Verbal Behavior,* 5, 549–560.

Kintsch, W., & van Dijk, T.A. (1978). Toward a model of text comprehension and production. *Psychologial Review, 85,* 363–394.

Klein, W. (1982). Deixis and spatial orientation in route directions. In R. J. Jarvella & W. Klein (Eds.), *Speech, place, and action.* New York: Wiley.

Labov, W. (1972). *Language in the inner city.* Philadelphia, PA: University of Pennsylvania Press.

Lesgold, A.M., Roth, S.F., & Curtis, M.E. (1979). Foregrounding effects in discourse comprehension. *Journal of Verbal Learning & Verbal Behavior, 18,* 291–308.

Linde, C. (1979). Focus of attention and the choice of pronouns in discourse. In T. Givon (Ed.), *Syntax and semantics. Vol. 12: Discourse and syntax* (pp. 337–354). New York: Academic Press.

Linde, C., & Goguen, J.A. (1978). Structure of planning discourse. *Journal of Social & Biological Structures, 1,* 219–251.

Linde, C., & Labov, W. (1975) Spatial networks as a site for the study of language and thought. *Language, 51,* 924–939.

MacWhinney, B. (1077). Starting points. *Language, 53,* 152–168.

MacWhinney, B. (1984). Grammatical devices for sharing points. In R.L. Schiefelbusch & J. Pickar (Eds.), *The acquisition of communicative competence* (pp. 325–374), New York: University Park Press.

Mani, K., & Johnson-Laird, P.N. (1982). The mental representation of spatial descriptions. *Memory & Cognition, 10,* 181–187.

Marslen-Wilson, W., Levy, E., & Tyler, L.K. (1982). Producing interpretable discourse: The establishment and maintenance of reference. In R.J. Jarvella & W. Klein (Eds.), *Language, place, and action* (pp. 339–378). New York: Wiley.

Minsky, M. (1975). A framework for representing knowledge. In P. Winston (Ed.), *The psychology of computer vision* (pp. 211–277). New York: McGraw Hill.

Morgan, J.L., & Sellner, M.B. (1980). Discourse and linguistic theory. In R.J. Spiro, B.C. Bruce, & W.F. Brewer (Eds.), *Theoretical issues in reading comprehension: Perspectives from cognitive psychology, linguistics, artificial intelligence, and education* (pp. 165–200). Hillsdale, NJ: Erlbaum.

Morrow, D.G., (1982). *The role of prepositions and verb aspect in foregrounding and backgrounding narrative content.* Unpublished doctoral dissertation, University of California at Berkeley.

Morrow, D.G., (1985a). Prominent characters and events organize narrative understanding. *Journal of Memory & Language, 24,* 304–319.

Morrow, D.G., (1985b). Prepositions and verb aspect in narrative understanding. *Journal of Memory & Language, 24,* 390–404.

Morrow, D.G., (1986a). Grammatical morphemes and conceptual structure in discourse processing. *Cognitive Science, 10,* 423–455.

Morrow, D.G., (1986b). Places as referents in discourse. *Journal of Memory & Language, 25, 676–690.*

Morrow, D.G., (1988). *Spatial Models, prepositions, and verb aspect markers.* Submitted for publication

Morrow, D.G., Greenspan, S.L., & Bower, G.H. (1987). Accessibility and situation models in narrative comprehension. *Journal of Memory & Language, 26,* 165–187.

Morrow, D.G., Greenspan, S.L., & Bower, G.H. (1988). *Protagonists in narrative comprehension.* Unpublished manuscript.

Olson, D.R., & Filby, N. (1972). On the comprehension of active and passive sentences. *Cognitive Psychology, 3,* 361–381.

Omanson, R.C. (1982). An analysis of narratives: Identifying central, supportive, and distracting content. *Discourse Processes, 5,* 195–224.

Reichman, R. (1978). Conversational coherency. *Cognitive Science, 2,* 283–327.

Reinhart, T. (1982). *Pragmatics and linguistics: An analysis of sentence topics.* Bloomington, IN: Indiana University Linguistics Club.

Robinson, A.E. (1981). Determining verb phrase referents in dialogs. *American Journal of Computational Linguistics, 7,* 1–16.

Sachs, J.S. (1967). Recognition memory for syntactic and semantic aspects of connected discourse. *Perception & Psychophysics, 2,* 437–442.

Sanford, A.J., & Garrod, S.C. (1981). *Understanding written language: explorations of comprehension beyond the sentence.* New York: Wiley.

Schank, R., & Abelson, R. (1977). *Scripts, plans, goals and understanding.* Hillsdale, NJ: Erlbaum.

Segal, E.M., Bruder, G.A., & Daniels, J. (1985, November). *Deictic centers in narrative understanding.* Paper presented at the twenty-fifth annual meeting of the Psychonomics Society, San Antonio, TX.

Sharkey, N.E., & Mitchell, D. (1985). Word recognition in a functional context: The use of scripts in reading. *Journal of Memory & Language, 24*, 253–270.

Sharkey, N.E., & Sharkey, A.J.C. (1987). *Journal of Memory & Language, 26*, 255–276.

Sidner, C.L. (1984). Focusing in the comprehension of definite anaphora. In M. Brady & R.C. Berwick (Eds.) *Computational models of discourse* (pp. 267–330). Cambridge, MA: MIT Press.

Talmy, L (1978). The relation of cognition to grammar. In D. Waltz (Ed.), *Theoretical issues in natural language processing II.* (pp. 1–11) Washington, DC: Association for Computational Linguistics.

Talmy, L. (1983). How language structures space. In H. Pick & L. Acredolo (Eds.), *Spatial orientation: Theory, research, and application* (pp. 225–282). New York: Plenum Press.

Trabasso, T.D., Secco, T., & van den Broek, P. (1983). Causal cohesion and story coherence. In H. Mandl, N. Stein, & T. Trabasso (Eds.), *Learning and comprehension of text* (pp. 83–112). Hillsdale, NJ: Erlbaum.

Winograd, T. (1972). *Understanding natural language.* New York: Academic Press.

Wunderlich, D., & Reinelt, R. (1982). How to get there from here. In R.J. Jarvella & W. Klein (Eds.), *Speech, place, and action* (pp. 183–201). New York: Wiley.

Yekovich, F.R. Walker, C.H., & Blackman, H.S. (1979). The role of presupposed and focal information in integrating sentences. *Journal of Verbal Learning & Verbal Behavior, 18*, 535–548.

Zubin, D.A. (1979). Discourse function of morphology. The focus system in German. In T. Givon (Ed.), *Syntax and semantics, Vol. 12: Discourse and syntax* (pp. 469–504). New York: Academic Press.

CHAPTER 4

The Everyday Use of Anaphoric Expressions: Implications for the 'Mental Models' Theory of Text Comprehension*

ALAN GARNHAM
Centre for Research on Perception and Cognition
Laboratory of Experimental Psychology
University of Sussex

JANE OAKHILL
MRC Perceptual and Cognitive Performance Unit
Laboratory of Experimental Psychology
University of Sussex

INTRODUCTION

As part of its recent campaign to disseminate information about AIDS, the British Government distributed a leaflet to every household in the country. This leaflet was intended to reach the widest possible audience. It was, therefore, presumably written in such a way as to be readily comprehensible. The following extracts are taken from this leaflet:

> . . . many teenagers do have sex and some may experiment with drugs. Even if you think your children *don't* . . . they may have friends who encourage them *to*.

*We would like to thank Nicola Yuill for adding some interesting anaphors to our collection.

> Anyone who misuses drugs should not inject. If you ever *do*, never share equipment. . . . You could be injecting the virus straight into your blood stream. *It* is extremely dangerous.
>
> . . . go direct to a clinic for sexually transmitted diseases . . . *they'll* let you know. . . .
>
> But the virus is spreading. And as it *does*. . . .

The italicized expressions in these extracts, which were not, of course, italicized in the original, are *anaphoric expressions* of various types. In context the intended meaning of each of these expressions is perfectly clear, though it may be momentarily uncertain, until all the relevant contextual information has been considered. Nevertheless, from both a linguistic and a psycholinguistic perspective, these expressions are problematic, for reasons that will become clear as this chapter progresses. Linguistic and psycholinguistic theories are best suited to accounting for more straightforward uses of anaphoric expressions, such as:

> John came into the room. *He* sat down.

In this chapter we will discuss the problems that certain uses of anaphoric expressions cause to the comprehension system, and attempt to show how the mental models theory of discourse comprehension (e.g. Garnham, 1981; Johnson-Laird, 1983) can be extended to handle these problems. We will illustrate these difficulties using examples taken from similar sources to the ones above. Almost all the examples will be from published material.

We begin the chapter by discussing some of the linguistic literature on anaphora that is relevant to our current concerns. We then outline the mental models of theory of text comprehension, as it has been developed to date. We emphasize its strengths and point to places in which it needs further development if it is to account for the data we will present later in the chapter. We follow this discussion with an account of how we collected a corpus of anaphoric uses from published text, and we examine in detail some examples from this corpus. We then consider the kinds of extension to the mental models theory that these examples suggest. We also discuss some experiments that examine the online processing of problematic anaphoric expressions.

ANAPHORA—SOME LINGUISTIC BACKGROUND

The linguistic literature on anaphora is vast, and most of it is not directly relevant to our current concerns. We will not, therefore, try to give a detailed account of how linguists have attempted to account for anaphoric reference. A brief overview is given by Garnham (1987a). What we will do, however, is to provide that part of the linguistic background to anaphoric reference that is relevant.

Model-Interpretive Anaphora and Ellipses

Hankamer and Sag (1976) drew a distinction between *deep* and *surface* anaphors. In more recent works (Sag, 1979; Sag and Hankamer, 1984) they have revised the distinction, and renamed the two types of anaphor *model-interpretive anaphora* and *ellipses*.

Model-interpretive anaphors include definite pronouns (*he*, *she*, *it*, *they*), indefinite pronouns (*one*, *some*), full definite NP anaphors (. . . *a man* . . . . *the man* . . .), sentential *it* (John's a fool, but no one knows *it*) and *do it*. Ellipses include verb phrase ellipsis (VPE), *so*, *do so*, and the constructions known as sluicing (*someone's trying to wreck my plans, but I don't know who*), and gapping (*a magpie landed on the fence and a jay on the lawn*).

Model-interpretative anaphors and ellipses differ from each other in two main ways First, deep anaphors often take their meaning from some aspect of the nonlinguistic context in which they are uttered. *She* can refer to a person that speaker and hearer can both identify, even if she has not previously been mentioned. If I'm alone with a friend in a room, and we both notice someone coming in at the door, I can say without further ado, *she's my doctor*. Ellipses, on the other hand must have *linguistic* antecedents. If someone drops and breaks a cup, I cannot say *So did Bill*. Such a sentence is only appropriate in a context such as

> John broke a cup. So did Bill.

Hankamer and Sag formulate this distinction by saying that model-interpretative anaphors can be either linguistically or pragmatically *controlled*, but ellipses must be linguistically controlled.

Second, not only must surface anaphors be linguistically controlled, but the linguistic *form* of the antecedent must be (almost) identical to that of the ellipsed material. This restriction does not, however, apply to model-interpretive anaphors, even when they have linguistic antecedents. Consider, for example, the model-interpretive anaphor *do it*. In a sentence such as:

> The floor was swept by Bill, because Sarah wouldn't do it.

this anaphor is perfectly acceptable, even though *do it* means *sweep the floor* and there is no expression of this *form* in the first clause. However, with verb phrase ellipsis, the corresponding nonparallel construction is much worse. The sentence:

> The floor was swept by Bill, because Sarah wouldn't.

is unacceptable. It must be:

> Bill swept the floor, because Sarah wouldn't.

In this second version the ellipsed material is identical in form to the antecedent (apart from the inflection on the verb, which is one reason why *identical* had to be qualified by *almost*).

These two properties of ellipses have a number of consequences for the way such expressions are interpreted. The most important is that, because an ellipsis must have a syntactically parallel antecedent, an elliptical clause may be interpreted differently when it follows sentences with the same meaning. This phenomenon can be illustrated using verb-phrase ellipsis as an example. Consider the following passages.

> The moose had chased the squirrel.
> The beaver had, too.

> The squirrel had been chased by the moose.
> The beaver had, too.

The first sentences in the two passages describe the same situation, and the second has exactly the same form in each. Nevertheless, the first passage says that the beaver had chased the squirrel, and the second that the moose had chased the beaver. Therefore, in order to interpret an ellipsis correctly it may be necessary to remember, not only the meaning of the antecedent-containing clause, but also its exact wording. It follows that ellipses are likely to be difficult to understand, because, in normal circumstances, people are not very good at remembering the exact form of sentences. They remember the meaning of what they were told, but not how it was phrased, unless the phrasing affects the message conveyed, as it does in jokes (cf. Keenan, MacWhinney, & Mayhew, 1977). Indeed, memory for the exact wording of a message deteriorates within a few sentences or less (Sachs, 1967).

This observation suggests that ellipses are likely to be misinterpreted unless they either occur very close to their antecedents or there are other cues to their interpretation.. Often such cues are available. It is well known that comparatively difficult syntactic constructions are easier to interpret if semantic cues force, or strongly suggest, one reading. For example, Slobin (1966) found that passives such as:

> The flowers were watered by the girl.

are easier to interpret that ones such as:

> The girl was followed by the boy.

because flowers cannot (or do not usually) water girls, but girls can follow boys.

Similarly, in the passage:

> It had been a busy morning at the hospital.
> The elderly patient had been examined by the doctor
> during the ward round.
> The child had, too.

The ellipsis is easy to interpret because it is natural for a doctor to examine a child during a ward round, but not natural for a child to examine an elderly patient. However, if the elliptical sentence had been *the nurse had, too*, the linguistically correct interpretation (that the nurse had been examined by the doctor during the ward round) would have been less likely than the 'suggested' interpretation that the nurse had examined the elderly patient. This interpretation is suggested because the passage describes a doctor examining a patient. This event could have been described in the sentence.

> The doctor had examined the elderly patient during
> the ward round.

in which case the antecedent for the elliptical verb phrase would have been *examined the elderly patient*.

Garnham and Oakhill (1987) showed that passages of the second type were frequently misinterpreted, even when the phrase intervening between the antecedent and the ellipsis (*during the ward round*, in the example above) was left out. Furthermore, the elliptical clauses took a long time to interpret in these passages, especially with the intervening phrase included. It seems that people are trying to produce the correct interpretation—they are not just selecting the most plausible interpretation as quickly as they can. However, they have more difficulty retrieving the antecedent of an elliptical verb phrase when the two are separated by even a short distance. We will discuss in more detail the question of how ellipses are interpreted later in this chapter.

Anaphoric Islands

Generally speaking, an anaphoric expression cannot take its meaning from a component of the meaning of its antecedent that has no overt manifestation in the form of that antecedent. For example, although an orphan is, by definition, a child with no parents, it is not possible for *they* following *orphan* to refer to the orphan's parents simply by virtue of the fact that the word *orphan* has been used. Compare:

> Max's parents are dead and he deeply misses them.
> Max is an orphan and he deeply misses them. (Postal, 1969, example 3)

This observation does not seem very surprising. However, in the late 1960s and early 1970s, a linguistic theory known as generative semantics was in vogue and this theory had difficulty explaining such facts. Generative semantics postulated very abstract underlying structures for sentences. For example, in the representation underlying the 'orphan' sentence, there would be no single element corresponding to the word *orphan*. Instead, its components, including PARENTS, were present, waiting to be collected together and *lexicalized* at some point in the derivation of the sentence from its underlying structure. Generative semantics inherited from its predecessors the idea that pronouns in the surface form of sentences were produced by a transformation that replaced one of two identical items in the underlying structure of a sentence by a pronoun. But if *orphan* in surface structure comes from, among other things, PARENTS in underlying structure, the underlying forms of *Max is an orphan* and *Max deeply misses his parents* should be combined by (two applications of) the pronominalization transformation to produce: *Max is an orphan and he deeply misses them.*

Ross (1967) put forward an idea that remains important in syntactic theory to this day, namely that elements in certain structures cannot be affected by transformations. Those structures are *islands* that transformations cannot reach into. To explain the 'orphan' phenomenon, Postal (1969) claimed that lexical items are *anaphoric islands*—no matter what their internal structure, they cannot contain antecedents for anaphors. This *anaphoric island constraint* must apply comparatively late in the derivation of a sentence, after lexicalization has occurred.

Postal extended this constraint to what he called *derivatives*, which he also claimed to be anaphoric islands. Derivatives, unlike words such as *orphan*, do betray their underlying origins. So, to take one of Postal's examples, *McCarthyites* is a derivative of *McCarthy*. Nevertheless, the Anaphoric Island Constraint rules out *his* taking its meaning from the *McCarthy* in *McCarthyites* in the second of the sentences below:

Followers of McCarthy are now puzzled by his intentions.
McCarthyites are now puzzled by his intentions. (Postal, 1969, example 42)

However, despite Postal's strong intuitions about his own idiolect, anaphoric reference into derivatives is possible for many people, though acceptability judgements are reported to vary widely from speaker to speaker. Corum (1973) suggests the following alternative to the anaphoric island constraint.

1. If the lexical item (i.e., the putative anaphoric island AG/JO) and the antecedent are not morphologically related, the sentence is unacceptable.
2. If the lexical item and the antecedent are morphologically related and if the lexical item commands the pronoun, the sentence is judged as deviant, but not to the same extent as starred (i.e., ungrammatical AG/JO) sentences.
3. If the lexical item and the antecedent are morphologically related and the lexical item does not command the pronoun, the sentence is either of moderately questionable acceptability or fully grammatical, depending on the idiolect of the speaker.

She supports this claim with examples of the following kind, which many people find perfectly acceptable.

Sergeant Pepper claims not to drink, but I have seen him take one now and then.

Jim reviewed that book and it will be published in 'Linguistic Inquiry'.

Shakespearean imitators usually fail to capture his style.

I speak French fluently because I lived there for eight years.

Barbara Walters interviewed Howard Hughes, but it won't be broadcast until Saturday.

Corum, extending the geographical metaphor, refers to dialects in which such sentences are acceptable as *peninsular dialects*, since peninsulas, unlike islands, can be reached directly from the mainland. We will present a number of examples from our corpus similar to Corum's later in this chapter.

Since the demise of generative semantics in the mid-1970s, linguists have lost interest in anaphoric islands. Indeed, one might argue that, as with many of the phenomena with which 'late' generative semantics concerned itself, the questions posed by anaphoric islands and peninsulas are primarily of psychological interest. However, psychologists have not investigated anaphoric islands or anaphoric peninsulas. Nor have they solved the problems that such constructions pose for theories of comprehension. We will present the results of some preliminary studies of our own below.

THE MENTAL MODELS THEORY OF TEXT COMPREHENSION

A number of theories of text comprehension have been inspired, either directly or indirectly, by the computational metaphor for the mind. Among the most promising of these accounts is Johnson-Laird's (1983) theory of mental models, which is closely related to developments in the neighboring disciplines of artificial intelligence (e.g., Webber, 1979), linguistics (e.g., Seuren, 1985) and formal semantics (e.g., Kamp, 1981; Heim, 1983; Barwise & Perry, 1984). The theory of mental models, as developed by Johnson-Laird is, as the subtitle of his book, *Mental Models*, indicates, a general theory about 'language, inference and consciousness.' Indeed, the most thoroughly worked out application of the theory is not to discourse understanding, but to syllogistic reasoning (Johnson-Laird, 1983, chap. 5; Johnson-Laird & Bara, 1984).

The principal idea of the theory is that thinking and language processing require the construction and manipulation of mental analogues of parts of the real or an imaginary world. In particular, the theory of mental models follows Bransford (e.g., Bransford & Franks, 1971; Bransford, Barclay, & Franks, 1972) and other earlier psycholinguists in rejecting the idea that the mental representation of the content of a text corresponds to any of its linguistic representations. Instead, it proposes that the representation is similar in form to those constructed from perceptual inputs. The theory of mental models tries to improve on earlier accounts by providing a detailed account of how information from different parts of a text is integrated into a coherent mental representation. Bransford, for example, had comparatively little to say about the *processes* that link information from different sentences.

The application of mental models to discourse comprehension has already been outlined in several places (e.g., Garnham, 1981; Garnham, Oakhill & Johnson-Laird, 1982; Johnson-Laird, 1983, chap. 14). The theory claims that each discourse or text is represented in its own mental model, which corresponds to the part of the real or imaginary world that the text is about. Mental models are constructed incrementally. Each new sentence adds to the model, and the model so far provides (part of) the context for interpreting the next sentence. Models are easier to construct if the text is coherent, and one of the main ways that a text can be coherent is by exhibiting *referential continuity*. The prototypical case of a referentially continuous text is one in which the present sentence or clause always contains a reference to something mentioned in the previous sentence or clause. Referentially continuous texts are presumably easier to understand because they place fewer demands on working memory (cf. Kintsch & van Dijk, 1978), though the details of this proposal remain to be worked out. Evidence that referentially continuous texts are easier to understand than discontinuous ones is provided by Garnham, Oakhill, and Johnson-Laird (1982), Ehrlich and Johnson-Laird (1982), and Oakhill and Garnham (1985).

Coherent, referentially continuous texts are not necessarily plausible. They may describe bizarre sequences of events. Plausibility as well as referential continuity contributes to the ease with which a text can be understood (Garnham, Oakhill, & Johnson-Laird, 1982; see also Johnson-Laird, 1983, pp. 375–377; Black, Freeman, & Johnson-Laird, 1986). The plausibility of texts derives primarily from their fit with how things usually happen in the world. Indeed, we claim that the structure that story grammarians find in text mirrors the structure of the way people think about events in the world. The use of that structure in text comprehension is, therefore, simply another example of the use of knowledge about the world in understanding.

Johnson-Laird (1983) closes his chapter on the coherence of discourse with brief discussions of the use of mental models to interpret definite and indefinite descriptions (see also, Garnham, 1982), the contrast between referential and attributive uses of definite descriptions (cf. Johnson-Laird & Garnham, 1980), and the use of mental models and propositional representations to interpret pronouns and other anaphors. This discussion introduces many of the problems that will be taken up later in this chapter, but it is based largely on ideas from linguistics, formal semantics, and artificial intelligence, rather than on psychological evidence about the interpretation of anaphors. We believe that, if the theory is to be successfully extended, it must be modified in the light of such evidence.

In a further chapter on the nature of mental models (1983, chap. 15) Johnson-Laird introduces some additional assumptions about these representations. First he proposes three general principles governing the construction of mental models. Models are computable, they are finite and their structure represents the structure of states of affairs. A more specific idea is that models obey a principle of economy. This principle states that one model will be constructed even if the description in a text is incomplete or ambiguous. Indeterminacies are, therefore, represented only if they can be incorporated within a single model. Otherwise a model may never be constructed. This idea is supported by the experimental results of Mani and Johnson-Laird (1982), who showed that the verbatim form of indeterminate descriptions of spatial layouts was better remembered than that of determinate descriptions. This finding suggests that indeterminate descriptions are held in propositional form, rather than being translated into a model.

From the general idea that mental models parallel the structure of the part of the world that they represent, Johnson-Laird concludes that such models are different from the sorts of propositional representation embodied in, for example, semantic network theories. He illustrates this idea by comparing the mental models representation of quantified sentences, which derives from the mental models theory of syllogistic reasoning, with one influential semantic network representation of such sentences, Hendrix's (1979) system of partitioned networks. For simpler sentences there is no need to make such a distinction. Garnham (1982), for example, illustrates some aspects of the mental models theory using a network-like representation.

The reason for insisting that mental models should have the same structure as part of the world is that these models are Janus-like entities, with two roles to play. First, they are *representations* and, as such, in need of an interpretation. Second, they have to play the role of models in model theory and provide an interpretation for expressions in natural languages. The idea of making the structure of models parallel that of the world is to make *their* semantics perfectly transparent so they can be treated as if they were the set-theoretic constructs that they represent (see also, Garnham, 1982, who draws the analogy between the representational system of mental models and that employed in mathematical writings on model theory). Where the semantics of semantic networks is transparent, they are on a par with mental models, but the semantics of partitioned nets is far from transparent.

The same duality of function—representation and model—must be attributed to the *discourse representation structures* in Hans Kamp's (1981) discourse representation theory. Spencer-Smith (1987, p. 4) makes this point particularly clearly by describing the language of discourse representation structures as "a logically perfect language, i.e. one in which the surface form of expressions corresponds to their semantic form." Any extension to the mental models theory should preserve this property of semantic transparency.

THE CORPUS OF ANAPHORIC EXPRESSIONS

Over a period of about a year we have been collecting examples of potentially interesting uses of anaphoric expressions. One way of making such a collection—a way that we have not followed—would be to interrupt speakers who use anaphoric expressions oddly and ask them what they really meant to say and what information they meant to convey. This technique has been widely used in the study of speech errors. We decided not to use it, because monitoring for such uses of anaphoric expressions is harder than monitoring for speech errors. As we will show, unusual uses of anaphoric expressions do not usually cause a misperception of the message that the speaker intended to convey—they are not so noticeable as speech errors.

We decided, therefore, to collect anaphors that we noticed in the course of our ordinary reading. In fact, our examples are taken mainly, though not exclusively, from written material intended for a general readership. Much of this material is journalistic and presumably prepared under some time pressure. Nevertheless, it has gone through several stages of editing. It was written by people whose trade is writing. It has, therefore, been produced and checked over by people who know how to present information so that it can easily be understood. Furthermore, it should also have been checked over by editors, who should correct any errors or unclarities that they notice. In less formal material—spontaneous speech or private letters, for example—people frequently say or write things that they realize are incorrect, or would judge to be incorrect if they could contemplate them at their leisure. However, in informal material these 'mistakes' are not always corrected, or even noticed, for various reasons. In a conversation the information content may not be of primary importance, or speakers may be unable to remember the exact form of an antecedent that they are trying to refer back to elliptically.

We are *not* claiming that all the examples that we cite are correct English, or even that they are all *grammatical* in the sense in which linguists use that term. Indeed, many ordinary readers can be led to claim that they are unsatisfactory. Nor are we claiming that every one of them is readily comprehensible, despite the efforts of their authors and editors, though most of them are. What we do believe, however, is that *most* readers of *most* of the examples we cite would not notice anything unusual about them if they were reading normally and not deliberately searching for examples of bad writing. Our language understanding systems cope perfectly well with these linguistic oddities, and we shall try to explain how they might do so.

In this exploratory stage of the project we have not attempted to be systematic, for three main reasons. First, until we started the corpus, we did not have a very clear idea of what we expected to find. We simply wanted to get some idea of the ways that anaphoric expressions are used in published texts. Second, anaphoric expressions are very common. If we had undertaken a systematic analysis of all anaphoric expressions, we would have been restricted to a few texts and our findings may have reflected the idiosyncrasies of a small number of writers. To be systematic, we would have had to consider all anaphors, because our ideas about what might make the interpretation of an anaphor difficult were very broad. Third, we would, in any case, have had to exercise extreme caution in any claim to be systematic. Most anaphors are so readily understood, that problems they cause to the language understanding system may go unnoticed. Just as in the case of speech errors, certain types of anaphoric oddity are easier to spot than others (see Cutler, 1981).

Fortunately, the arguments that we will make in this chapter are all of a kind that parallel what Cutler calls *some-errors* arguments about speech errors. A some-errors argument says that at least one speech error of a certain kind is well attested and that, therefore, the way that type of error arises must be explained. Some-errors arguments are relatively safe, since they depend only on the unambiguous identification of one error of a certain type, though more than one is obviously preferable. The other two main types of arguments identified by Cutler are *more-errors* arguments and *no-errors* arguments. A more-errors argument claims that one kind of error is more common than another. In making such arguments, particularly on the basis of an ad hoc collection of errors, one has to be aware of the possibility that one type of error may be easier to spot than another. A no-errors argument, which claims that errors of a certain type are impossible, risks refutation by a single counterexample. Despite their risky nature, no-errors arguments, if they are justified, can be important in testing models of the language processor. If there is something that the processor cannot do, that fact imposes constraints on its architecture. Cutler concludes that some-errors arguments are "without question the theoretically safest kind to make on the basis of error data collected from everyday speech" (1981, p. 578). This conclusion applies equally well to our corpus of anaphors from everyday text, though we are not committed to the view that our examples should be thought of as errors.

Two informal observations about the collection of our corpus are relevant to the way anaphoric expressions like those in our examples are usually interpreted. First, the number of examples we found increased with the amount of effort we put into finding them. We believe that many potentially troublesome anaphors go unnoticed by people who are reading for content. This observation, suggests, though it does not prove, that such anaphors do not require any out-of-the-ordinary processing. Second, even when we noticed a potentially interesting anaphor, it was not always immediately apparent how it should be characterized. It was sometimes hard work explaining just what was unusual about the anaphor. Again, this observation suggests that the anaphor was originally understood by a highly automatic process, typical of ordinary text comprehension, and not available to conscious awareness.

We can illustrate these points, and some others, with an example from the corpus. The following extract is from J.G. Farrell's novel *A Girl in the Head*.

> On his return from Boscobel he noticed a handbag on the hall table and a strange perfume on the stairs. Just for a moment he thought it might be Inez . . . (Farrell, 1981, p. 33)

When one of the authors (AG) first saw this example out of context, he could not understand why it had been included in the corpus. He assumed that

Inez was a perfume and that the interpretation of *it* was, therefore, straightforward. However, readers of the novel know that Inez is one of the female characters. They never consciously consider the possibility that it might be the perfume, but interpret that word as referring to something like 'the female presence that he detected in the house.' Nevertheless, if they are asked to spell out explicitly what *it* refers to, they may have some difficulty, as they might in explaining how they derived that interpretation.

It is impossible to explain the interpretation of such anaphors on the basis on their linguistic properties and those of the surrounding text. We will argue that they can only be understood with reference to a mental model of the situation that the text is about, a model that is, or can be, elaborated using relevant knowledge about the world. We now turn to the question of how the mental models theory must be extended and elaborated to account for the comprehension of anaphors in real texts.

SOME EXTENSIONS TO THE MENTAL MODELS THEORY

Surface Anaphora

As Johnson-Laird argues, texts are initially represented in a relatively superficial form, which he refers to as propositional, that encodes (some of) the verbatim detail of the sentences. Obviously, superficial details of the text have to be encoded as a step to working out its meaning, but is there any reason why such an encoding should be more than transitory? One reason is suggested by Mani and Johnson-Laird's (1982) finding, mentioned above, that *indeterminate* descriptions of spatial layouts—those compatible with more than one class of layout—tended to be encoded as 'propositions' rather than as a model. In normal text this propositional encoding might be retained in the hope that diambiguating information will be forthcoming. Mani and Johnson-Laird's further finding that determinate descriptions were easier to recall overall is consistent with Stenning's (1978, p. 193) suggestion that texts compatible with only one model are 'psychologically privileged.' People prefer to make models, but if indeterminacy make it too difficult, they try to retain surface form.

Hankamer and Sag's (1976) notion of surface anaphora, or ellipsis, suggests another role for memory for the surface form of a text. If the interpretation of an ellipsis depends on the exact form of a previous sentence or clause, that form must be remembered if the ellipsis is to be interpreted correctly. Although readers can always look back to the previous text, if they are skilled they usually do not. Listeners often have no choice. In producing discourse, writers and speakers must retain information about what they have just written or spoken, if they are to use surface anaphors felicitously. Writers, like readers, can always look back. However, as we shall see shortly, even writers use surface anaphors in a way proscribed by Hankamer and Sag's criteria. So the naive theory that listeners and readers retain a representation of the surface form of the text they have just heard or read, so that they can interpret surface anaphors when they come across them, is probably incorrect in many cases.

Murphy (1985) suggests that, particularly when surface anaphors are well separated from their antecedents, meaning-based representations contribute to their interpretation. We will return to this idea below. First, we will consider another claim of Murphy's, that, when a surface anaphor follows its antecedent closely, it may be interpreted by a process of *copying* its antecedent.

As Murphy himself would concede, this procedure cannot work in general, if the idea of copying is taken literally. Sag and Hankamer (1984) describe two classes of cases that are problematic for a copying account and, indeed, problematic for their original transformational account of surface anaphora.

First, as Sag (1976) and Williams (1977) have noted, the occurrence of quantifiers in the antecedent of an elliptical verb phrase complicates its interpretation in a way that is hard to explain in terms of copying theory or a transformation theory. Consider the following example (Sag & Hankamer's example 11, 1984, p. 330).

> Sandy thinks someone loves everyone.
> Chris does, too.
> Chris thinks someone does, too.

Sag and Hankamer claim that the ambiguity of the embedded clause *someone loves everyone* is preserved under the first 'deletion,' but not under the second. They go on to explain this fact, and others, on the assumption that verb phrase ellipsis depends not on identity of *surface forms* but on identity of *logical forms*, in the sense in which that term is used in Chomsky's (1981) theory of Government and Binding. Sag and Hankamer claim that this idea is compatible with recent theories of parsing in which the parser recovers binding relations as well as surface structure information. If the parser returns a surface representation that also indicates binding relations, it produces just the right information for the correct interpretation of surface anaphors.

Second, as Partee (1975) pointed out, indexical terms such as *I*, *you*, *here*, *there*, *now*, *then* behave in ellipses in a way that is incompatible with the copy theory:

> Please scratch my back.
> I will (i.e., scratch *your* back).

These facts do not follow from the identity of logical form theory of verb phrase ellipsis, but can be explained by a further principle that is compatible with that theory.

It would appear that Sag and Hankamer have fulfilled the promise of the title of their paper, to take us a step 'toward a theory of anaphoric processing.' The human sentence parsing mechanism computes logical forms (i.e., surface structures plus binding relations). An ellipsis takes its interpretation from a nearby element with a logical form that would be identical (up to alphabetic variation) to that of elided material, if it were there. Logical forms are used to construct mental models, though the details of this process are not provided by Sag and Hankamer. Deep anaphors are interpreted by relating them directly to a mental model.

The problem with this theory is that it is, at the same time, much better and much worse at interpreting anaphors than people are. It is better at classifying ellipses, in particular, as correct or incorrect. But it provides no mechanism for the rapid interpretation of ellipses that violate the condition on logical form. Of course, Sag and Hankamer could appeal to a problem solving mechanism to work out an interpretation for such ellipses, but we believe that they are interpreted quite naturally as part of the normal process of language understanding.

We would like to distinguish two abilities that people have. The first of these abilities is the ability to make judgements about the grammaticality, acceptability, or whatever one likes to call it, of sentences containing ellipses. The second is the ability to produce and understand ellipses in the everyday use of language. Sag and Hankamer implicitly assume that these two abilities are closely related. Whether they are is an empirical question that psycholinguists must try to answer.

We are in no sense challenging Sag and Hankamer's judgements about the felicity and possible meanings of ellipses. Most readers of the literature on ellipses, ourselves included, agree with most of the judgements in that literature. Sag and Hankamer's theory may well provide an account of the linguistic competence underlying this ability. What we are questioning is, first, whether the majority of native speakers of English have this ability, and, second, whether those that have it use it to interpret ellipses rather than to make judgements about them. On the first count, we believe that the ability to make reliable judgements about ellipses is strongly associated with a high degree of literacy. But even those people who might be expected, on these grounds, to have this ability to a high degree (professional writers) often produce apparently infelicitous ellipses. Casual observation suggests that infelicitous ellipses are even more common in speech. On the second count, the number of ellipses for which a simple copying strategy, or even an identity of logical form strategy, fails to produce an interpretation, but which seem entirely unproblematic to the naive reader, suggests that other factors help to determine the meaning of such ellipses. Furthermore, experimental evidence suggests that ellipses are not simply interpreted by copying pieces of representations of surface form or logical form.

The observation that we are making here is a special case of one that has long puzzled psycholinguists attempting to incorporate linguistic insights into a theory of language processing. A generative grammar can be thought of as a device for classifying strings of words as sentences of a language or not, and for assigning structures to those that are. However, the primary use of language is to convey information, and utterances used for this purpose may be grammatical or ungrammatical. The ability to make grammaticality judgements is a secondary one and one that is only well developed among the literate. So this ability is probably unnecessary for language processing, and the way we assign meanings to utterances may be very different from the way we might go about deciding if they are grammatical.

Before discussing some experimental results on the interpretation of ellipses, we will illustrate some the ways that ellipses are used in texts with some examples from our corpus.

Some ellipses from the corpus. 'Real' ellipses often violate the conditions laid down by Hankamer and Sag, namely that all ellipses should have linguistic antecedents and that those antecedents should have a parallel form to the elided material. It is comparatively rare to find an ellipsis without a linguistic antecedent. Indeed, our only clear case, which we have borrowed from Johnson-Laird (1983, p. 394), is a deliberate violation of Hankamer and Sag's constraint, designed to produce a particular effect:

> It is a spottily hilarious and technically brilliant cavort through art, using actors and animated pictures and starting with the Arnolfini family by Van Eyck. Oh yes you do. The one where the bride wears her wedding dress so oddly bunchy over her stomach. (Nancy Banks Smith, the *Guardian*, September 5, 1980)

Here the author anticipates her audience saying that they do not know the Van Eyck picture she is talking about. Her 'oh yes you do' is a retort to this anticipated response. Understanding this quotation requires processing that is more like problem solving than ordinary language understanding. We believe, therefore, that the language processor expects an ellipsis to have some linguistic antecedent, though not necessarily one with a parallel form.

The condition that the antecedent of an ellipsis should have a parallel form is indeed frequently violated. The simplest cases are those in which the ellipsis differs inflectionally from its antecedent. Although the occasional fervent generative grammarian has tried to deny it, these cases are fully grammatical. In such cases the ellipsis clearly differs in surface form from its antecedent on a literal interpretation of surface form. How the two differ in logical form is not clear, because generative grammarians have had little to say about the representation of tense and aspect. What is true, however, is that such inflectional mismatches usually go unnoticed and provide no phenomenological evidence for difficulty of understanding.

For example, in the following example of the *gapping* construction, the gap would be filled by *are*, though the antecedent is *is*.

> The wisdom is theirs, the errors mine. (Ward, 1986, p. 4)

Similarly, only a pedant would complain about the following instruction from a bottle of shampoo, where the anaphoric *do* in the simple present tense has as its antecedent a present participle (*-ing*) form.

> Avoid getting into eyes. If it does rinse with warm water.

One further example:

> There's nothing magic about sailing a ship. Columbus did.
> (*Sunday Times*, November 5, 1986)

Another kind of mismatch between antecedent and surface anaphor arises when the antecedent is affirmative and the anaphor negative, or vice versa. In itself this observation is of little interest, since it can be accommodated by appropriate assumptions about the position of the negative in surface form or logical form. However, a further complication arises when so-called *negative polarity* items appear. Two examples from the corpus have 'not . . . any . . .' in the antecedent, but the affirmative ellipsis must be read as *some*, not *any*:

> The sculpture does not represent any particular observer's-point-of-view, but the photograph does. (Fillmore, 1975, Lecture 2)

> If you see me walk on level ground, you might not notice any irregularities, but I do. (Heller & Vogel, 1986, extracted in *The Sunday Times*, August 24, 1986)

Again, this observation is inconsistent with a simple copying theory. It could be made compatible with an identity of logical form theory by a suitable assumption about the representation of negative polarity items at the level of logical form.

The kind of mismatch between antecedent and anaphor that Hankamer and Sag particularly wished to rule out for ellipses was the kind of gross syntactic mismatch exemplified by, for example, an active ellipsis with a passive antecedent. Yet examples of this type are found in our corpus:

> The Annan Committee recommended that the concession be dropped on the grounds that the BBC should not be expected to finance social benefits from licence revenue, but it is unlikely that any government would do so. (*New Society*, May 30, 1986)

Do so is a surface anaphor and its meaning in this example is *drop the concession*. Its antecedent in the complement clause of the first sentence of the extract is passive. Admittedly, this passage is a comparatively difficult one, but it would be hard to argue that the source of the difficulty was the nonparallelism between antecedent and anaphor, rather than the content.

Another transformational nonparallelism is illustrated in the following passage, in which Gore Vidal uses a quote by Tennessee Williams as the source of the antecedent for his own anaphor. Williams has topicalized the phrase *for love*, taking it out of the main verb phrase. Yet that phrase is included in Vidal's ellipsis.

> 'For love I make characters in plays,' he wrote; and did. (*Sunday Times Magazine*, August 3, 1986)

Vidal is saying that Williams made characters in plays for love, not just that he made characters in plays.

There can also be examples in which the ellipsis is transformed, but not the antecedent. The following example from the corpus illustrate this possibility. However, it is one of the less acceptable of our examples.

> The cotton knitwear felt good on and you didn't have to be a size 8 to wear it. The slinky jersey dress that you did, looked wonderfully sexy. (*Sunday Times*, February 15, 1987)

In this example the ellipsis is inside a relative clause, so the words missing (or replaced by *did*) are *had to be a size 8 to wear*. There is no overt direct object for *wear*, other than the relative clause's head noun phrase. The antecedent, of course, has the explicit direct object *it* following *wear*.

We move now to some examples that even the mildly pedantic would probably, at least after considered thought, complain about. Yet we have found that we often miss these 'monstrosities' on first reading and have to think quite hard about exactly what is wrong with them. We even go through phases of not noticing them at all, and only start to pick them up when we are trying to expand our corpus. These examples are easiest to spot when the antecedent and anaphor are close together and in comparatively simple 'sentences'. The following, from a book club catalogue, positively leaps out of the page.

> THE IRRESISTIBLE GARFIELD So many of you obviously couldn't!

Others are less obvious.

> great baggy shorts for those of us with big bottoms and tight jersey skirts for those who don't (*Sunday Times*, February 15, 1987)

In this example the interchangeability of *person with a big bottom* and *person who has a big bottom* has conspired with our concern for meaning rather than surface form to disguise, at least to some extent, an elliptical verb phrase with a prepositional phrase as its antecedent.

In the next example the close connection between the existence of a theory and someone producing it suggests an antecedent for *do* that the text does not provide.

> in his very good extended entry on personality, Reber says there are no coherent behavioural accounts, but that is exactly what Lundin tried to do. (Foss, 1986, p. 250)

Other examples seem willful, and are perhaps intended to create an amusing turn of phrase.

> and one [bedroom] with a four-poster bath built for a previous owner of this Queen Anne house who did little else. (*Sunday Times*, January 11, 1987)

The anaphors in our corpus suggest that lack of parallelism between surface anaphors and their antecedents need not be a major hindrance to comprehension. We suspect that, in spoken language, lack of parallelism is even more common. If speakers are to ensure parallelism, they must remember surface details of what they have just said. But people are not very good at remembering such details and are probably even poorer at doing so when they are concurrently trying to plan their next utterance. A more important conclusion is that, if ellipses are not parallel in form to their antecedents, readers and listeners cannot rely on a superficial representation of the preceding text to interpret them. Presumably they must be making use of a meaning-based representation, such as a mental model.

We turn now to some experimental evidence on the interpretation of surface anaphors in an attempt to clarify the way they might be understood.

Experiments on surface anaphora A number of studies (Garnham, 1987b; Garnham and Oakhill, 1987; Malt, 1985; Murphy, 1985; Tanenhaus, Carlson, & Seidenberg, 1985) have shown that, as the distance between a surface anaphor and its antecedent increases, the anaphor takes longer to interpret. This finding is consistent with the well-established fact that memory for the verbatim detail of a text is short-lived and begins to fade rapidly after the end of the current clause (cf. Caplan, 1972; Jarvella, 1979). However, the finding must be interpreted with caution, for two reasons. First, such distance effects have been reported for deep anaphors, whose interpretation does not depend on memory for surface form. These effects are often really change-of-topic effects that disappear when the intervening material is altered so that the topic does not change (see Garnham, 1987a, for discussion). Some distance effects for surface anaphors appear to be of this kind. Second, the involvement of memory for surface form in the interpretation of surface anaphors does not exclude the use of other information, as some of Malt's results show (see below).

Tanenhaus and Carlson (1985, 1986b) have investigated the effect of parallel and nonparallel antecedents in the interpretation of both deep and surface anaphors. They used a judgement task in which subjects were asked to read a context sentence and then judge whether a further target sentence made sense in that context. They found that parallelism regularly affected the number of 'makes sense' judgements for surface anaphors, but not for deep anaphors. Typically, the number of such judgements was around 90% for all deep anaphors and for surface anaphors with parallel antecedents. It fell to around 70% for surface anaphors with nonparallel antecedents. This figure is still fairly high, in accord with the idea that people are not greatly affected by nonparallelism, even for surface anaphors. However, too much should not be made of a particular numerical value, since it undoubtedly depends on the kind of items that were included in the experiment to produce some clear 'no' responses. The time to make the (positive) judgements was always affected by parallelism for both deep and surface anaphors, perhaps for different reasons. Nonparallelism not only breaks the fit between a surface anaphor and its antecedent, it changes the topic structure. For example, an antecedent may be made nonparallel by passivizing the sentence containing it. Passivization affects the person or thing on which attention is focused, which may in turn affect the availability of the antecedent of a deep anaphor. It any case, as some of the examples from the corpus suggest, nonparallelism does not have the disastrous effect on the interpretation of surface anaphors that a naive interpretation of Hankamer and Sag's theory might suggest.

The involvement of processes other than matching to the antecedent in the interpretation of surface anaphors is shown by Malt's (1985) finding that elliptical verb phrases are easier to interpret if they are part of an answer to a question containing their antecedent. Malt also shows that a change of focus between antecedent and elliptical verb phrase can make that verb phrase more difficult to understand. So, for example, the clause containing the ellipsis was read more quickly in the first of the following passages.

'I liked the Monet exhibit,' Heather remarked.
'It was really very nice.'
'I did, too,' Richard replied.

'I liked the Monet exhibit,' Heather remarked.
'Renoir is my favorite, though.'
'I did, too,' Richard replied.

However, this result may not reflect the interpretation of the ellipsis itself, so much as the fact that some of the people and things referred to within the elliptical VP, in this case the Monet exhibit, have become defocused.

The use of meaning-based representations in the interpretation of ellipses is also demonstrated by the findings of Garnham and Oakhill (1987) discussed earlier in this chapter. Garnham and Oakhill showed that elliptical verb phrases were often given plausible, but linguistically incorrect, interpretations, if those interpretations were suggested by context. The effect of distance, which was present when the plausible and linguistically correct meanings coincided, was greatly enhanced when the correct meaning was implausible. It seemed as though people were trying to interpret the ellipses correctly, but that, when the surface form of the preceding text was difficult to retrieve, it sometimes lost out in a prolonged battle with plausibility to assign a meaning to the elliptical verb phrase. If plausibility can override the correct meaning in this way, it can surely produce an interpretation when there is no linguistically suitable antecedent.

The Interpretation of Surface Anaphors—Conclusions

Surface anaphors are often readily interpreted even though their antecedents are not identical in surface form or in logical form. If such interpretation occurred only with difficulty, and appeared to be a reasoning-out process rather than a normal process of text understanding, it would have only indirect relevance to a theory of text comprehension. This idea may be correct in the case of those rare surface anaphors that have *no* linguistic antecedent. However, the fact that lack of parallelism between surface anaphor and antecedent often goes unremarked, both by language producers and by those trying to understand them, suggests that the understanding of ellipses without direct reference to surface or logical form is a normal part of comprehension.

A number of questions remain to be answered. In particular, we have not yet found any examples in real texts in which a surface anaphor whose linguistically correct antecedent is implausible has a plausible interpretation suggested by context. It may be that writers and speakers avoid anaphors of this type, although they are not consciously aware that they are doing so.

The other side of this coin might be that readers and listeners assume that they will not encounter these kinds of anaphors and that this assumption provides a partial explanation for why the elliptical verb phrases in our experiment (Garnham & Oakhill, 1987) were quite frequently misinterpreted.

Deep Anaphora

Sag and Hankamer (1984) contrast ellipsis with *model-interpretive anaphora* (their former deep anaphora). Model-interpretive anaphors do not, in any sense, require parallel antecedents, since they do not require linguistic antecedents at all. The problems that they pose to the language understanding system are, therefore, rather different from those posed by surface anaphors.

Sag and Hankamer (1984, p. 335) claim that model-interpretive anaphors are understood by 'direct reference to constructs of the understander's discourse model.' On this account model-interpretive anaphors are interpreted using only a representation of content. More superficial representations have no role in how they are understood. We saw above that meaning-based representations are sometimes used to interpret ellipses, even when the use of those representations leads to such anaphors being misinterpreted. We may, therefore, ask whether superficial representations play any role in the interpretation of model-interpretive anaphors, even though their use is unnecessary. One reason to think they might was mentioned above. Distance effects have often been reported for deep anaphors. The further the anaphor is from its antecedent, the longer it takes to understand. Since there is no reason to think that a mental model encodes the temporal order in which it was constructed, the existence of these distance effects appears to suggest a role for superficial representations in the interpretation of deep anaphors.

Garnham (1987a), in reviewing these distance effects, concluded that many, if not all, of them could be explained by a change of topic or focus between the anaphor and its antecedent. When changes of focus are eliminated, distance effects often disappear as well (e.g., Lesgold, Roth, & Curtis, 1979). However, it remains uncertain whether all distance effects with deep anaphors can be explained away in this manner. So, a representation of the verbatim detail of a text may be implicated in the resolution of deep anaphors.

The question of focus is a vexed one. Clearly, items do become more or less easy to refer back to, depending on the direction of the conversation or text. However, in what sense readers and listeners have to keep track of sets of focused items and how focusing determines what kind of reference is appropriate are questions that remain to be answered. One view is that only focused items can be pronominalized (e.g., Sanford & Garrod, 1981). However, this view is in danger of becoming circular if the only test for whether an item is focus is to try to refer to it using a pronoun. Some examples from our corpus show that pronouns are often used where there is unlikely to be any misunderstanding but where there has been some digression between them and their antecedents. For example, in the following passage there is no doubt that *they* refers to the authors of the book under review. However, the use of a pronoun is not altogether felicitous:

> The authors then discuss in chapter three the nature of analogies and how they are used to build an explanatory science. The fourth chapter explores the limitations of cognitive psychology, in particular the 'modularity hypothesis' or the notion than many psychological processes are sufficiently explained by the operation of separate information-processing modules within the minds of individuals. Naturally, they reject this approach. (Aitkenhead, 1986, p. 419)

Further examples in which a pronoun is used after what appears to be a change of focus, with no danger of misunderstanding, are:

> The latest in this line is Professor David Premack's star pupil, a chimpanzee called Sarah, who thanks to her ability to manipulate symbols representing words has become the heroine of many a psychological textbook. In *Gavangi!* he concludes that . . . (Sutherland, 1986)

> I can also let Leader readers into another Men Only secret. Mandy T is 18-year-old Luci Lewis. . . . The pictures were taken by Brighton photographer Jim Deans, who runs glamour sessions for amateur photographers. She was pictured in the Leader last July. (*Brighton and Hove Leader*, January 12, 1985)

The following example is, perhaps, more surprising:

> They said good-bye, and Doctor Copeland watched them from the dark front porch as they went together up the street. Their footsteps as they walked away had a lonesome sound and he felt weak and tired. When they were a block away Willie began playing his harmonica again. The music was sad and empty. He stayed on the front porch until he could neither see nor hear them any longer. (McCullers, 1943 p. 77)

Here one might expect *he* to refer to Willie, the last mentioned male character. Yet the content of the sentence makes it clear that it refers to Doctor Copeland. In fact it is difficult to decide whether Willie has become focused in this passage and whether Doctor Copeland has become defocused—presumably it was the latter who found Willie's music sad and empty. This example, therefore, points to difficulties in the notion of focus and questions whether it has a role in a theory of anaphor interpretation.

The idea that deep anaphors are interpreted directly in mental models, which is one version of the mental models theory of text comprehension, leaves two questions unanswered. First, what are the 'constructs of the understander's discourse model' (Sag & Hankamer, 1984, p. 335)? What are the elements of a discourse model that are waiting to be referred to by model-interpretative anaphors? Second, it is really true, as Sag and Hankamer's way of talking suggests, that every element that might be referred to by a model-interpretative anaphor is already in the model? Might not some only be inserted into the model in response to subsequent expressions that cannot otherwise be understood?

The first question is more difficult to answer than it initially appears to be. The simplest cases are easily accommodated. If a story begins by talking about *a man*, an element standing for that man is introduced into the discourse model and, provided that certain conditions are satisfied, a subsequent *he* will be taken to refer to that man.

> A man entered the room.
> He started rummaging through the drawers.

However, even this text contains some model-interpretive anaphors that are less easy to explain. What justifies a definite reference to *the room* in the first sentence, or to *the drawers* in the second?

The answer is different in the two cases. There is a sense in which *the room* should have been *a room*. This is the first mention of the room, and an indefinite reference is appropriate. However, definites can be interpreted as introductory references if there is no element in the model for them to refer to. The choice of a definite reference when a person or thing is first introduced suggests a familiarity with that person or thing that an indefinite reference would not. Such a choice can, therefore, be used as a literary device to create this effect of familiarity.

The drawers in the second sentence, on the other hand, is justified by the fact that rooms are quite likely to have drawers in them. A reference to drawers following a reference to a room will usually be taken to designate the drawers in the room. The use of the indefinite *some drawers* would, in this case, suggest either that the man looked in some but not all of the drawers or that this was not the kind of room that one would expect to find drawers in. Establishing that the drawers are drawers in the room is an example of what Clark (1977) calls a *bridging inference*. Haviland and Clark (1974), in a classic psycholinguistic experiment, established that bridging inferences take time and they they are made in a backwards direction. So, when people read:

> We checked the picnic supplies.
> The beer was warm.

They take longer to read the second sentence of this passage than when they read the same sentence in:

> We got some beer out of the trunk.
> The beer was warm.

Only when they read about beer, and cannot find a referent for it, do they introduce beer into their mental model. This finding shows that it is not, in general, true that anaphoric expressions have referents that are already in the mental model. Sometimes those referents must be put there using knowledge-based inferences, for example the inference that picnic supplies may well include beer.

The general problem of how world knowledge in used in text comprehension is a very difficult one, and one that cannot be solved here. However, there are other, more tractable, questions about the kinds of elements that expressions of different types introduce into discourse models and make available for subsequent reference. Webber (1979) provides an answer to some of these questions. For example, noun phrases may introduce individuals, sets, stuff, generics, and prototypes. A single noun phrase may introduce more than one possible referent. For example, *three men* introduces both a set of three men considered as a set and that set considered as three individuals. So,:

> Three men lifted a piano.

may mean that they lifted one piano together or that each one of them lifted a piano separately.

Violations of the anaphoric island constraint. In the rest of this section we will be concerned primarily with violations of the anaphoric island constraint, which was introduced earlier in this chapter, and their implications for anaphoric processing. Pronominal references that violate the anaphoric island constraint are well represented in our corpus. A typical example is:

> 'Snodland is a hotbed of socialism,' he said. 'I've been fighting them since 1945 and I will never give up.' (*Sunday Times*, June 29, 1986)

Them are the socialists whose existence is implied by the mention of socialism in Snodland. Another example, which does not involve direct speech, is:

> The unemployed are traditionally apathetic and Murphy does not deny it exists, but he sees the dangers of harping on the fact. (*New Society*, April 18, 1986)

In this case the use of *it* to refer to the apathy implied by the fact that the unemployed are apathetic seems a little strained.

Pronominal references of this kind pose two main problems for a theory of text comprehension. First, they further complicate the question of what items an expression makes available for future reference, and whether those items are computed prospectively or retrospectively. Second, they raise the question of when a syntactically non-matching antecedent is preferred to a matching one. In almost any text there will be some possible matching antecedent for a pronoun, provided that pronoun is far enough into the text.

A convenient fiction, maintained in much of the linguistic literature, is that definite pronouns agree syntactically with their antecedents. Well-known classes of exceptions, such as referring to cars as *she*, are regarded as aberrant in some way. However, as Weise (1983) and others have argued, even in languages such as French and German, in which nouns have syntactic gender, syntactic agreement is regularly violated. Examples from our corpus include one from an advertisement for a bedding centre:

> If you want a bed quick we have got them.

and one from a Sainsbury's leaflet on milk:

> Protein: These are our body's building bricks.

Such examples inevitably complicate a theory of text comprehension. It is often assumed that, when a pronoun is encountered, the possible antecedents that are considered first are those that were introduced by expressions that have the same syntactic number and gender. This assumption would explain why pronouns are understood more quickly when there is a gender cue to their interpretation. For example, Garnham and Oakhill (1985) showed that the second clause of:

> Alan lent a pen to Jill, because she wanted to write a letter.

is understood more quickly than that of:

> Vicky lent a pen to Jill, because she wanted to write a letter

However, the idea that items in a mental model are marked with information about the syntactic number and gender of expressions that introduced them is not comparable with the basic assumption of mental models theory, namely that mental models are structurally similar to parts of the world, not to linguistic representations. It may turn out, therefore, that the results of Garnham and Oakhill (1985) should be explained in terms of semantic, not syntactic, number and gender. Furthermore, in a broader context plausibility also plays a role in the interpretation of pronouns, often overriding purely syntactic considerations.

What happens, for example, when the first matching antecedent for a pronoun is a long way back, and perhaps out of focus? In such cases a nonmatching antecedent for a pronoun may be preferred to a matching one. This possibility is illustrated in the following passage:

> I've had all that and no longer care about the social scene. I don't *need* parties. Besides, I'm quite fanatical about smoking and won't go anywhere where I have to inhale it. (*Sunday Times Magazine*, January 18, 1987).

It is 'obviously' smoke. Its nearest matching antecedent is *the social scene*, but that cannot be the referent. A more interesting example is:

> He made a considerable mark as Information Technology Minister, a post which ideally suited his interests as one who appreciated and became involved in the whole word processor revolution while he was still in opposition during the 1970s and long before most people had heard of them. (*Times Higher Education Supplement*, May 30, 1986).

In this example, *them* cannot be most people (it would have to be *themselves*), nor does it mean 'the 1970s' or 'his interests.' The case of *the 1970s* is the most interesting one, since that expression both agrees syntactically with the pronoun and is nearer to *them* than the anaphoric island *word processor revolution* that actually provides the antecedent. General knowledge tells us that most people who might be reading the *Times Higher Educational Supplement* in 1986 have heard of the 1970s. Such examples, therefore, rule out the idea that, if a syntactically matching antecedent is found, it is automatically accepted.

Our corpus also contains examples that are similar to those that violate the anaphoric island constraint, but in which there is no single word or compound that can be identified as the island. In these examples the antecedent is almost mentioned, but not quite. It can be derived by a simple inference. For example, in the following passage:

> I bedded a student for a while (a female one) but dared not tell you because you would not accept its simple goodness. (Bannister, 1983, p. 52)

there is no explicit antecedent for the sentential *it*, which means, roughly, 'my bedding of that student' or 'the fact that I bedded a student.' A similar example comes from one of Van Gogh's letters.

> I have sometimes wondered why I was not more of a colourist, because my temperament decidedly seems to indicate it—but up to now it has developed very poorly. (Bernard, 1985, p. 48)

Here the meaning of it is 'my being a colourist.' Finally, a slightly different kind of inference is needed to find the antecedent for *it* in the following passage about Harold MacMillan.

> In everything his sense of humour was unfailing, and of a particular quality. Often black, it also contained a special delight in the ridiculous. As with courage, he greatly prized it in others.

It can hardly be 'his sense of humour,' but must be 'a sense of humour.'

Reference into an anaphoric island, a phenomenon in which linguists apparently lost interest after the demise of generative semantics, raises a large number of empirical questions about text comprehension. One set of questions concern whether reference into an anaphoric island is more difficult than a corresponding reference to a matching antecedent and, if it is, whether that difficulty arises because of the relative unacceptability of reference into an anaphoric island. We recently explored these questions in two experiments (Garnham & Oakhill, in press). In these experiments subjects read passages such as:

> Young Toby gets into fights every playtime.
> but they never lead to injuries.
>
> Young Toby fights every playtime,
> but they never lead to injuries.

In the first of these passages, the pronoun *they* has an explicit antecedent *fights*. In the second passage the pronoun has only an implicit antecedent. The activity of fighting is mentioned, but the fights themselves are not. According to Postal, *fights*, the verb, should be an anaphoric island in which *they* cannot find an antecedent.

In an attempt to vary the acceptability of the passages, we used either present tense verbs, as above, chosen so that the third person singular always had the same form as the corresponding plural nouns (*fights-fights*) or past tense verbs for which this parallel was broken (*fought-fights*). This manipulation was based on an empirical observation, rather than on any theoretical considerations. Our own judgements about the acceptability of examples in our corpus and of those discussed by authors such as Postal (1969) and Corum (1973) suggested that parallel forms were less likely to jar. We, therefore, hoped to determine whether any extra difficulty in understanding the second type of passage could be explained in terms of acceptability.

We used a timed judgement task similar to that used by Tanenhaus and Carlson (1985, 1986b) and, as Table 1 shows, our tense manipulation did affect acceptability. The passages with explicit antecedents were almost always judged acceptable, whereas those with implicit antecedents within anaphoric islands were not. However, when the verbal anaphoric island was not identical to the corresponding plural noun, the proportion of negative judgements was higher. One possible explanation of this result is that a *superficial* representation of the sentences is influencing the interpretation of *deep* anaphors. A very low-level representation, such as a purely phonological one, would not distinguish between *fights* the plural noun and *fights* the verb. So the language processor might be misled by finding *fights* in that representation into thinking that *them* has an explicit noun antecedent.

Table 2 shows that times taken to make the judgements. Reference into an anaphoric island always took longer to resolve than reference to an explicit antecedent. However, the difficulty was no greater for the passages with nonparallel verbal antecedents, which were much less acceptable overall. These results suggest that the relative slowness of the judgements when reference is into an anaphoric island cannot be explained by the relative unacceptability of the passages. It follows that the mention of an activity, such as fighting, does not automatically make the corresponding entity, fights in this case, available for anaphoric reference. Just as beer has to be connected with picnic supplies by an inference when beer is subsequently mentioned, it appears that fights are only inferred from the activity of fighting when there is a subsequent reference to them.

Table 1. Percentage of YES and NO Judgements for Each Type of Passage in the 'Anaphoric Island' Experiments

	Type of Passage		
	Explicit Noun Antecedent	Implicit Verb Antecedent	
(a) Verbal antecedent identical to nominal (fights-fights)			
YES	92	54	73
NO	8	46	27
(b) Verbal and nominal antecendents nonidentical (fought-fights)			
YES	93	29	61
NO	7	71	39

Table 2. Times (in milliseconds) to Make Positive and Negative Judgments in 'Anaphoric Island' Experiments (Table 1 shows the percentage of judgements of each kind)

	Type of Passage	
	Explicit Noun Antecedent	Implicit Verb Antecedent
(a) Verbal antecedent identical to nominal (fights-fights)		
YES	1871	2281
NO	2577	2422
ALL	1926	2347
(b) Verbal and nominal antecendents nonidentical (fought-fights)		
YES	2225	2640
NO	2913	2583
ALL	2268	2600

CONCLUSIONS

Our long-term goal is to extend the mental models theory to produce a more adequate theory of text comprehension. The theory emphasizes that mental representations of the context of texts are constructed incrementally and that the representation constructed so far provides (part of) the context for the interpretation of the next sentence. One of the principal devices that links the information in different sentences of a text is anaphoric reference. So part of a goal is to provide a satisfactory account of how anaphoric expressions are interpreted.

Some previous accounts, whose proponents were, perhaps, led astray by the phenomenological ease with which anaphors are understood, have regarded anaphoric reference as, essentially, a trivial problem. Indeed, mental models theory is important not so much because the general principles it embodies are new, but because it stresses that there are hard problems to solve in formulating a theory of anaphoric processing. Bransford, for example, pointed out the integrative and constructive aspects of text processing, but he never went on to produce a detailed theory of integration and construction.

A more recent trend—one that has tended until now to inform developments in mental models theory is to look to accounts of anaphor interpretation in linguistics and, to a lesser extent, AI as the basis for a psychological account. This approach has already paid dividends, for example in recognition that there are two broad classes of anaphoric expression, deep (model-interpretive) anaphors and surface anaphors (ellipses). However, as we have seen, Sag and Hankamer's account of these two types of expression is more likely to form the basis of a theory about considered judgements on the acceptability or felicity of anaphoric references than one of how anaphoric references are resolved during text comprehension. Empirical evidence suggests that ellipses are not interpreted solely from a superficial representation, nor does the interpretation of deep anaphors depend only on what is in a mental model.

Our own approach to constructing a psychological theory has been to look at the kinds of anaphoric expression that are found in published texts. We have not taken the simplistic view that any published anaphor must be one that is readily interpreted. Nor have we claimed that all the anaphors in our corpus are fully acceptable. Good writers would certainly want to reformulate some of our examples, though others were produced by established literary figures, such as Vidal, McCullers and Farrell. It is fair to say, however, that our examples have all got past the censors of, admittedly sometimes rather rushed, writers and editors. Most of these oddities do not prevent readers from deriving the intended messages from the passage, even if they do, as our experimental evidence suggests, slow readers down by a few hundred milliseconds.

We regard our corpus as a source of ideas about the way people usually process anaphors. However, we are keenly aware of the limitations imposed by the way the examples were collected. We are not suggesting that ideas suggested by those examples should be accepted without further experimental evidence. Indeed, those ideas are being or will be put to experimental test.

One example of such an experimental test, our work on anaphoric islands, was reported at the end of the chapter. We have shown that a number of lines of evidence converge on the conclusion that, although linguistically deviant in some way, pronouns with implicit antecedents supplied by verbs are readily understood. They appear with some regularity in published prose; they are, at least in some cases, sanctioned by linguists; in our experiment, they are, in the majority of cases, allowed as sensible continuations of preceding material. Nevertheless, they do cause problems to the language processing system. They are judged as nonsensical far more often than pronouns with explicit noun phrase antecedents and, even when they are judged to be acceptable, the judgement takes a comparatively long time to make. It appears, therefore, that implicit antecedents are not incorporated into mental models until they are needed to interpret a subsequent expression. The conditions under which they are readily introduced and the circumstances under which lack of an explicit antecedent does not produce a linguistic anomaly both require further investigation.

In this chapter we have only discussed some kinds of example from our corpus. We will close by briefly mentioning two more. Many of the examples of deep anaphors illustrate the role of knowledge of the world in text comprehension, perhaps the most difficult aspect of text comprehension to elucidate. One particularly interesting example illustrates the power of knowledge—in this case, specific information provided by context—to override what ought to be the linguistic meaning of a text. Under the subheading 'How does it feel to be a parent?' the editor of *Practical Parenting* writes:

> Actually, I already know, I have two.

Well, yes, one is tempted to say, most people do have two parents. However, the passage continues:

> The ones shown in the picture with me.

The accompanying picture shows the editor with Ben, aged 5, and Miranda, 18 months. So the message she is conveying is not that she has two parents from whom she has found out what it is like to be a parent but, rather:

> I already know how it feels to be a parent, I have two children.

The context provided by the picture leaves no doubt that this message is the intended one, despite the linguistic form of the text.

A second kind of example involves surface anaphors, rather than deep anaphors. A single sentence may contain two ellipses. If it does, the question arises as to how the order of the ellipses relates to the order of their antecedents. Our corpus contains only two such examples, but in both cases the dependency is crossed. The antecedent of the first ellipsis occurs earlier than the antecedent of the second.

> She thought 'If this boy finds and gives me a cornelian I shall marry him', which he did and she did. (*Sunday Times*, March 8, 1987)

> *That's* the good life, not everyone is capable of it, not everyone has the courage. We are, and we have. (Murdoch, 1986, p. 91)

While it would be dangerous to generalize on the basis of two examples, if this preference for crossed rather than nested dependencies continues to hold as more examples are collected, it may have implications for a theory of language processing. It may, on the other hand, turn out that the interpretation of double ellipses depends on plausibility, in which case such ellipses would provide further evidence for the use of meaning-based representations (mental models) in the interpretation of surface anaphors.

REFERENCES

Aitkenhead, M. (1986). Review of R. Harre, D. Clark, and N. De Carlo, *Motives and Mechanisms. Bulletin of the British Psychological Society*, 39, 419–420.

Bannister, D. (1983). *Burning leaves.* London: Pan.

Barwise, J., & Perry, J. (1984). *Situations and attitudes.* Cambridge, MA: MIT Press.

Bernard, B. (Ed.). (1985). *Vincent, by himself.* London: Orbis.

Black, A., Freeman, P., & Johnson-Laird, P.N. (1986) Plausibility and the comprehension of text. *British Journal of Psychology, 77*, 51–62.

Bransford, J.D., Barclay, J.R., & Franks, J.J. (1972). Sentence memory: A constructive versus interpretive approach. *Cognitive Psychology, 3*, 193–209.

Bransford, J.D., & Franks, J.J. (1971). The abstraction of linguistic ideas. *Cognitive Psychology, 3*, 331–350.

Caplan, D. (1972). Clause boundaries and recognition latencies for words in sentences. *Perception and Psychophysics, 12*, 73–76.

Chomsky, N. (1981). *Lectures on government and binding.* Dordrecht, Netherlands: Foris.

Clark, H.H. (1977). Bridging. In P.N. Johnson-Laird & P.C. Wason (Eds.), *Thinking: Readings in cognitive science.* Cambridge, England: Cambridge University Press.

Corum, C. (1973). Anaphoric peninsulas. *Chicago Linguistics Society, 9*, 89–97.

Cutler, A. (1981). Guest editorial: The reliability of speech error data. *Linguistics, 19*, 561–582.

Ehrlich, K. & Johnson-Laird, P.N. (1982). Spatial descriptions and referential continuity. *Journal of Verbal Learning and Verbal Behavior, 21*, 296–306.

Farrell, J.G. (1981). *A girl in the head.* Glasgow, Scotland: Fontana.

Fillmore, C. (1975). *Santa Cruz lectures on deixis.* Bloomington, IN: Indiana University Linguistics Club.

Foss, B. (1986). Review of Reber, *Dictionary of psychology. Bulletin of the British Psychological Society, 39*, 250–251.

Garnham, A. (1981). Mental models as representations of text. *Memory and Cognition, 9*, 560–565.

Garnham, A. (1982). *On-line construction of representations of the content of texts.* Bloomington, IN: Indiana University Linguistics Club.

Garnham, A. (1987a). Understanding anaphora. In A.W. Ellis (ed.), *Progress in the psychology of language* (vol. 3). London: Erlbaum.

Garnham, A. (1987b). Effects of antecedent distance and intervening text structure in the interpretation of ellipses. *Language and Speech, 30*, 59–68.

Garnham, A., & Oakhill, J.V. (1985). On-line resolution of anaphoric pronouns: Effects of inference making and verb semantics. *British Journal of Psychology, 76*, 385–393.

Garnham, A., & Oakhill, J.V. (1987). Interpreting elliptical verb phrases. *Quarterly Journal of Experimental Psychology, 39A*, 611–628.

Garnham, A., & Oakhill, J.V. (in press). 'Anaphoric islands' revisited. *Quarterly Journal of Experimental Psychology.*

Garnham, A., & Oakhill, J.V., & Johnson-Laird, P.N. (1982). Referential continuity and the coherence of discourse. *Cognition, 11*, 29–46.

Hankamer, J., & Sag, I.A. (1976). Deep and surface anaphora. *Linguistic Inquiry, 7*, 391–428.

Haviland, S.E., & Clark, H.H. (1974). What's new? Acquiring new information as a process in comprehension. *Journal of Verbal Learning and Verbal Behavior, 13*, 512–521.

Heim, I. (1983). File change semantics and the familiarity theory of definiteness. In R. Bauerle, C. Schwarze, & A. von Stechow (Eds.), *Meaning, use, and interpretation of language*. Berlin: de Gruyter.

Heller, J., & Vogel, S. (1986). *No laughing matter*. London: Jonathan Cape.

Hendrix, G.G. (1979). Encoding knowledge in partitioned networks. In N.V. Findler (Ed.), *Associative networks: Representation and use of knowledge by computers*. New York: Academic Press.

Jarvella, R.J. (1979). Immediate memory and discourse processing. In G.H. Bower (ed.), *The psychology of learning and motivation, 13*. New York: Academic Press.

Johnson-Laird, P.N. (1983). *Mental models: Towards a cognitive science of language, inference, and consciousness*. Cambridge, England: Cambridge University Press.

Johnson-Laird, P.N., & Bara, B. (1984). Syllogistic inference. *Cognition, 16*, 1–61.

Johnson-Laird, P.N., & Garnham, A. (1980). Descriptions and discourse models. *Linguistics and Philosophy, 3*, 371–393.

Kamp, H. (1981). A theory of truth and semantic representation. In J. Groenendijk, T. Janssen, & M. Stokhof (Eds.), *Formal methods in the study of language*. Amsterdam, Netherlands: Mathematical Centre Tracts.

Keenan, J.N., MacWhinney, B., & Mayhew, D. (1977). Pragmatics in memory: A study of natural conversation. *Journal of Verbal Learning and Verbal Behavior, 16*, 549–560.

Kintsch, W., & van Dijk, T.A. (1978). Towards a model of text comprehension and reproduction. *Psychological Review, 85*, 363–394.

Lesgold, A., Roth, S., & Curtis, M. (1979). Foregrounding effects in discourse comprehension. *Journal of Verbal Learning and Verbal Behavior, 18*, 668–682.

McCullers, C. (1943). *The heart is a lonely hunter*. London: The Cresset Press.

Malt, B.C. (1985). The role of discourse structure in understanding anaphora. *Journal of Memory and Language, 24*, 271–289.

Mani, K., & Johnson-Laird, P.N. (1982). The mental representation of spatial descriptions. *Memory and Cognition, 10*, 181–187.

Murdoch, I. (1986). *The good apprentice*. Harmondsworth, England: Penguin.

Murphy, G.L. (1985). Processes of understanding anaphora. *Journal of Memory and Language, 24*, 271–289.

Oakhill, J.V. & Garnham, A. (1985) Referential continuity, transitivity and the retention of relational descriptions. *Language and Cognitive Processes, 1*, 149–162.

Partee, B.H. (1975). Deletion and variable binding. In E.L. Keenan (Ed.), *Formal semantics of natural language*. Cambridge, England: Cambridge University Press.

Postal, P.M. (1969). Anaphoric islands. *Chicago Linguistics Society, 5*, 205–239.
Ross, J.R. (1967). *Constraints on variables in syntax.* Unpublished doctoral dissertation, MIT.
Sachs, J.S. (1967). Recognition memory for syntacticc and semantic aspects of connected discourse. *Perception and Psychophysics, 2*, 437–442.
Sag, I.A. (1976). A logical theory of verb phrase deletion. *Chicago Linguistics Society, 12*, 533–547.
Sag, I.A. (1979). The nonunity of anaphora. *Linguistic Inquiry, 10*, 152–164.
Sag, I.A., & Hankamer, J. (1984). Toward a theory of anaphoric processing. *Linguistics and Philosophy, 7*, 325–345.
Sanford, A.J., & Garrod, S.C. (1981). *Understanding written language: Explorations in comprehension beyond the sentence.* Chichester, England: Wiley.
Seuren, P. (1985). *Discourse semantics.* Oxford, England: Blackwell.
Slobin, D.I. (1966). Grammatical transformations and sentence comprehension in childhood and adulthood. *Journal of Verbal Learning and Verbal Behavior, 5*, 219–227.
Spencer-Smith, R. (1987). Semantics and discourse representation. *Mind and Language, 2*, 1–26.
Stenning, K. (1978). Anaphora as a approach to pragmatics. In M. Halle, J.W. Bresnan, & G.A. Miller (Eds.), *Linguistics theory and psychological reality.* Cambridge, MA: MIT Press.
Sutherland, N.S. (1986, June 27). Animal magic: A review of D. Premack, *Gavagai! Or the future history of the animal language controversy. The Times Higher Educational Supplement.*
Tanenhaus, M.K., & Carlson, G.N. (1985). Processing deep and surface anaphora. In *Proceedings of the Fourteenth Annual Meeting of the Northeastern Linguitic Society.* Amherst, MA: GLSA.
Tanenhaus, M.K., & Carlson, G.N. (1986). Processing verb phrase anaphors. In C. Clifton (Ed.), *Proceedings of the Eighth Annual Conference of the Cognitive Science Society.* Hillsdale, NJ: Erlbaum.
Tanenhaus, M.K., Carlson, G.N. & Seidenberg, M.S. (1985). Do listeners compute linguistic representations? In D.R. Dowty, L. Karttunen, & A.M. Zwicky (Eds.), *Natural language parsing.* Cambridge, England: Cambridge University Press.
Ward, C. (1986). *Chartres: The making of a miracle.* London: The Folio Society.
Webber, B.L. (1979). *A formal approach to discourse anaphora.* New York: Garland Publishing.
Weise, B. (1983). Anaphora by pronouns. *Linguistics, 21*, 373–417.
Williams, E. (1977). Discourse and logical form. *Linguistic Inquiry, 8*, 101–139.

CHAPTER 5

Component Processes of Reference Resolution in Discourse*

A.J. SANFORD
Department of Psychology
University of Glasgow
Scotland

For more than a decade, psychologists have been studying language understanding with a view to modelling the way in which understanding comes about (e.g., Sanford & Garrod, 1981). This chapter is essentially a report on some aspects of the more recent research which my colleagues and I have been carrying out, and relates our findings to the work of others. The work reported is dominated by two questions: What is the time course of processes leading to the cohesion of a discourse, and what kinds of thing constrain the prominence that different parts of a discourse play in cohesion. These two themes are highly related to one another, of course. What is important is likely to get priority in processing. Above and beyond the accounts of empirical research, we shall ask what type of devices(s) might be needed to support the kinds of results found, in a rather more general way. This part of the chapter will be more discursive and less tight than the earlier parts, which were designed to lay a good empirical foundation for the arguments. Finally, no attempt has been made to be exhaustive in introducing the work of others. The present chapter is to be understood more as an interim report than as a review of the literature.

---

*Most of the experimental work described here was carried out jointly with colleagues at the University of Glasgow. Special thanks should be given to Simon Garrod, together with whom I have had many fruitful discussions. Thanks are also especially due to Linda Moxey, with whom I had discussions about the present paper and who assisted in its preparation. Thanks are also due to the Science and Engineering Research Council, and the Economic and Social Research Council, both of which supported various aspects of the research reported here.

A. BACKGROUND

In our earlier work (Sanford & Garrod, 1981), we were concerned to explore a number of mechanisms supporting understanding. The framework which we developed rested upon the simple principle that a language processor is so arranged that it seeks the significance of an input as early as possible. A similar view has been put forward by other workers (e.g., Just & Carpenter, 1980). We further assumed that linguistic input serves as a partial match on structured knowledge within the reader; if a good match is made, the appropriate background knowledge becomes available for use. We used the term *implicit focus* for this activated knowledge, and the term *scenario* for its content. A scenario is similar to a script or some other schema structure. Once a particular scenario structure is in implicit focus, it is used to interpret subsequent input, precisely in the way Schank and Abelson (1977) supposed scripts to enable understanding. The utterances actually used we assumed had representations too; these we assumed were limited by working-memory limitations. As a text unfolded, both implicit and explicit focus might change: in other words, they were assumed to be *dynamic*.

We presented a variety of data in support of these claims, concentrating partly upon the idea that language retrieves situation-specific knowledge from memory. Elsewhere, this angle has been expanded upon considerably (see, e.g., Sanford, 1987, for an overview). Our other concern was to track down some of the detail of how linguistic expressions interact with the proposed structure of focus. We proposed that different types of reference address different aspects of focus, and produced the following formulations. *Definite reference* using pronouns was restricted to explicit focus; that is, a pronoun was thought of by us as a search directive to the processor to examine representations of things which had been explicitly mentioned. Pronouns, we argued, would only initiate a search of implicit focus if a search of explicit focus failed. In contrast, we supposed that full noun phrase reference could be used to address implicit focus. An illustration will help at this point.

(1) Erica unlocked the door with difficulty.
(1′) The key tended to stick.

As we have argued extensively (especially Garrod & Sanford, 1982), verbs like *unlock* have as part of their representation in implicit focus *key*. A noun phrase reference to *the key* in (1′) initiates a search in implicit focus leading to a mapping between the representation of *key*, which is part of the verb-schema, and *the key*, which has just been mentioned. The idea is that this can be done readily with a full noun-phrase, but not with a pronoun. We shall discuss this a little later.

In a similar way, indefinite expressions like *a waiter* were assumed not to result in a search for antecedents primarily, but to result in a new representation being put into explicit focus for that character. However, it was assumed that some sort of mapping takes place between an indefinite expression and the contents of focus. This makes sense with the following example:

(2) Ian picked up the menu in the cafe.
(2′) A waiter came to take his order.

At once, *a waiter* is a *new* individual, but derives his significance and raison d'etre from the cafe scenario in implicit focus. So indefinites do not preclude searches. Indeed, if they did, the following would not seem unusual:

(3) Steve went into the supermarket.
(3′) A waiter was standing near the fruit stall.

These and other expressions were thus thought of as initiating searches of different types. Differences in times for anaphoric reference resolution under different conditions in explicit focus were considered to be the result of differing degrees of *emphasis* of some sort in explicit focus—this we termed *foregrounding*, taking this term as denoting a continuum of activation.

A major assumption made in this work was that the processing of anaphora occurred as immediately as possible, and this claim is examined in some detail in the sections which follow. The other assumption, that anaphora processing can be construed as a search operation, is also scrutinized in some detail, utilizing the results of a number of different studies.

B. IMMEDIACY OF ANAPHORIC REFERENCE

Since anaphora is one of the major contributors to text cohesion (Halliday & Hasan, 1976), anaphor resolution is a good candidate for an *immediate* process. It helps the process of constructing a discourse model, if the reference of expressions can be decided early on. In this section, evidence will be reviewed for the claim that anaphoric reference establishment is a rapid, online process, and that it seems to have various *automatic* qualities. Let us begin with some relatively early data.

In an early exploration of anaphoric reference resolution with full noun phrases, Garrod and Sanford (1977) explored the role which the *strength* of a class membership relation could play in coreference between category and exemplar descriptions, as in

(tank)
(4) A (bus) came roaring around the corner.
(5) The vehicle almost flattened a pedestrian.

Of course, *tank* is a low frequency exemplar of a vehicle, while *bus* is a high frequency exemplar, and this fact is known to influence class membership decision times (Wilkins, 1971; Sanford, Garrod, & Boyle, 1977). Using a self-paced reading time task, Garrod and Sanford established that the time to read a target sentence (like (5)) was shorter following a high frequency antecedent (e.g., *bus*), an effect of about 90 msecs. This is consistent with the idea that retrieval of the necessary identity relation between the two forms of description depends upon the time to retrieve the class membership relation. But more importantly, the presence of such an effect (called a *conjoint frequency effect*) can be taken as an index of a matching check between the two entities taking place.

The conjoint frequency effect can be used to show how prominent and powerful such matching operations are. The autonomous nature of the process is revealed through examples of the following type:

(tank)
(6) The (bus) came trundling down the hill.
(6′) It nearly crushed a vehicle.

Sentence (6′) has a strange ring about it—it seems curiously truncated. What is interesting is that Garrod and Sanford found a small (47 msec) conjoint frequency effect for materials of this type. In other words, some process of matching *tank/bus* to *a vehicle* is taking place. This is happening although other cues militate against the sense of the process. For instance, since *It* has presumably been taken as coreferential with tank/bus, coreference with *a vehicle* is ruled out syntactically and pragmatically. Furthermore, the indefinite article in *a vehicle* is a further cue that there is no coreference. The results thus suggest an autonomous matching process based upon preexisting and text-independent semantic relations between *tank/bus* and *vehicle*. This result has been replicated in an unpublished study (Cuthill, Golombok, & Rana, 1978).

If there is an autonomous semantic process, then by ruling out coreference *semantically*, there should be no sign of a conjoint frequency effect. Garrod and Sanford used materials like the following in a test of this proposition:

(7) A bus/tank came trundling down the hill.
(7′) It nearly collided with a horse-drawn vehicle.

This time there was no conjoint frequency effect. The semantic matching operation was either not undertaken given the restricting adjectival phrase (i.e., was ruled out), or perhaps a fast check finding no overlap would result in a rejection of more detailed processing (in the manner suggested by Smith, Shoben, & Rips, 1974, for explicit category judgements). Either way, there is evidence for automatic checking for coreference, even in the presence of strong syntactic and pragmatic cues to the contrary. We shall return to this issue in the discussion.

These observations suggest that attempts at reference resolution may take place relatively automatically as soon as a noun phrase in encountered, a view central to Sanford and Garrod's (1981) original framework, and shared by other investigators. More direct tests of this proposition have been attempted using a variety of techniques, including priming test and spelling error detection. For instance, Dell, McKoon, and Ratcliff (1983) presented subjects with short texts one word at a time. At various points during the presentation a single word was exposed as a recognition test. By choosing the points in the text at which the recognition probe was used, these authors were able to show that anaphoric noun phrases speeded the recognition time for both the antecedent noun, and other nouns taken from the sentence in which it was mentioned. Their data supported the view that such information was recovered within 250 msecs of encountering the critical noun phrase. Such a result gives credibility to the idea that the sort of information associated with anaphoric resolution is somehow activated as soon as a potential anaphor is encountered. However, priming experiments of this sort might show that information relevant to anaphoric resolution is activated as soon as a potential anaphor is encountered, but they do not show that anaphor resolution actually takes place.

A somewhat different task was used by Garrod and Sanford (1985) to try to overcome some of these problems. Rather than use priming as an index of facilitation, we used the latency of detecting a simple spelling error during self-paced reading. This is a visual analog of the technique devised earlier by Cole and Jakimik (1978, 1980), in which they used mispronunciation detection latency in continuous speech as a index of the influence of context. In the present experiment, materials of the following type were used:

(8) Elizabeth was a very inexperienced swimmer, and wouldn't have gone into the pool if the lifeguard hadn't been nearby. But as soon as she was out of her depth, she started to panic and wave her hands about in a frenzy. Within seconds Elizabeth/the lifeguard jumped/sank in the pool . . .

The passage continues, but the snippet here takes us up to the important point. In the last sentence there are two alternative character references—one to Elizabeth, the main character, and the other to the lifeguard, a secondary character. Then there are two alternative verbs, *jumped* or *sank*. In the context of what went before, *jumped* is obviously most appropriate for *the lifeguard*, while *sank* is most appropriate for *Elizabeth*, given her prior behaviour and state.

Table 1. Results of the Error Detection Study, Taken from Garrod and Sanford (1985). Times in Msecs. Section (a) Shows the Results for Noun Phrase Anphors, Section (b) the Results With Pronouns.

	Anaphor	Target Verb: Consistent	Target Verb: Inconsistent	Difference
(a)	Proper name (main character)	2442	3182	740
	Role description (secondary character)	2800	3372	572
(b)	Main character pronoun	1926	2477	551
	Secondary character pronoun	2212	2053	−159

Our reasoning was that, if an appropriate verb is somehow enhanced in accessibility over an inappropriate verb, as a result of prior context, then we should be able to detect this through a spelling error detection task with the error on the verbs. With the present example, *jumped* was translated to *jimped*, and *sank* to *senk*. We hypothesized that, if verb predictability is influenced by prior context, then the error detection latency for *senk* would be faster following *Elizabeth* than *the lifeguard*, and that the latency for *jimped* would show the opposite pattern. Twenty-eight passages of this sort were used in the experiment.

The results, shown in Table 1(a), demonstrate a clear effect of verb predictability as a function of context. Subsidiary tests, based on controls, suggest that the spelling errors are being detected as soon as the readers encounter them (see Garrod & Sanford, 1985, for details). This observation has quite strong implications for the timing of a variety of processes, and for a general account of interpretation. First, note that there is nothing to *intrinsically* associate sinking with a swimmer and diving with a lifeguard. Certainly, over the whole set of materials used, there is no intrinsic association between verbs and characters within passages. So the predictability effect in this experiment is *entirely* due to bringing the current situations of the characters in the particular passage to bear on verb recognition; it is not due to some generalized associative priming. Secondly, the enhanced recognition speed due to local context could *only* be brought to bear on line after the character descriptions has been read. This is so because it is only through the character description that the *relevant* contextual background could be accessed. So the process of resolving the anaphoric reference must be immediate rather than deferred, and be a process enabling the weightings of particular verbs to be very rapidly changed.

We believe the results of this study to be important in that they not only suggest immediate reference resolution to occur with simple noun phrases, but show how rapidly quite complex and extensive knowledge structures can be utilized. Clearly, it would be a good think to have more evidence on this issue, and doubtless it will be forthcoming. But the evidence strongly supports immediate resolution of noun-phrase anaphors.

C. IMMEDIACY, CHARACTER-TYPE, AND PRONOUNS

When one turns to pronominal anaphora, the question of immediacy can again be asked. Indeed, the problem has received considerable attention from a number of investigators (e.g., Shillcock, 1982; Tyler & Marslen-Wilson, 1982; Vonk, 1985). We will look at some of these points shortly, but first I shall report the results of the second part of the Garrod and Sanford (1985) study described above. With some minor adjustments to the materials, it is possible to substitute *he* or *she* for the full noun phrases in the passages of the spelling error study, with the pronouns making unambiguous reference back to one or other of the characters. Then, precisely the same question could be asked as was asked before. Is the spelling error on the verb detected more quickly as a result of being made predictable through the character referred to by the pronoun? When this experiment was carried out, exactly the same results as those previously obtained held for the character which was given a proper name (the main character of the story), but there was no effect of predictability on error recognition time for the character given a *role* description.

These results are shown in Table 1(b). Garrod and Sanford (1985) argued that, while pronouns referring to *main* characters receive an immediate interpretation, those referring to secondary characters do not, and this lack of interpretation is why there is no verb predictability effect with the secondary character. There is thus a difference between the results using full noun phrases and those using pronouns, which must be attributable to the different functions of these forms of description.

The difference is consistent with our differentiation of explicit and implicit focus. As we have argued, full noun phrases serve (typically) to introduce individuals or reintroduce individuals which have a low priority in explicit focus. Pronouns address entities in explicit focus (typically), and they are therefore sensitive to the priority of items within explicit focus (i.e., degree of foregrounding). While this goes some way toward an explanation, it is obviously incomplete. Why and how is resolution deferred in the case of the secondary character? To understand why, we need to look a little deeper at what it is that is special about main characters, and to look at other evidence relating to the main/secondary distinction. This we shall do in the next section of the chapter. So far as the process which differentiates resolution in the cases of the two character types is concerned, we shall turn now to other evidence. First, in spontaneous *speech*, he/she pronouns at the start of the sentences tend to be *used* for referring to the thematic subjects of the discourse, which is usually the "main character" of the discourse (Karmiloff-Smith; 1980; Marslen-Wilson, Levy, & Tyler, 1982). So pronouns seem to be used for maintaining reference to an important character. This is not so true of pronouns in later positions, where various prior features of the sentence may serve to cue other interpretations. But with pronouns right at the beginning cannot be influenced by such factors, if they are processed immediately. And our claim is that in the case of *main characters*, they are processed immediately. So it would seem that we have to think of main-characterhood as operating in a top-down way. When a pronoun is encountered and its interpretation is not otherwise restricted, a test is done between the pronoun and the main character to check for a match. If no *other* information is available, then the only data which can be used for a matching process are number and gender. Now what is particularly interesting is that in our experiment, some of the subjects reported a curious effect. They reported feeling as though the *wrong pronoun* had been used, and perhaps mispelt themselves, either by adding an *s* to give he->she, or omitting one. Such reports are certainly compatible with the view that the pronoun is rapidly interpreted as referring to the main character, but if this is erroneous, then verb information will be used to alter interpretation to the secondary character. On this line of reasoning, it should always be the verb relevant to the main character which is primed (hence producing a faster recognition time for the spelling error). Although it is not statistically reliable, there is a negative effect of − 159 msecs for verb predictability with the secondary character, which is otherwise difficult to explain unless one assumes that it is merely noise. The conclusion, then, is that main characters capture sentence-initial pronouns for interpretation. With full noun phrases, of course, this is precluded, since whichever character is being described is directly signalled.

With different types of materials in which two characters interact, and where the text is very short, it becomes difficult to use the intuitive concept of main character in any convincing way. For instance, who is the main character in (9)?

(9) Jack apologized profusely to Joss.

Elsewhere (Sanford & Garrod, 1981) we have argued that, for single active sentences, the initial noun phrase is likely to be highlighted over other noun phrases. This can be shown using reading time techniques, and by using the frequency with which a particular noun phrase is used in a task where subjects are invited to continue writing a passage on the basis of an initial sentence. The percentage of times the first noun phrase is chosen over any other noun phrase can be considered an index of the degree to which the referent of that noun phrase is highlighted. However, Newman and Canham (1986) recently investigated the pattern of continuations resulting from four sentence types (active, passive, cleft, and pseudocleft). The choice of topic for continuation was most frequently the *agent* of the stimulus sentence, regardless of sentence position. Thus it would seem to be agency rather than sentence position which is the determiner of highlighting.

A recent study by Hudson, Tanenhaus, and Dell (1986) looked at the resolution of anaphoric references to agents and nonagents in simple two-person interactions. An example of their materials is:

(10) Jack apologized profusely to Josh.
(10a) He had been rude to Josh yesterday.
(10b) He had been offended by Jack's comment.

In (10) Jack is the agent, and (10a) refers to Jack. This they call a *centered* target (the second sentence is the target sentence). The alternative target, (10b), refers to Josh and is called *noncentered*. In one study, they asked people to *judge* whether the second sentence fitted sensibly with the first, as quickly as possible. Centered targets were accepted as sensible much more quickly than noncentered targets (almost 500 msec faster). This is interesting, especially when the following targets to (10) are considered.

(10c) Jack had been rude to Josh yesterday.
(10d) Josh had been offended by Jack's comment.

Target (10c), although *centered*, was processed considerably more slowly than the pronoun in version (10a). Furthermore, with the fuller noun phrase, there was no effect of centering at all. The investigator's interpretation was that the target sentence pronouns would be *immediately* interpreted as referring to the discourse center (agent). If it turned out not to, then readers would have been garden-pathed and have to assign the pronoun, leading to longer judgement times. In addition, the authors predicted that the agent-centered interpretation pattern would be blocked if a full noun phrase was used. This happened. A further experiment supported the results of the first.

Thus we have evidence from two quite different types of experiment which produces convergent results. Pronouns in initial positions in sentences seem to be automatically interpreted as referring to the currently

Table 2. Typical Material Used in the Anderson et al. Study.

In the Restaurant
The Browns were eating a meal in the restaurant.
The waiter was hovering around the table.
This restaurant was well known for its food.
Five hours/forty minutes later the restaurant was empty.
They) (eating
He) had enjoyed (serving all the good food. (target)

[Parentheses denote options in presentation]

most important entity in the discourse (main-character; discourse center). Full noun phrases are not interpreted in this way. There is thus good evidence for a powerful top-down constraint on the interpretation of pronouns, to which we shall return later.

D. THEMATIC SUBJECTS: ADDITIONAL FINDINGS

Apart from the pronoun/noun phrase difference, the effects described above hinge upon a difference in the way in which processing activities relate to type of character. This in turn raises the question of how we might define type of character. For the moment, let us accept the intuitive distinction between main and secondary characters, and note some additional ways in which they differ in terms of processing. The first evidence we obtained was reported in Anderson, Garrod, and Sanford (1983). It was noted that, while the availability and accessibility of representations of main characters was unaffected by shifts of episode in a story, both the availability and accessibility of secondary character was reduced. This result has been amply discussed elsewhere, so it will only be briefly outlined here. Passages like the one in Table 2 were used.

Intuitively, *the Browns* are the main characters in this vignette; the waiter is secondary, owing his presence in the story to it being set in a restaurant.

The fourth sentence in the passage indicates a time-shift. The argument was that the 5-hour time reference would cue the end of an episode. The background scenario in implicit focus (basically a restaurant script) would be no longer relevant at this point, and we predicted that it would no longer be active as implicit focus. In contrast, the 40-minute shift does not so cue the end of an episode.

Our predictions were that subsequent anaphoric references to the secondary character (*the waiter*, here) would be influenced by time shift. Using a reading time task, if the final sentence of the passage (the target sentence) refers back to *the waiter*, then it is read *more slowly* after the large time shift than after the small time shift. In complete contrast, with main character references (*The Browns*), time shift had no effect. So one characteristic of secondary characters seems to be that, when the background shifts so that their normal setting is shifted out of focus, they become less accessible. A further result was that the time to read sentences referring to secondary characters was generally slower than that to read sentences referring to main characters, even if the secondary character had been referred to more recently.

Main and secondary character representations seem to be associated with different patterns of inferential activity, quite apart from the control of the immediacy of pronominal reference resolution discussed in the previous section. In particular, certain types of inference seem to be more readily made with respect to main characters than secondary characters. To illustrate, consider the following illustrative example:

(11) Mr. Brown went into the local cinema to see the latest spy movie.
(11′) He was shown to the only remaining seat by the usherette.
(11″) The air felt hot and sticky.

Sentence (11″) is what might be termed a *psychological atmosphere statement*. It refers to some aspect of the setting has which is not directly attributed to a persons perception, but which is the kind of thing that is mediated by perception. Other examples would be "The city seemed busy that night" or "visibility was rather poor." Returning to (11″), how might such a sentence be interpreted? Either Mr. Brown or *the usherette* might well be affected by the air feeling hot and sticky. But, intuitively, it seems more likely that the statement will be seen as more relevant to the main character. Thus, in some sense, (13) seems to be easier to answer than (12):

(12) Did the usherette find the air hot and sticky?
(13) Did Mr. Brown find the air hot and sticky?

Al-Ahmar and I (unpublished research; Sanford Garrod, & Al-Ahmar, 1985) surmised that the first of these questions suggests a *no* or *don't know* answer, while the second suggests a *yes* answer. Using a variety of materials of this type, Sanford and Al-Ahmar found that there were more *yes* responses to such questions asked of main characters, while for secondary characters the responses *no* and *don't know* were more prominent, as shown in Table 3. By way of caveat, it should be pointed out that this effect only emerged under self-paced reading conditions with a requirement to choose a response category as quickly as possible. In a pencil-and-paper version, with indefinite time to deliberate, the category *don't know* predominated. This is scarcely surprising, since, strictly speaking, the answers to the questions are indeterminate. Speeded responding precludes careful, logical analysis.

Table 3. Frequency of Choosing *Yes*, *No*, or *Don't Know* As a Response to Atmosphere Questions Asked About Main and Secondary Characters.

	Yes	No	Don't Know
Main	.85	.08	.08
Secondary	.51	.27	.22

The results support the proposition that the content of a psychological atmosphere statement is more easily judged as relevant to main characters than to secondary characters. Possibly, atmosphere statements are interpreted as reflecting what is *perceived* by the main character, and obtain their textual coherence in that way. Alternatively, the results might simply reflect plausibility judgements made at the time of answering the questions, rather than inferences made at the time of interpretation. In order to cast more light on the problem, a further study was carried out (Garrod & Sanford, 1988).

A typical material from this experiment is shown in Table 4. Following the title, the first two sequences introduce the main and secondary characters. The third sentence is the optional atmosphere sentence. The next sentence is used to reintroduce one of the characters, which will be the one mentioned in the final sentence. The final sentence can be about the main or the secondary character also. This is the sentence of interest, since what it depicts is an action which is a natural consequence of the atmosphere being hot and sticky. Now, if the inference had been drawn that "Alistair found the atmosphere hot and sticky," then the reading time for *He mopped his brow* would be faster than it is if the atmosphere sentence was absent. And if, as we suspect, a similar inference is not drawn with respect to *the waitress*, then there should be no difference between the condition where the atmosphere statement is present, and where it is absent, for the sentence *she mopped her brow*.

Table 4. Example of Materials in the Reading Time Study of Inferences Relating to Psychological Atmosphere Statements (after Garrod & Sanford, 1988).

Lunch at the Cafeteria

Alistari hung up his coat and picked up a tray.
The waitress smiled as she poured the coffee.
The atmosphere was hot and sticky.
He took)
She offered) the cup.
He) (his
She) mopped (her brow.

[Parentheses denote options in presentation]

Table 5. Reading Times for Target Sentences in the Atmosphere Sentence Study. The Presence or Absence of an Atmosphere Sentence Affects the Reading Times for Main Character Sentences, but not for Secondary Character Sentences (from Garrod & Sanford, 1988). Times in Msecs.

	Presence of atmosphere statement	
	Present	Absent
Main character	1379	1650
Secondary character	1430	1463

Using materials of this type, the influence of the presence or absence of atmosphere sentences on target-sentence reading times was tested for main and secondary character conditions. The results were shown in Table 5. Two things are notable.

First, for the main character condition, target sentence reading time is considerably shorter in the presence of the atmosphere statement. In the secondary character condition, there is no real difference. This is more or less as hoped for, and shows that atmosphere statements certainly have an effect on processing in relation to main characters but not on secondary characters. But the second point complicates the picture. If anything, the effect on main character seems to result from an *increase* in reading time in the absence of an atmosphere statement. So, while it is possible that the attributional inference is drawn during reading (Garrod & Sanford, 1988), producing an enhancement, there is also a strong negative effect.

One possibility is that on encountering the description of an action undertaken by a main character, the processor attempts to fit this action into a causal sequence. If the atmosphere statement had been made, then that statement provides the *reason* for the main characters action, so the statement is easily interpreted. In the absence of an atmosphere statement, the goal of setting the main character's actions in a causal chain cannot be fulfilled so readily. Presumably, time is spent while the process trying to forge causal connections is failing to find a satisfactory outcome and this explains the extended reading time. In the case of *secondary* characters, however, the processor is not disrupted by the absence of a reason for the action, suggesting that the same attempt to locate the action in an unambiguous causal chain is not being made. So once again, we end up in the position that when a main character is mentioned, a different kind of activity is initiated in order to establish coherence.

In story-schema analyses, stories are usually portrayed as being about the activities of a single main character or set of characters (e.g., Grimes, 1975; Mandler & Johnson, 1977; Rumelhart, 1975). The events which make up the episodes of a story are usually based on the goals and plans of the character(s). The processor spends time looking for antecedents to unexplained actions when the main character is the actor. But this aspect of story structure is not enough to explain the results described above, since the texts are far too brief to allow the frequency of goals associated with a character to be the determiner of main-characterhood. Indeed, it is not even frequency of mention which determines it in the example above. It is an open question how these character-based constraints on processing are initiated, although, for very simple texts, we have recently investigated some possible factors. First of all, there is primacy of mention. If a character is introduced first into a short snippet of discourse, it is possible that this would serve to give that individual special status in processing. Intuitively this does not seem to be very compelling:

(14) The manager dictated very quickly.
(14′) The secretary had trouble taking it all down.

Neither *the manager* nor *the secretary* seem to be particularly good candidates for main-characterhood. In contrast, another potential cue is the use of a *proper name*, often used in short narratives to introduce main characters. This seems much more potent:

(15) Mr. O'Donnell dictated very quickly.
(15′) The secretary had trouble taking it all down.

This seems to highlight Mr. O'Donnell, if anything, while, in the example below, it is Sheena who is highlighted:

(16) The manager dictated very quickly.
(16′) Sheena had trouble taking it all down.

An empirical test of these contentions is to invite people to write the next line of such two-line passages. Presumably, the character they continue with corresponds to the main character (Sanford, Moar, & Garrod, in preparation). Indeed, in a study of this type, it was established that giving characters names caused those characters to be mentioned most frequently in continuations. If both characters have proper names, or both are given role descriptions, then each will be mentioned about equally often. Primacy or recency had no observable effect. Thus the use of a proper name serves to highlight a character, and, once a character is used in a continuation, it is likely to figure in a longer continuation, thus being a main character in the sense of being mentioned frequently.

Before leaving this issue, it is worth noting that it is possible to carry out a reading-time version of this experiment. When this is done, it is found that sentences containing pronominal anaphoric references to characters given proper names are read more rapidly than the same sentences referring to characters described by roles. This corresponds to earlier results found for main-versus-secondary characters (Anderson, Garrod & Sanford, 1983). However, in keeping with the distinction made between the role of pronouns and full noun phrases, the advantage of named characters disappears when the anaphors are repeat names or role-descriptions.

E. OTHER EVIDENCE FOR IMMEDIACY WITH PRONOUN PROCESSING: BONDING

In this section I shall introduce a rather different set of observations suggesting automatic, rapid processing with pronouns, this time particularly in relation to the pronoun *it*. This is the king of pronouns, or course. It can refer to practically anything:

(17) John went to town to buy a car he had seen.
It was a good place to shop. (it = town)
It was an old Jowet Javelin. (it = car)
It was the height of extravagance. (it = buying the car)
It upset his girlfriend. (it = his entire action)

and so on. This pronoun is a prime candidate for ambiguity. The examples shown above clearly reveal the hazards of immediate interpretations of the sentence-initial *It*. Small wonder it is a happy hunting ground for discourse psychologists.

During normal speaking and normal writing, we are all prone to use this pronoun infelicitously on occasion. This is what Kantor (1977) calls "strained anaphora." We can usually get the intended referent in the end, but an utterance can still seem to be strained in some way. And sometimes it has a humorous side to it, as in the following example, from Hirst (1979):

(18) If an incendiary bomb lands near you, don't lose you head. Put it in a bucket and cover it with sand.

This amusing example is extremely interesting. The effect, which I call "false bonding" (Sanford, 1985), is easily recognized, but knowledge-based coherence rules out the false bond as an intended mapping. A slightly weaker example is the following, from Wilks (1975):

(19) John left the window and drank the wine on the table. It was brown and round.

Here it sounds as though it is the wine which was brown and round, although, of course, the reader will normally come to the conclusion that it is not.

Such false bonding effects are certainly not random, and are due to emphasis or foregrounding effects. For instance, in the case of (18), the if-then structure of the first sentence presumably foregrounds that action *don't lose your head*, in which the main object is *head*. Although there is a need for a thorough analysis of the foregrounding structure of if-then sentences, it is relatively easy to make up examples parallel to (18). In the case of (19), *the wine* is the direct object, which is being manipulated by the actor. It seems likely that entities being directly manipulated by actors will be foregrounded, although more evidence is required here to separate such a possibility from straightforward syntactic effects.

A study of false bonding was undertaken by Sanford, Garrod, Lucas, and Henderson (1983) in an attempt to study several aspects of pronominal reference resolution. First, we contended that pronouns *automatically* bond to representations of entities in explicit focus; pronouns are only used to refer to entities in implicit focus as a secondary option. Earlier, Sanford and Garrod (1981) had noticed that false bonding seemed to occur with examples where pronouns were used to refer to implied entities rather than explicit ones:

(20) Mary dressed the baby.
(20′) They were made of pink wool.

and

(21) Sailing to Ireland was eventful for Harry.
(21′) It sank without trace.

Intuitively, in (20′) *They* seems to refer to *Mary and the baby*, while in (21′) *It* seems to refer to Ireland. Although one does get a correct reading, the incorrect bonding is prominent too, coexisting with the final correct reading. Notice that in example (21), the *event* is put in the initial position, but even this does not make the implied means of transport predominate over the simple explicit description *Ireland*.

The false bonding effect here vanishes in the presence of an explicit intended antecedent:

(22) Mary put the clothes on the baby.
(22′) They were made of pink wool.

and

(23) Taking the boat to Ireland was eventful for Harry.
(23′) It sank without trace.

This too seems to suggest that, if there *is* a suitable *explicit* entity, then the pronoun becomes bonded to it; if it is relevant, all is well. If not, false bonding ensues. To test whether false bonding actually led to a disruption of processing, materials were presented to subjects in a self-paced reading time setting. We predicted a slowing due to bonding in the *absence* of an appropriate explicit antecedent. With an appropriate antecedent, we predicted a relatively fast reading time. As a control, we included an antecedent present/absent contrast with materials which were *bond enabling*. For instance:

(24) Being arrested)
Being arrested by the police) was extremely embarassing for Andy.
(24′) They took him to the station in a van.

There is no explicit plural antecedent onto which the pronoun *they* could falsely bond. The anticipated result was that the reading time for sentence (24′) should be longer in the absence of an explicit antecedent, but not so long as that obtained for the bond-enabling materials.

The results of this study are shown in Table 6. There was a massive increase in target sentence reading time for the conditions supporting false bonding. This emerges as a huge interaction between antecedent status and whether the materials were bond-enabling or not.

In subsequent experiments, it was established that, in the absence of an explicit antecedent, even under bond-enabling conditions, there is a small increase in the reading times for the target sentence over that obtained in the presence of an explicit antecedent. This is consistent with the idea that pronouns have explicit focus as a primary locus for reference resolution. As a result of these studies, it seems that *event-based* bonding does not occur, or if it does, it must be extremely difficult to engineer. An example of an event-based bond-enabling material is the following:

Table 6. Reading Times for Target Sentences as a Function of Antecedent Sentence Type (bond-enabling or bond-excluding), and the Presence or Absence of an Explicit "Intended" Antecedent. Data from Sanford et al. (1983).

	Antecedent	
	Explicit	Not Explicit
Bond-enabling	2259	3163
Bond-excluding	2317	2456

(25) Driving was very tiring.
(25′) It had extremely poor air conditioning.

The pronoun *it* in (25′) *could* refer to the event depicted in (25) unless the predicate information is used to rule out the possibility (as it ultimately is, of course). In contrast, with the following, the target could not possibly refer to the event:

(26) Marrying his second wife made Tony happy.
(26′) She is certainly compatible with him.

Using sets of materials with this sort of contrast, there was no observable difference between the explicit antecedent/no explicit antecedent conditions. Seemingly, then, in the absence of an explicit antecedent *entity*, the pronoun *it* does not automatically map onto the whole event. This is an interesting limitation, since sentences beginning with *It* frequently refer to entire events or actions.

In summary, the basic false-bonding effect suggests the following. First, when a pronoun occurs at the start of a sentence, a matching operation is brought about between the pronoun and likely antecedents. The match is made against antecedents in *explicit focus*, and is presumably based on number and gender matching. (As argued, degree of foregrounding in explicit focus is also an important factor, but this is not well understood in the present context.) This process is automatic and occurs straight away. Subsequently, information in the rest of the sentence may run counter to the match, in which case false bonding has occurred. This, of course, is very much the same idea as that discussed in the previous section, only we have approached it from a different angle. Furthermore, the idea that rapid "temporary assignment" of pronouns to referents occurs, only to be later changed if necessary, receives support from other investigators, using different techniques, (e.g., Vonk, 1985).

There are problems with the formulation given above, however, and these are interesting from the point of view of the notion of "temporary assignment." Sanford (1985) observed that, when the pronoun *it* is used nonanaphorically, whether in cleft constructions or in cataphora, there is no impression of false bonding. For instance, contrast (27) and (28):

(27) Sailing to Ireland was eventful for Jim.
(27′) It sank half way.

versus

(28) Sailing to Ireland was eventful for Jim.
(28′) It was a really windy day.

There is no trace of a bonding effect with the second of these. Now, if we want to argue that matching of pronouns to antecedents is immediate and automatic, as perhaps we need to in order to explain the false bonding phenomena, then we have to assume that the predicate information is not being used at the point (e.g., Ireland doesn't sink half way). Yet in order to explain away the lack of false bonding with (28′), it is necessary to suppose that the predicate information was used to recognise the different function of the pronoun. So there is something of a dilemma, on the face of it, especially in view of the fact that, in a reading-time study, no trace of any false-bonding could be found using materials of this type (Sanford, 1985). So the absence of an effect is not restricted to phenomenal experience; there is simply no evidence for backtracking.

There are two major ways of going about solving the problem. One is to suppose that certain verbs, like the copula *was*, predisposes the processor towards a nonanaphoric reading of *It*, as in:

(29) It was a beautiful day.

or

(30) It was Dr. Kazminsky who gave Mary aspirin.

However, although *was* is the most common verb used in narratives where these constructions are concerned, it is not the *only* one, and *was* is frequently used when the pronoun is referential. So the line of argument that verb information is used to rule out referential usage (hence blocking false bonding) requires care. We would have to say that *sank*, for instance, is a verb that not only indicates a referential use of *it*, but that it is potentially applicable to Ireland. This seems unlikely.

An alternative is to suppose that, in order to achieve *genuine* reference resolution, a number of processes converge. First, a fast match between the pronoun and an entity in explicit focus begins. This happens first because the pronoun is encountered first, and should be complete before the rest of the sentence is processed to any substantial degree. As the rest of the sentence is processed, the bond between the pronoun and the entity in focus is used as a *locus* for testing various semantic relations, if needed. So, if the predicate clearly signals a referential usage, then the locus is tested as a site for the *coreference relation* (or a cosignificance relation, again depending on the predicate). In the case of a genuine and intended coreference relation, everything should be fine. In the case of false bonding, the coreference relation, signalled by the rest of the sentence, leads to a state of affairs which simply does not mesh with the general knowledge (e.g., it is silly to assert that Ireland sank half way during a boat journey). So a process of reinterpretation has to take place as a secondary process. Now consider the cataphora and cleft sentence cases. Here the predicate rules out coreference at the locus of the bond. There is no reassignment; there is simply no need to test for coreference.

The explanation given above removes the dilemma by supposing that bonding is different from the establishment of a full semantic relationship (as required for reference *resolution*). Rather, a bond simply defines a locus for the establishment of a full semantic relationship if needed. Although conditions for treating *it* as though it will be used anaphorically are set up, in no sense has that *decision* been taken. The bond is set up as early as possible on this account, rather like the immediate mapping effect found with the spelling error detection task. The evidence for rapid, automatic mapping is thus extremely strong.

F. DISCUSSION

In the preceding sections, a number of illustrations of issues surrounding the question of immediacy of reference resolution have been given. The claim that the processing of reference takes place as quickly as possible has been refined somewhat, and the discussion will begin with a summary.

While there is evidence that processes supporting reference resolution are initiated immediately a noun phrase or a pronoun is encountered, a number of important qualifications must be stated. First, with full definite noun phrases, referring to characters in narrative discourse, there seems to be good evidence for immediate resolution. The data presented in section B show, in fact, that the resolution process allows for changes in the predictability of verbs as a result of the state of implicit goals of the character identified. However, with pronoun reference under precisely the same conditions, there is evidence that immediate resolution only occurs with main-character references. With secondary characters, resolution seems to be *deferred*. This picture is complicated by the fact that the processor seems to behave as though sentence-initial pronouns are going to refer to main characters. Independent research by other investigators was reported which was consistent with this view.

Additional research in which main and secondary characters are contrasted reveals a number of differences in the way they influence processing. In particular, evidence was presented to show that setting information (psychological atmosphere statements) seems to be classed as relevant to main characters, but not to secondary ones. This too suggests a heavy top-down involvement of the kind of character being currently referred to on the likely course of processing. It was additionally noted that simple cues, like the use of proper names, might serve to bring about these processing differences, or lead toward them occurring.

In a final section, research on pronoun bonding was introduced. Resting upon some curious rhetorical effects, this phenomenon also indicates early initiation of processes supporting the interpretation of pronouns, this time especially in relation to the pronoun *it*. This is an interesting outcome, worthy of much more detailed study, since this pronoun can be used to refer to any singular thing whatsoever. In general, psychological research on this most important pronoun has been somewhat restricted. An investigation of false bonding in relation to nonanaphoric uses of the pronoun produced an interesting question: How is predicate information in (*it*) pronoun initial sentences used to assist reference resolution? One suggestion is that, rather than establishing a full semantic relation, a major part of the resolution process for the pronoun *it* is to establish a locus, or pipeline, between what follows the use of the pronoun and a major aspect of the prior representation. Coreference, codesignation, or any other form of referential link is only specified at that locus if forced by what follows in the discourse. At that point, a semantic value of some sort can be assigned, and the pipeline used for inference passing, and so on.

This potted summary of some aspects of reference resolution is one part of the message of this chapter. Much of it is probably contentious as it is. Now I want to move on to use the data in a different way, and ask about the general characteristics of some of these processes. Inevitably, much of what is said is discursive, speculative, and preliminary.

Autonomy and Veto

Many of the processes described above have the character of independent mechanisms which are seemingly uninfluenced by information which is pertinent to interpretation. This is suggestive of two things: automatic triggering of the processes when the conditions are right, and relative impermeability of the processes to veto. These characteristics are illustrated in the following list:

1. *Noun phrase anaphor checks (section B).* If two noun phrases bear a class member/class relation, then they are automatically checked for potential coreference, even when there are syntactic cues which clearly rule out this possibility. Veto of this process appears to be restricted to the semantic level (e.g., an adjectival qualification ruling it out). Within limits which are still being explored, there is evidence for an automatic matching process which has the appearance of being autonomous over a range of conditions.

2. *The immediacy of noun phrase initiated anaphor resolution (section B).* An immediate anaphor resolution process results in a change in the accessibility of verbs relevant to the goals of the character in question. The presence of a definite noun phrase seems to be sufficient for this to be initiated.
3. *Theme-driven pronominal mapping (sections C and D).* Main characters and agents seem to drive the reference resolution process to the extent of overriding gender mismatches, at least in the short term. The impression of a referential problem seems to remain even after the "correct" interpretation has been given. The verb-accessibility results with pronouns suggest that pronouns are matched against representations which relate to the main character. So when the pronoun matches, resolution appears to be immediate. When it does not match, the process appears to be delayed. This is a clear top-down effect.
4. *Pronoun bonding effects (section E).* Bonding seems to be an immediate and automatic process, which cannot be vetoed by later information, in that there is a clear impression of backtracking. Yet the final interpretation fits what is intended. Bonding may be irrelevant to later interpretation.

These processes capture some of the elements which have to enter into any psychologically valid account of the reference resolution process. Some of them clearly have the appearance of being relatively autonomous. At one extreme, autonomous mechanisms are potential candidates for devices which operate independently of any input except for the conditions which trigger them. For instance, the occurrence of a definite noun phrase may be enough to trigger a search of some sort for potential antecedents. Such a process could occur in parallel with other processes relevant to the process of interpretation. Alternatively, the appearance of relative modularity could be revealed through relatively stable processing patterns in more richly interactive structures, such as connectionist networks (Anderson and Hinton, 1981; McClelland and Rumelhart, 1986). Any question of autonomy is slightly complicated by an observation which can be made on the basis of automatic coreference checking, and the bonding effect: although final interpretations successfully fit the facts of the world, there remains a strong "sounds-like" effect in each case. For instance, when false bonding occurs, as in example (18), for example, there is an impression that *it* equates with *head* rather than *bucket*, the "sounds-like" effect. Such phenomena are quite widespread, and raise an important issue: Under what conditions are conflicting cues processed in such a way as to remain unconscious, and under what conditions do they lead to a "sounds-like" effect? An understanding of such effects is clearly essential to a proper model of how a writer comes to choose expressions which minimize them in normal discourse production. From a process model point of view, it raises the question of how "sounds-like" phenomena and proper reference resolution co-occur and coexist in our examples. Clearly, automomy of outcome and interaction of process leading to these outcomes can co-occur.

A problem which is related to reference resolution further complicates matters. Component processes which must interact in order to produce coherence under an intended interpretation may not interact, or may not be prominent enough to enter into an interaction, leading to incompleteness of processing. Some illustrations are presented below.

Incompleteness and Coherence

It is trivially true that all of the indefinitely large number of inferences which could be made on the basis of a discourse are not made. Indeed quite frequently, interpretations will result which appear to be *misinterpretations* from the producer's point of view, but which result from incomplete processing. Let us begin with a simple illustration of a limit on coherence testing:

(31) There was a tourist flight on its way from Vienna to Barcelona. On the lost leg of the journey, it developed severe engine trouble over the Pyrenees. The pilot last control and it crashed, right on the border. Wreckage was equally strewn in France and Spain, and one question facing the local people was where the survivors should be buried.

Many people find that there is a definite delay before they realise the stupidity of the passage; some people don't see the problem at all, debating locations for burial like France, Spain, or "back home." The problem is that, while one does not bury survivors, the use of the term *survivors* in the present context can go unnoticed as leading to an anomaly. Perhaps the verb *buried* requires a patient, and the term *survivors* has a lexical semantics sufficiently similar to *the victims* or *the dead* to produce a good enough match on the requirements of a patient. To my knowledge, there is no systematic empirical treatment of this type of problem, although it does show how incomplete processing can be.

Another example comes from my own experience with reference problems that break gender expectations. While most people get a distinct feeling that the following example poses a processing problem, others have said to me that there is *no* problem:

(32) The manager called the secretary in for dictation.
(32′) She dictated too quickly for him.

The way in which it was thought to be "no problem" was based on failing to notice that the manager was a woman and the secretary was a man, a reversal of the usual roles (Sanford, McDougall, & Simmons, unpublished data). While most people notice this, some do not. What might have happened to prevent people from noticing it? In fact, to interpret the whole thing, a number of quite complex processes must have occurred. For instance, managers *dictate*, so *the manager* in (32) would have to be mapped onto the agent slot dictate in (32′).

Similarly, *secretary* would be mapped into the patient slot. Now, *manager* is the agent of (32), so it should be treated as the referent of the initial pronoun in (32′). By parallel function (Sheldon, 1975), *secretary* should map onto the object pronoun. If these mappings are fast enough, they could occur before gender had entered into the process. The representations would still be *coherent*, but the gender mismatch would not be noticed. Another possibility is that when *she* appears in (32′), a bottom-up process is initiated that creates a match with *secretary*. Obviously, there are a lot of processes and an equally large number of possible outcomes. Although the account given above is entirely speculative, it raises an important possibility: with a number of autonomous, presumably parallel, processes going on to check coherence, small changes in wording, or emphasis, or structure, should exert a huge influence on the pattern of processing. Perhaps the person who can generate particular effects in writing is simply manipulating the balance of these processes in terms of timing and priority. So far as incompleteness like that discussed above is concerned, it seems likely that in *normal* processing, incompleteness occurs, but it simply doesn't matter from the point of view of being able to use a message. This obviously deserves investigation.

Easily Modified levels of Activity

An important characteristic suggested by the work cited in this chapter is that levels of prominence or activity of background knowledge must be easily modified. In terms of Sanford and Garrod's (1981) scenario theory, not only were linguistic descriptions supposed to activate appropriate background knowledge, but certain specific cues seem to deactivate that knowledge (section D; Anderson, Garrod & Sanford, 1983; Sanford & Garrod, 1985). This obviously requires a very fast search or match facility which must, equally obviously, be a parallel process. In terms of activating appropriate knowledge in the first place, there are two possibilities: one is that certain key conceptions need to be mentioned in order to bring about retrieval of an appropriate scenario (cf. Schank & Abelson's 1977 discussion of script activation). The other is that activation is more continuous: Scenario representations are made more available the more cues there are that are relevant to them. This raises the question of whether there *are* modular situation-based structures in memory. Fortunately, the notion of modularity is relative. Actions, etc., which belong together (scripts) will tend to be highly related to one another, while incompatible actions can be mutually inhibitory.

The deactivation of a knowledge structure (scene shifting) can seemingly be precipitated by using control expressions like *after that*, or *later*, or *after he had finished*. Whatever memory structure had been active must be rapidly deactivated at this point. For independent evidence of this claim, see Sharkey and Mitchell (1985). Recently, Sharkey and Sharkey (1986) have shown how it is possible for such a process to take place within a connectionist model of background knowledge representation. The problems for any scheme of activation are immense, however. In the spelling error detection task, rapid shifts in the activation levels of verbs as a function of discovering the actors state poses a problem. Whole fields of potential verbs presumably have their activation levels changed very quickly.

In Conclusion

The bulk of this paper has been concerned with the identification of dominant processes which underlie reference resolution and associated phenomena of coherence establishment. The main argument is for a resolution process which is initiated as soon as definite expression is encountered. However, especially in the case of pronouns, there is a strong top-down component which exerts a strong bias on interpretation in favor of thematic subjects. Full noun phrases, to some extent in contrast, seem to result in a relatively bottom-up dominated process which can rapidly influence the accessibility (or activation) of verbs relevant to the goal structures to which the individual designated by the noun-phrase is bound. The pattern matching operations underlying these processes seem to go on to completion, even when they are treated as cues and integrated with other data to produce the interpretation intended by the writer. When this happens, a "sounds-like" effect is produced, in addition to an appreciation of intended meaning.

It seems distinctly possible that many of the phenomena described here could be modelled within a connectionist framework, which would naturally handle the rapid change in verb accessibility, and has a good potential to handle the coexistence of interpretations and "sounds-like" effects. Of course, it is possible to account for the findings in other frameworks too. The real test will be whether finer predictions about interaction patterns drop naturally from network or alternative models.

REFERENCES

Anderson, A., Garrod, S.C., & Sanford, A.J. (1983). The accessibility of pronominal antecedents as a function of episode shifts in narrative text. *Quarterly Journal of Experimental Psychology, 35a*, 427–40.

Anderson, J.A., & Hinton, G.E. (1981), Models of information processing in the brain. In G.E. Hinton & J.A. Anderson (Eds.), *Parallel models of associative memory*, (pp. 9–48). Hillsdale, NJ: Erlbaum.

Cole, R.A., & Jakimik, J. (1978). Understanding speech: how words are heard. In G. Underwood (Ed.), *Strategies of information processing*. London: Academic Press.

Cole, R.A., & Jakimik, J. (1980). A model of speech perception. In R.A. Cole (Ed.), *Perception and production of fluent speech*. Hillsdale, NJ: Erlbaum.

Cuthill, A.P., Golombok, S.E., & Rana, R.K. (1978), *An investigation of the integration of semantic information while reading using co-referential and noncoreferential sentences*. Unpublished project report, Psychology department, University of Glasgow.

Dell, G.S. , McKoon, G., & Ratcliff, R. (1983). The activation of antecedent information during the process of anaphoric reference in reading. *Journal of Verbal Learning and Verbal Behavior*, *22*, 121–132.

Garrod, S.C., & Sanford, A.J. (1977). Interpreting anaphoric relations: The integration of semantic information while reading. *Journal of Verbal Learning and Verbal Behavior*, *16*, 77–90.

Garrod, S.C., & Sanford, A.J. (1982). Bridging inferences and the extended domain of reference, In J. Long & A.D. Baddeley (Eds.), *Attention and performance IX* (pp. 311–346).

Garrod, S.C., & Sanford, A.J. (1983). Topic dependent effects in language processing. In G.B. Flores d'Arcais & R.J. Jarvella (Eds.), *The process of language understanding*. (pp. 271–298). Chichester, England: John Wiley & Sons.

Garrod, S.C., & Sanford, A.J. (1985). On the real-time character of interpretation during reading, *Language and Cognitive Processes*, *1*, 43–61.

Garrod, S.C., & Sanford, A.J. (1988). Thematic subjecthood and cognitive constraints on discourse structure. *Journal of Pragmatics*, *12*, 357–372.

Grimes, J.E. (1975). *The thread of discourse*, the Hague, Netherlands: Mouton, Janua Linguarum.

Halliday, M.A.K., & Hasan, R. (1976). *Cohesion in English*. London: Longman.

Hirst, G. (1979). *Anaphora in natural language understanding: A survey*. Technical report 79-2, Department of Computer science, The University of British Columbia.

Hudson, S.B., Tanenhaus, M.K., & Dell, G.S. (1986), The effect of the discourse center on the local coherence of a discourse. In C. Clifton (Ed.), *Proceedings of the Eighth annual Conference of the Cognitive Science Society, 1986* (pp. 96–101). Hillsdale, NJ: Erlbaum.

Just, M.A., & Carpenter, P.A. (1980), A theory of reading: from eye-movements fixations to comprehension. *Psychological Review*, *87*, 329–355.

Kantor, R.N. (1977). *The management and comprehension of discourse connection by pronouns in English*. Unpublished doctoral dissertation, Ohio State University.

Karmiloff-Smith, A. (1980). Psychological processes underlying pronominalisation and nonpronominalisation in children's connected discourse. In J. Kreiman & A.E. Ojeda (Eds.), *Papers in the parasession on pronouns and anaphora* (pp. 231–250) Chicago, IL: Chicago Linguistic Society.

Mandler, J.M., & Johnson, N.S. (1977). Remembrance of things parsed: Story structure and recall. *Cognitive Psychology*, 9, 111–151.

Marslen-Wilson, W.D., Levy, E., & Tyler, L.K. (1982). Producing interpretable discourse: The establishment and maintenance of reference. In R.J. Jarvella & W. Klein (Eds.), *Speech, place, and action* (pp. 339–378). Chichester, England: John Wiley.

McClelland, J.L., & Rumelhart, D.E. (1986). *Parallel distributed processing: Explorations in the microstructure of cognition.* London: The MIT Press.

Moar, K., Sanford, A.J., & Garrod, S.C. (in preparation), Proper names and the control of focus in discourse.

Newman, J.E., & Canham, L. (1986, April). *Focus and topic selection in a sentence production task.* Paper presented at Annual meeting of the southwestern Psychological Association, Austin, Texas. Report from the Department of Psychology, University of New Mexico at Albuquerque.

Rumelhart, D.E. (1975). Notes on a schema for stories. In D.G. Bobrow & A. Collins (Eds.), *Representing and understanding: Studies in cognitive science.* New York: Academic Press.

Sanford, A.J. (1985). Aspects of pronoun interpretation: An evaluation of search formulations of inference. In G. Rickheit & R. Strohner (Eds.), *Inferences in text processing* (pp. 183–204). Amsterdam, Netherlands: North-Holland.

Sanford, A.J. (1987). *The mind of man: Models of human understanding.* Brighton, England: The Harvester Press.

Sanford, A.J., & Garrod, S.C. (1981). *Understanding written language.* Chichester, England: John Wiley.

Sanford, A.J., & Garrod, S.C. (1985). The role of background knowledge in psychological account of text comprehension. In J. Allwood & E. Hjelmquist (Eds.), *Foregrounding background*, (pp. 223–234). Lund, Sweden: Doxa AB.

Sanford, A.J., Garrod, S.C., & Al-Ahmar, 'H.F. (1985), *The role of character perspectives in discourse comprehension.* End-of-project report to the Economic and Social Research Council, project number C00232009.

Sanford, A.J., Garrod, S.C., & Boyle, J.M. (1977). An independence of mechanism in the origin of reading and classification-based semantic distance effects. *Memory and Cognition*, 5, 214–230.

Sanford, A.J., Garrod, S.C., Lucas, A., & Henderson, R.J. (1983). Pronouns without explicit antecedents? *Journal of Semantics*, 2, 303–318.

Schank, R., & Abelson, R. (1977), *Scripts, plans, goals and understanding: An enquiry into human knowledge structures*, Hillsdale, NJ: Erlbaum.

Sharkey, N.E., & Mitchell, D.C. (1985). Word recognition in a functional context: The use of scripts in reading, *Journal of Memory and Language*, 24, 253–270.

Sharkey, N.E., & Sharkey, A.J.C. (1986). KAN: A Knowledge Access Network Model. In R. Reilly (Ed.), *Communication failure in dialogue*, North-Holland, Netherlands: Elsevier.

Sheldon, A. (1975). The role of parallel function in the acquisition of relative clauses in English, *Journal of verbal learning and verbal behavior*, 13, 272–281.

Shillcock, R. (1982), The on-line resolution of pronominal anaphora. *Language and Speech, 24*, 4.

Smith, E.E., Shoben, E.J., & Rips, L.V. (1974). Structure and process in semantic memory: A featural model for semantic decision. *Psychological Review, 81*, 214–41.

Tyler, L.K., & Marslen-Wilson, W. (1982). Processing utterances in discourse contexts: On-line resolution of anaphors. *Journal of Verbal Learning and Behaviour, 18*, 545–569.

Vonk, W. (1985). The immediacy of inferences in the understanding of pronouns. In G. Rickheit & H. Strohner (Eds.), *Inferences in text processing.* (pp. 205–218). North Holland, Netherlands: Elsevier BV.

Wilkins, A.J. (1971). Conjoint frequency, category size and categorization time. *Journal of Verbal Learning and Verbal Behavior, 10*, 383–5.

Wilks, Y. (1975). A preferential pattern-seeking semantics for natural language inference. *Artificial Intelligence, 6*, 53–74.

Winograd, T. (1972). *Understanding natural language.* London: Academic Press.

CHAPTER 6

Extending Marker-Passing to Microfeatures

JAMES A. HENDLER
Department of Computer Science and Systems Research Center
University of Maryland

1. INTRODUCTION

Connectionism has recently become a growing force in the cognitive science community. The spectrum of connectionist models is quite large, with systems ranging from distributed models of weighted networks with learning algorithms to local models in which each node corresponds to a discrete concept. Despite the wide diversity of such systems, they all fall within a new paradigm for cognitive science, that of *subsymbolic* systems (Smolensky, 1988).

One type of system that falls within the subsymbolic approach is that which replaces a system in which the lowest level of representation is a discrete system of nodes each corresponding to a high level concept with a distributed memory in which lower levels of representation, usually corresponding to perceptual features, are used. Thus, instead of a discrete symbolic representation, for example a frame, corresponding to "grandmother," a subsymbolic model might augment this by having a set of perceptual features that jointly imply an activation of this node. These lower level concepts, sometimes called *microfeatures*, have little meaning on their own, it is only in conglomeration that they take on discrete meaning.

Using a microfeature based approach seems to have many advantages in terms of cognitive science. One of these is the correspondence with results in the spreading activation literature; another is the massively parallel nature of computations performed on these networks and the potential neurobiological plausibility of such systems.[1] A major problem with these models,

[1] For more discussion of such models see, for example, McClelland, Rumelhart, and Hinton (1986) or Feldman and Ballard (1982).

however, is the difficulty of linking them with present systems. The basic control structures for such computing are significantly different from those used in present models of parsers, planners, etc. Thus, using present techniques, one is forced to choose between two seemingly uncompatible computing techniques.

One alternative to a connectionist approach is that of using an activation spreading algorithm which passes symbolic information through an associative network in which links representing symbolic information, rather than link strengths or the like, are present. These *marker-passing algorithms* can pass very simple information (cf. Fahlman, 1979) or complex marks which include various amounts of control information for use in the marker-passing process (cf. Quillian, 1966; Charniak, 1983).[2] These systems have been shown to integrate gracefully with the standard AI technologies of parsing (Charniak, 1983, 1986; Hirst, 1983; Norvig, 1986) and planning (Hendler, 1987).

A concrete example of the difference between a marker-passing system and a connectionist model can be seen by contrasting the word sense disambiguation systems of Hirst (1983) and Cottrell (1985a). Hirst's model approaches ambiguity by setting up a set of frames that put word senses in one to one correspondence with semantic meanings. Each sense of a given word competes for recognition based on two criteria: the syntactic information available and the connectedness of the words to each other. This connectedness was found via a marker-passing system.

The form of marker-passing used by Hirst is based on a model proposed by Charniak (1983). This model involves passing symbolic information through an associative network to find paths between marked nodes. To understand this model of marker-passing we propose the following analogy: Consider an associative network as a bunch of hockey pucks connected to one another by springs hanging in midair. If one of these hockey pucks is struck, it will start to vibrate. This in turn will start the hockey pucks next to it vibrating, and so on and so on and so on. Because of friction, each puck would vibrate a little less than the puck before, until we reached some limit where the vibrations were relatively imperceptible. Assume that, at this point, we struck another hockey puck elsewhere in the entangled web. This puck would also start a set of pucks vibrating.

At some point, a hockey puck already vibrating from the first node might be shaken from the second. The collection of those pucks vibrating from both nodes forms a set with an interesting property: Informally, it is the set of all nodes for which some path connects the two first nodes. In a system where the associative network corresponds to the search space for some

[2] A detailed discussion of the various forms of these algorithms can be found in (Hendler, 1986.)

problem to be solved (as in a planning system), each path found corresponds to some way of traversing the space between the two nodes.

As an example of this use of marker-passing, consider how Hirst's system would handle the word *ball* in the phrases:

1. The rubber ball
2. The formal ball

The dictionary entry for *ball* would contain two word senses, one corresponding to the frame *sphere* and one corresponding to *dance*. When phrase 1 was encountered, the system would pass markers between *rubber* and *ball* and a path such as

rubber → (isa rubber material) → material →
(composed-of physobj material) → physobj →
(isa sphere physobj) → sphere →
(world-sense ball sphere) → ball

would be found. This would cause the preference of *sphere* over *dance*. If, on the other hand, phase 2 was being analyzed, a path like

formal → (isa formal social-event-type) →
social-event-type → (has social-event social-event-type) →
social-event → (isa dance social-event) →
dance → (word-sense ball dance) → ball

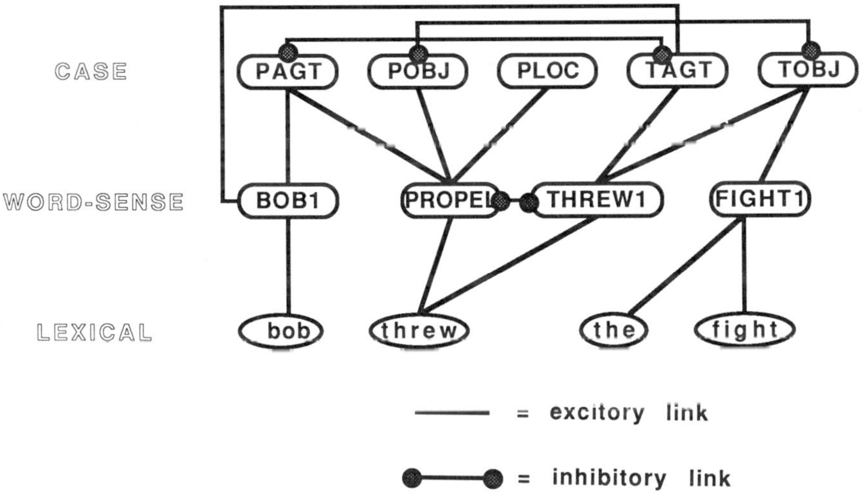

Figure 1. Bob threw the fight

would be found and the *dance* meaning used. When paths were found to more than one meaning of a word, Hirst's system used a path strength heuristic (based mainly on the number of intermediate nodes) to determine which meaning to prefer.

As a contrast to this, we consider the local connectionist word sense disambiguation system developed by Cottrell (1985a). Cottrell's word sense disambiguation system is designed to interact with a word recognition system, rather than a parsing system. His system uses a massively parallel, numeric activation spreading algorithm as more than a single part of a word sense disambiguation system. It uses a local connectionist network to take into account not only the effects Hirst handles, but also, at the same time, to perform case-slot assignment. As an example, Cottrell's system would use the network of Figure 1 (weights on the arcs are omitted) to handle the disambiguation of *threw* in the sentence

Bob threw the fight.

The correct interpretation of the sentence would be found at the end of a period of numeric computation by examining the nodes remaining activated in a stable configuration. In the above example, these would be *bob1* as *tagt*, *threw1*, and *fight1* as *tobj*.

Further, Cottrell's system is designed to dovetail with a system such as the Rumelhart and McClelland (1986) system for doing the recognition of words from letters which are in turn found at the microfeature level with letter orthography features.[3] Thus, a full system could conceivably go from the perceptual features of the letters through to case slot assignments.

From the above examples we see that what Hirst accomplishes with marker-passing is a subset of what Cottrell achieves with the network. On the other hand, Hirst's system tied into a traditional AI Syntactic parser, which Cottrell's system couldn't. Hirst was therefore able to use syntactic clues in performing disambiguation. For example, the word *tire* in the sentence

The left front tire is flat.

can only be a noun, not a verb, and Hirst's system could use this information to ignore the *verb* senses. Cottrell's system, however, could not take advantage of such information. For Cottrell to produce a syntactic parse to provide such information to the disambiguator would require the design of a syntactic parsing scheme within the massively parallel framework. This may not be

[3] For example, horizontal line, vertical slope left to right, and other more-or-less perceptual features of the letters occurring in words.

impossible, in fact Cottrell (1985b) discusses some steps in exactly such a direction, but at present connectionist schemes are not having tremendous success at producing such results.

It is our belief that the integration of the two approaches, the symbolic and the subsymbolic as it were, holds tremendous promise if it can be performed gracefully. Such a hybrid system could take advantage of some of the virtues of a distributed memory, particularly that of their ability to handle similarity and generalization (Hinton, McClelland, & Rumelhart, 1986), without losing the traditional controlled inferencing of traditional AI systems. In this paper we discuss a new implementation of a marker-passing system, similar to that used by Hirst, that is able to make some use of microfeatures. The system is an extension of the SCRAPS system (Hendler, 1987), which uses this type of marker-passing to augment the choice mechanism in a planning system. In the rest of this paper we will describe the SCRAPS system as previously implemented, show where the use of a microfeature-like approach is needed for such a system, and describe a new version of the SCRAPS planner which can handle such cases.

2. SCRAPS

To demonstrate the integration of a local connectionist model into a marker-passing system, we will describe an extension to an existing system. This system, called SCRAPS, is designed to improve the behavior of a planning algorithm in the presence of certain types of information. In particular, a major problem faced by problem-solving systems today is that of making a choice in the presence of multiple alternatives. It is often the case that the system has access to knowledge that would lead to the optimal solution, or that could avoid a conflict, but this knowledge is not used.

Consider the case of a system trying to solve a goal such as *Satisfy hunger*. It must make a choice between alternatives such as *Go to restaurant* and *Eat at home*. In the absence of other knowledge it may not matter which path is chose, but what if we had already asserted *You have no money* to this system? At this point there is a conflict down the *restaurant* path, but none down the *eat at home* path. Thus the system should choose the latter. Unfortunately, most present day systems are unable to take advantage of this sort of information and would make the choice at random. At a later point in the problem solving these systems encounter the error (assuming the *restaurant* path is taken) and are forced to backtrack. Avoiding backtracking has been a primary goal of problem-solving researchers.

The SCRAPS system presented a new approach to the issue of choosing paths during problem solving. We showed that a suitably structured marker-passer could be used to aid the choicen mechanism used by a problem solver. In this section we will briefly describe the implementation of SCRAPS and how the marker-passer was used. We will only present enough of SCRAPS for the sake of our later discussion of extending the marker-passer. The details of SCRAPS, the marker-passing algorithm, cognitive implications of the work, and a comparison between our view of marker-passing and the work in connectionism can be found in (Hendler, 1987).

2.1. Inside SCRAPS

Most planning systems work by making step-wise refinements on subplans until primitive actions are reached. Thus *Restaurant* would be broken down into *enter*, *order*, *eat*, *pay*, and *leave*. *Order* could be broken down into *read menu*, *decide*, and *tell waitress*, etc. Steps such as *read menu* would be primitive and the final plan is comprised of a partial ordering of these operations.

It is this step-wise refinement that causes the problem with planning choice as discussed previously. At the time we decide which plan to use we do not yet know what steps it will eventually decompose into. However, it is when we try to perform these primitives that the conflicts will most often manifest themselves. It is clear that, if a mechanism could be developed that would examine all these primitive actions looking for possible conflicts and identifying them prior to the making of a choice then the problem solver would benefit. This mechanism would need to find all possible bindings for the variables being passed through the various levels of substeps checking for conflicts.

Unfortunately, such a mechanism does not presently exist. To be computationally viable, this mechanism would need to perform deduction extremely quickly. Further, since all possible bindings of any variables need to be examined, large numbers (susceptible to combinatorial explosion) of low-level deductions[4] would need to be done. Parallelism would improve the situation, but not solve it, due to inherent limitations on the efficiency of such deductions and the combinatorics of multiple possibilities of variable bindings.

It is possible, however, to approximate this mechanism with a system that can examine the primitive actions quickly while looking for special features. Our version of a marker-passing system is one such mechanism. SCRAPS was aimed at showing that merging a marker-passer with a planning system could provide substantial improvement to the choices made by the planner.

---

[4] Depending on the type of problem solver being used, these computations could involve unification, pattern-matching, script instantiation, etc. In this paper all of these types of tasks will be referred to as *low-level deductions* for want of a better term.

2.1.1. *What do we mean by Marker-Passing?*

There are many different uses of the term *marker-passing* in the literature. One use refers to the spreading of binary information in parallel to find various types of intersections (see Fahlman, 1979). This is not the way we are using the term. Instead, we use the term as proposed by Charniak (1983). This model proposes marker-passing in a manner similar to Quillian's (1966) spreading activation model. Quillian's model involves the marking of nodes adjacent to some node in memory with information that allows a trace back to that node, and then marking all nodes adjacent to those, etc. Thus, for example, in a traditional semantic net system, if we marked *car* we would then mark *wheels, vehicle,* and all those concepts directly related to *car*. Following this we would then mark *rubber, round* and all those concepts relating to *wheel,* as well as *transportation* and those concepts relating to *vehicle*. This process would proceed until either nothing was left to mark or the process was called to a halt. Each of the nodes reached (*wheels, vehicle, rubber, round, transportation,* etc.) would thus be marked with information providing a path back to the node *car*.

By passing marks from first one node and then another, we can see how these nodes are related by examining the paths found through those nodes marked from both. Thus, in this example from Charniak (1983):

> John wanted to commit suicide. He got a rope.

we would start passing markers at *suicide* and then pass markers starting at *rope*. This would find an intersection at the node *noose* which would be found both from *suicide* (the instrument of a *hang*) and from *rope* (as a material relation).

One difference between the marker-passer defined by Charniak and the binary marking method of Fahlman is that the former will return the path found, rather than reporting the node of intersection (as would the latter). Thus, in the example above, the marker-passer would return:

> *suicide* -> to do suicide *kill* self -> *kill* -> *hang* is a type of *kill* -> *hang* -> instrument of *hang* is *noose* -> *noose* -> *noose* is made of *rope* -> *rope*

Notice that the marker-passer does not perform any deductions during its run. If we had asserted that the agent and patient of a *hang* had to be different this path would still be found. This would not be a valid path, since the definition of *suicide* requires that the agent and patient must be the same. It is this blind behavior that enables us to implement marker-passing efficiently, at the expense of finding some false paths. The issues of evaluating false paths and avoiding their generation are omitted from this paper. Again, the interested reader should see (Hendler, 1987).

2.1. Inside SCRAPS (continued)

The flow of control used in SCRAPS is shown in Figure 2. In general the following occurs: To start, the problem-solver is invoked. In attempting to solve problems, it passes markers throughout the knowledge base. When these markers encounter other markers, an intersection is found. The paths meeting at this intersection are then returned to a path evaluation mechanism. This mechanism checks each path for information which can be used to help guide the problem-solver. Heuristics are used to determine if the path proposes a solution to an existing problem, causes a conflict, or suggests a modification to an existing plan. If one of these is identified the system takes appropriate action choosing, rejecting, or modifying the plan as necessary. If no such information is found, the path is ignored.

As an example consider the case of the use of SCRAPS for helping us identify a correct solution to a problem with several alternatives. For example, if we are attempting to plan

Trying to commit suicide while holding a gun.

the system needs to choose between each of the possible ways of performing the suicide plan. SCRAPS starts by passing markers to each of the alternatives; these in turn mark their subplans, etc. While this goes on we are also passing markers starting with the frame[5] *gun*. Since this marks *gun*, and since *shoot* has, as a precondition, the possession of a *gun*, we find an intersection. The path

suicide -> *to do suicide*(×) *kill* (x,x) -> *kill* -> method of *kill*(y,x) is *shoot*(y,x) -> *shoot* -> precondition of *shoot*(y,x) is possess (y,*gun*) -> *gun*

is reported to the path evaluator. Since this path presents a potential solution to the task which has been posed, committing suicide, the path evaluator can cause the problem solver to prefer the *shoot* plan for achieving the *suicide* goal.

A more interesting use of SCRAPS is the ability of the marker-passing system to identify conflicts between parts of plans. As an example consider the case of

Buying a cleaver while on a vacation.

The system starts by attempting to plan the buying of the cleaver. As such

[5] We will use *frame* here as a generic word for a planning structure. The user may replace it with *script*, *schema*, *plan* or other such term.

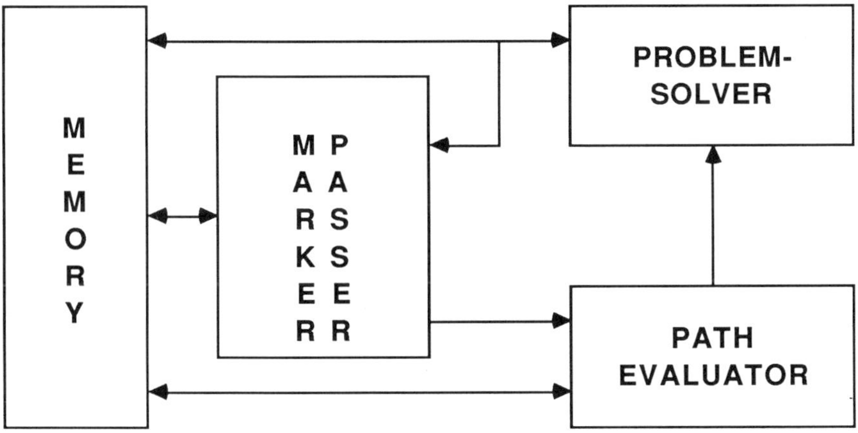

Figure 2. Flow of control in SCRAPS

it passes markers from the nodes *Cleaver* and *Buying*. From *cleaver* it finds the information that cleavers are commonly used as *kitchen tools* and as *weapons*, and it marks these nodes.

The system also marks the information about *buying* which includes the step of taking home that which is bought. Via the substeps of substeps the marker-passer eventually notices that one way to go home is to *fly* and that within *flying* one must pass through a *weapons check*. However, *weapons check* marks *weapon* and finds that it is already marked. Thus, an intersection has been reached and the information:

> *buying* → *take-trip* → *plane* → *boarding* → *weapons-check* →
> *if* you go through *weapons-check* with a *weapon* you get arrested[6] →
> *weapon* → *cleaver* → *cleaver*-1

The path evaluator examines this path and discovers that the intersection involves carrying a weapon through a weapons check, which leads to being arrested, an undesirable state.

The path evaluator can now use a heuristic which states that *IF a path causes one to reach an undesirable state THEN rule out the plan during which this state is enocuntered*. Since *Flying* is the plan which dominates the *weapons check*, we can rule it out as the method of getting the cleaver home.[7]

---

[6] The predicates connecting the nodes in paths will be left out for brevity's sake except where necessary to our discussion, as in this case, or where non-obvious.

[7] SCRAPS did no replanning when a potential error was found. The use of marker-passing paths for debugging such errors is described briefly in (Hendler, 1986) and is a research interest of ours at present.

2.1.2. Some relevant implementation details

SCRAPS is implemented using the problem solver NASL (McDermott, 1978) on a logic-based semantics system, FRAIL (Charniak, Gavin, & Hendler, 1983). Built into FRAIL is an implementation of a marker-passer. FRAIL and NASL are both implemented in the NISP dialect of Lisp (McDermott, 1983) and run on several different types of machines.

One important aspect of a marker-passing system has to do with the overall organization of the frame knowledge database being searched. For the purposes of performing marker passing, this database is viewed as a network of nodes, each representing a frame, that are interrelated via links along which markers may be passed. Each frame must be capable of both receiving and passing markers to neighboring frames through these links, which correspond to directed arcs relating frames to each other.

The FRAIL language provides such a system. Frames are translated into logical predicates which are stored in a LISP database as properties on atoms. These atoms serve as frames and the logical predicates serve as links. Thus, FRAIL provides both a logical language for deduction and a semantic-net representation useful for marker-passing.

The NASL problem solver uses FRAIL to retrieve the possible plans usable for satisfying a goal. When more than one plan can be applicable, FRAIL uses a simple choice mechanism that checks for plans which are ruled in or out by the present situation. SCRAPS uses this mechanism as the primary communication between the plan evaluator and the problem solver. When the marker-passer finds a path that the path evaluator recognizes as a potential solution that path is given priority over other paths. Similarly, when the path evaluator recognizes a problem, it simply rules out that plan which would cause the contradiction. Synchronization is provided by having the problem solver's choice mechanism wait until a message is received from the path evaluator allowing it to proceed. Modifications to plans are done by having the path evaluator make assertions directly into the FRAIL database.

As an example, consider once again the problem of being hungry but having no money. The system is started with the assertion that no money is available, and thus the *paying* frame is marked as unachievable via a forward chaining rule. We then ask the problem solver to solve the task of *eating* with some specific agent, *agent1*. NASL calls upon FRAIL to retrieve the potential plans and finds both *restaurant* and *eat at home*. As FRAIL retrieves these plans markers are passed. The path from *restaurant* to *money* is found and reported to the path evaluator. The path evaluator recognizes the conflict in the restaurant plan and therefore generates a *rule out* statement. It then signals NASL to continue. NASL now uses its choice mechanism to examine the possibilities. Since the restaurant frame has been ruled out, it uses the only other possibility, eat at home, and continues on.[8]

---

[8] Readers wanting more details about the format of the rules used by the choice mechanism in FRAIL and NASL are directed to Charniak, Gavin, and Hendler (1983).

In the SCRAPS system the path evaluator is composed of a set of *ad hoc* heuristics that can examine a path and see if any useful information can be gleaned. These rules serve two purposes: first, we wish to quickly reject paths which are not going to yield useful information, and second, we wish to be able to extract information from those paths which are useful. This is done by passing the path through a series of rules in a cascaded manner. The early rules are designed to reject paths known not to be useful; the later rules are designed to quickly check for a certain feature and pass the rule on to the next heuristic if it is not found.

The early rules work by first checking the node of intersection of a path. If the node of intersection is found to be a "promiscuous" node, one with a very high outbranching factor, the path is probably not useful and can therefore be rejected. If the node of intersection is not of this form, the path is checked to see if certain features can be detected that will cause the rule to be rejected. For example, a rule checks to see if a path is an *ISA Plateau*. This would be the situation if we passed markers from, for example, *dog* and *cat* and found an intersection at *animal*. These sorts of paths are useful for the word sense disambiguation use of marker-passing, but not for problem solving.

Once the obviously useless paths have been rejected the path evaluator then uses the second type of heuristic to see if the path yields valuable data. This is done by having each heuristic designed to quickly examine the path for some feature. If that feature is not found the path evaluator immediately passes this path on to the next heuristic. If none of these heuristics find such a feature then the path is considered uninteresting and rejected. If, however, such a feature is found the path is subjected to closer inspection to see if useful information is yielded. Due to the underlying logical formalism of FRAIL unification is used as a first test in these situations. If items in the path do not unify, it can be rejected. Charniak (1986) has been examining the paths returned by the marker-passer and trying to formalize path checking as a resolution proof procedure.

As an example of the heuristics used in SCRAPS consider what the path evaluator would do with the "Eating with no money" example described earlier. With variables included, the path found is of the form:

> *goal eat* (*me*) -> *eat* -> To-do *eat*(x) use *restaurant*(x) -> *restaurant* -> step-of *restaurant*(x) is *paying* (x y) -> *paying* -> *paying*(*me*) is Marked as unachievable

We are also informed that the place where the marks intersect is on the node *Paying*. First we examine the node *Paying* to see if it is inherently uninteresting or "promiscuous." Since it isn't, we go on to check the form of the path to see if we have found an ISA plateau. Again, this test fails, so we do not reject this path. We now move on to the second type of heuristic, those designed to extract information. One of these rules checks to see if either of the end-points of the path is marked as unachievable. Since this is the case, the rule is now subjected to closer scrutiny. The unification algorithm is used to see if variables in the path match up. Since, in this case, the person with the *Eat* goal is the same as the person trying to do the *Paying* the unifier reports the match. This then causes SCRAPS to rule out using the restaurant plan.

3. WHAT SCRAPS WOULDN'T DO

As the reader can see, the SCRAPS system can appropriately handle certain classes of previously difficult planning problems without the need for backup. Further, the basic SCRAPS program can be extended to handle multiple goal problems by finding paths representing interactions between the goals being solved and extending the path evaluator to recognize such interactions (Hendler, 1987). Thus, we feel that as an experiment in using marker-passing SCRAPS has been quite successful. Unfortunately, we have come across certain problems that a system like SCRAPS should handle, but which the present system cannot.

One problem with the present formulation of SCRAPS comes about when the system is faced with the following situation:[9] assume that, instead of proposing to buy a *cleaver* while on a vacation, SCRAPS had been invoked with trying to buy *an ornate Egyptian letter-opener*. SCRAPS depended crucially on the assertion:

> (*common-use cleaver weapon*); a common use for a cleaver is as a weapon.

While it does not bother us to assume that a cleaver is often used as a weapon, we are quite reluctant to assert that an ornate, Egyptian letter-opener is. Yet, it would seem that a system like SCRAPS should notice that the letter-opener causes the same problem when the weapons-check action occurs. That is, we must notice that the letter-opener could function as a weapon and thus a problem will be encountered when we try to get past the weapons-check at the airport.

Noticing that the letter-opener could be used as a knife is a difficult problem for present AI systems. While some knowledge representation schemes have an ability to do classification, this classification is viewed as a separate control process. Thus, for example, Brachman's (1978) KL-ONE system provides a classifier which can be run to determine what class is the best fit for some particular configuration of objects. (An example of this use of Brach-

[9] Suggested by Lokendra Shastri, personal communication 5, 1986.

man's classifier is the language parser described in Brachman & Schmolze, 1984).

This notion of separate classification will not fit in with the marker-passing technique of SCRAPS. The decomposition of an object into components and the classification of a set of components into an object must happen during the marker-passing. An algorithm is needed that causes a marker to flow from letter-opener to knife. We could use a predicate such as

(*is-like letter-opener knife*)

to solve this problem, but such a solution is clearly lacking in elegance and, in fact, will not solve the more general case. Consider the case of recognizing that some random *long, thin, pointed metal object*, (call it *object-27*) could not be taken onto the plane. It would seem clearly wrong to assert

(*is-like object-27 knife*)

without some sort of analysis of the object in terms of its features.

What is needed here appears to be some sort of breakdown of the letter-opener into a set of features and a recognition that those features include the defining characteristics of a knife. These features correspond to the connectionist notion of microfeatures discussed in the introduction. We need to recognize that our letter-opener is comprised of a set of features such as *pointed, metallic, and thin*, and that a knife would have these as defining characteristics.[10]

It will not be enough, however, to perform a simple match on all these characteristics. The ornate letter-opener (or the sharp metal object) will also have features which are not shared by the knife. We must move towards the connectionist notion of using an ensemble of microfeatures and an activation level. Thus, our new version of SCRAPS must include a marker-passer which can take into account such partial matches.

Encoding the information about the sets of microfeatures which define a given object is fairly easy in our present framework. We define a set a of microfeatures and a special network link called a *defining characteristic* link. Thus, each of the objects in our network is linked to set of microfeatures which define it.[11] Our *letter-opener* frame might now look like:

---

[10] If we use concepts like *pointy*, *thin*, etc., it would seem that we are talking of items that could simply be considered symbolic features rather than some sort of subsymbolic microfeatures. However, as previously discussed, the term *microfeatures* in the literature is generally used to represent items which, by themselves, do not define some particular entity, but rather represent that entity when taken as an *ensemble*.

[11] As will be shown later, not all of these characteristics must be present at a given time for activation to occur. Thus, we are preserving the distributed nature of this representation.

```
(frame: Letter-opener
 slots: . . .
        (Defining-characteristics (pointy thin metallic))
)
```

The associative network represented thus gains a section which resembles a typical local connectionist distributed memory as shown in Figure 3.

There are still two primary differences between this section of our associative net and the typical microfeature network. As presented so far, our network does not provide for weighted activation between nodes and it doesn't provide for inhibitory links. For now we will assume that the right thing to do is to treat all the links as having equal, excitory links, however, we will return to these issues later in this paper.

4. EXTENDING THE MARKER-PASSER

In this section we present an explanation of how our present marker-passing algorithm is extended to allow a graceful interaction with a microfeature based representation. Before we do this, however, to demonstrate the need for extending the present algorithm let us consider what would happen if we simply added the defining-characteristic links and let the present marker-passer handle things. What happens when we try to handle the situation of buying the letter-opener?

SCRAPS would be invoked to perform the purchase act on the letter-

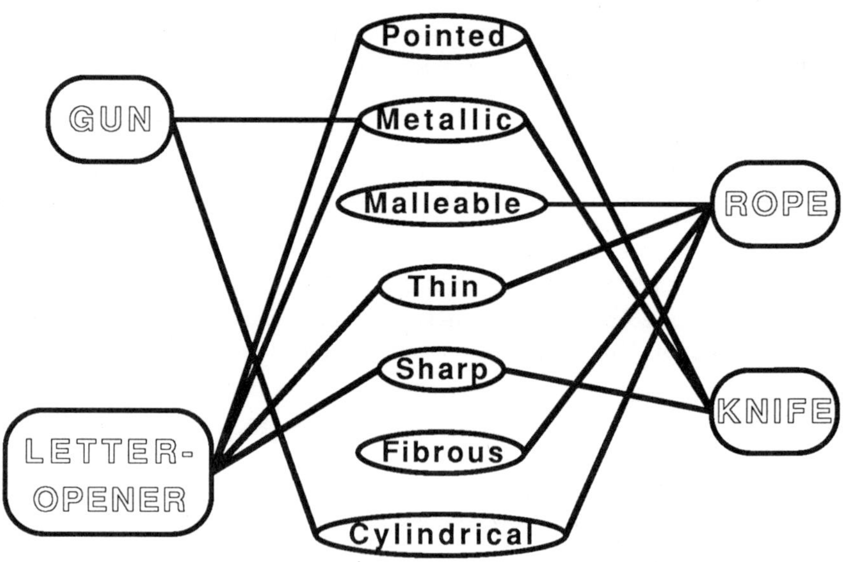

Figure 3. Definin-Characteristic links in a network

opener. As described previously for the cleaver, marks would now pass from the various objects in the initial situation: *buying*, *me*, and *letter-opener*. The marks from *letter-opener* would now pass down the defining-characteristic links. Using the network shown in Figure 3, this would mark each of the characteristics of the letter-opener: *pointed*, *metallic*, *thin*, and *sharp*. Since these are now marker they each mark their neighbors: *metallic* marks *gun* and *knife*, *thin* marks *rope*, *sharp* and *pointed* each mark *knife*. Marks would now flow as before, and barring any form of attenuation we would find the path:

> *buying* → *take-trip* → *plane* → *boarding* → *weapons check* →
> *if* you go through *weapons-check* with a *weapon* you get arrested →
> *weapon* → *knife* → *pointed* → *letter-opener*

Thus, it would seem that our present marker-passer works without any extention. Why must we change it?

The problem here comes because we don't use the distributed network appropriately. We do find the information that *letter-opener* and *knife* shared some information, but we also find many other paths that will cause us tremendous problems. For example, we will find:

> *buying* → *take-trip* → *plane* → *boarding* → *weapons-check* →
> *if* you go through *weapons-check* with a *weapon* you get arrested →
> *weapon* → *gun* → *metallic* → *letter-opener*

which would cause us to believe that we couldn't buy the letter-opener, since it was like a gun. Even worse, if we had started by trying to buy a rope we would have found a path such as:

> *buying* → *take-trip* → *plane* → *boarding* → *weapons-check* →
> *if* you go through *weapons-check* with a *weapon* you get arrested →
> *weapon* → *gun* → *metallic* → *letter-opener* → *thin* → *rope*

saying that we cannot buy the rope, since it shares a feature with a letter-opener, which shares a feature with a gun (which is a weapon and etc.)

For our present version of SCRAPS to be able to deal with these tasks, we would need to have our path heuristics examine the similarity between objects found to determine if the inferences really held. In the case of *letter-opener* and *knife* it would need to find them equivalent, while rejecting the similarity of *gun* and *letter-opener*, or *gun* and *rope*. But this is exactly the form of reasoning that traditional AI systems have trouble with. In fact, this difficulty is why we want to move to a distributed representation in the first place. Thus, if we do not extend the marker-passer, nothing is gained.

The problem, as described, is caused because we only perform half of the job. We correctly pass something akin to activation to each of the microfeatures of an object, but we do not combine activation from the microfeatures to see which, if any, other objects are strongly activated. Instead, we have simply marked *all* of the objects which share microfeatures. In many cases this will be most, if not all, of the objects in the memory. We must find a way to use the combination of activations through this distributed memory.

Intuitively, the answer to this problem is fairly obvious. We need to create a situation in which an object's being marked by the symbolic system corresponds to the same object being activated in the connectionist manner. This activation then spreads through the distributed memory, causing corresponding activations on the other objects in the network. These activations must then be returned to symbolic marks. In such a system, marking *letter-opener* would cause a large amount of activation on *knife*, with smaller activations on *gun* and *rope*. Adding some sort of threshold-like mechanism which would suppress marker-passing on the less activated nodes would complete the system. This situation, described in more detail in the following section, is shown in Figure 4.

Though conceptually simple, to implement such a mechanism requires blending two significantly different types of information. The symbolic marker-passer is passing discrete symbolic information that needs to propagate through the network, the distributed memory needs an activation strength that spreads via numerical combination. To integrate the two, a mechanism must be used which can provide such a numerical activation strength to the distributed memory. It turns out, however, that a mechanism already existing for other purposes in our system can be used for this purpose. We will describe this mechanism and then show how it is used for our present purposes.

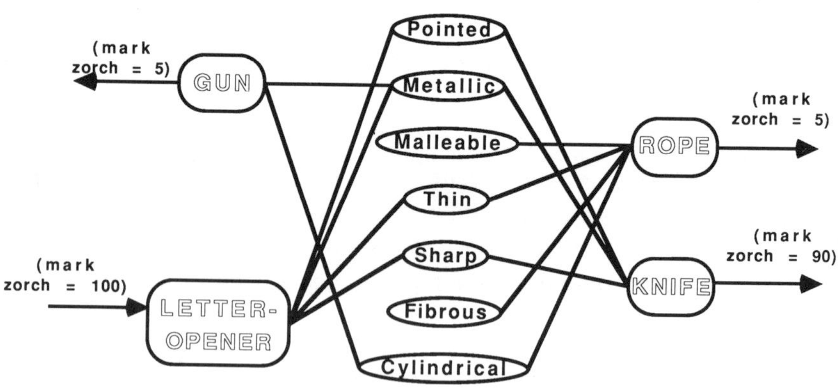

Figure 4. Zorch is divided as activation spreads through the defining-characteristic links

4.1. Zorch

One of the most important features in the design of a marker-passer is a mechanism which can limit the number of nodes marked and paths returned. If too many paths are found by the marker-passer, we lose efficiency, since the evaluator must examine all those paths which are reported. If too few paths are reported, the correct information can be missed. The marker-passer must be designed so as to constrain the number of paths reported.[12]

The simplest form of such an attenuation mechanism can be built into the algorithm in the form of a length limitation. In a complex network all the nodes are connected to each other. Some of these paths are long and tortuous—they require that a large number of inferences hold true. If one holds the breadth of activation spreading to a set limit, these long paths can be avoided. Most marker-passing systems to date have used such a limitation.

It turns out, however, that a limitation on length is not a sufficient mechanism since many paths that add no information are quite short. Consider the following paths from *suicide* to *rope*.

(1) *suicide* → *kill* → *hang* → *noose* → *rope*

(2) *suicide* → *action* → *hang* → *noose* → *rope*

These paths have the same length, and yet the first is more useful for inferencing than the second. The reason for this is that the node ACTION serves as the ISA hierarchy root for a large proportion of the frames in our system, and paths through such nodes as this are likely not to be useful. Further, allowing marks to flow through nodes such as ACTION will cause a tremendous proliferation of paths being found.

Charniak (1983) suggested dealing with this problem by checking the outbranching of nodes during marker-passing. Those promiscuous nodes which have an outbranching greater than some predefined constant will not propagate marks. They can be marked, but they cannot pass marks through. There are two major objections to this scheme.

The first drawback of this scheme is that, while it eliminates those paths that would be found by passing through a promiscuous node, it will not elim

[12] Another means of limiting the number of false paths is the removal of marks during processing. At present we have developed a model which degrades marker strength over time, depending on the original strength of the marking. When the strength is sufficiently decremented, the mark is removed. Other work, however, has suggested that the removal of marks is a more complex process than the one we have implemented. In this paper we will ignore the issue of removal of marks, since it does not effect the hybrid mechanism being discussed herein.

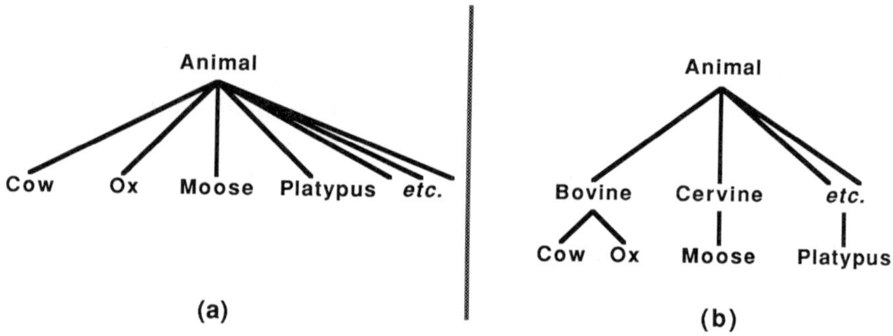

Figure 5. Networks varying in depth vs. breadth

inate those paths that meet at such a node. Thus, if we start marking from two frames that are both of type *object*, the first one will mark *object* and then stop. The second frame will do the same. However, at this point *object* will have been marked from each of the two nodes and thus will report a path. A solution to this is to amend Charniak's scheme so that promiscuous nodes not only serve as dead-ends to the marker-passer but also are not themselves marked.

The second drawback is shared by both the length and promiscuity limitations. When we use such limitations, our marking becomes very dependent on the form of the knowledge we are representing, not on the meaning. To see this, consider two ways to represent the relation between various frames concerning animals as shown in Figure 5. In the first case the paths from *cow* and *moose* to *animal* are short but go through a node with high branchout. In the second the paths are longer, but the nodes are less promiscuous. If we used a promiscuity limitation, the second path might be found and the first might not; if we used a length limitation, the reverse might be true.

Ideally, an attenuation mechanism should combine information about length and promiscuity so that longer paths are allowed if they involve little outbranching, and short paths that go through somewhat promiscuous nodes are also allowed. The attenuation mechanism used by SCRAPS is designed to limit both length and branchout of the paths found without violating the locality constraints needed for massively parallel computing. This is done by using a numerical constant, called *zorch*[13], to start marker-passing and dividing it by outbranching as we proceed. If we start at a highly branching node,

[13] The notion of attenuation is quite prevalent in discussion of spreading activation. The term *zorch* has often been used orally to describe this, although Hendler, (1987) appears to be the first use of the spreading activation literature. This use of the term appears to be based on an MIT use of the word to refer to "an amount of energy." More information on the MIT use of the word can be found in *The Hacker's Dictionary* (Steele et. al., 1983).

it and all its first set of descendants can be marked, but the zorch runs out quickly. If, however, we hit such a promiscuous node late in processing, its descendants are not marked, since we do not have enough zorch left.

It turns out that designing the attenuation mechanism in this way has many desirable features. In particular it can be shown that this mechanism fits nicely into a massively parallel implementation of marker-passing where it can serve as a feedback mechanism between the marker-passer and the planning component of SCRAPS. Further, by using a computation based on the attenuation numbers reaching a node of intersection, a path strength can be efficiently computed for use by the path evaluator. Finally, it appears that such an attenuation mechanism accounts for certain cognitive findings, particularly those relating to the fan effect, better than either a length or promiscuity limitation.[14]

4. EXTENDING THE MARKER-PASSER (CONTINUED)

At this point we have described all the mechanisms necessary for understanding the integration of our marker-passer with the distributed representation. Basically, the marker-passer is invoked as in the SCRAPS. Marks spread across the network carrying symbolic information. When a node with defining characteristics is encountered, the zorch of the mark is used as activation energy over the network defined by these defining-characteristic links. These links are used as a local connectionist network so that activation spreads in the analog manner over this subnetwork. As other nodes of this network gain activation they start passing marks with a zorch based on their activation. Marking then proceeds as normal, paths are found, and the path evaluator is invoked as normal.

As an example, let us now see how SCRAPS, extended with this new procedure, handles the *letter-opener* example. As in the standard version of SCRAPS marks are passed from the initial entities in consideration, including *letter-opener*. *Letter-opener*, as an origin of marking, is marked with a zorch equal to some initial constant. This zorch is now divided among the branches leading out from letter-opener. Some of this zorch goes through symbolic links to the symbolic properties of *letter-opener*, causing the marking of the neighbors of this node (perhaps *mail*). The rest of the activation is used as the constant starting the flow of activation through the microfeature network. Those nodes sharing microfeatures gain activation proportional to the number of microfeatures shared. The node *knife*, sharing multiple

[14] Details of this aspect of our mechanism can be found in Hendler (1987). A discussion of the fan effect is beyond the range of this chapter, Hayes-Roth (1977) provides a good discussion of the effect and related work in the psychological literature.

microfeatures with *letter-opener*, receives most of the activation. This enables marking to continue from *knife* with a zorch proportional to the amount of energy it received from *letter-opener*. This allows enough energy to reach the node *weapon*. We thus find the path:

> *buying* → *take-trip* → *plane* → *boarding* → *weapons-check* →
> *If* you go through *weapons-check* with a *weapon* you get arrested →
> *weapon* → *knife* → *letter-opener*

The path evaluator is invoked as usual and the system is able to rule out taking a plane for the flight home.

4.2. More details

Although the informal description presented so far serves to describe the mechanism, there are still some details we've yet to discuss. In particular, we have glossed over details of how symbolic information is carried through the distributed network.

To this point we have been describing marks as very simple entities. In point of fact, each marker in our system carries information about the origin of marking and the path that led to the node being marked, as well as information about the zorch. This information must be carried along with the activation energy that traverses the network and in our implementation this is exactly what is done. As the strength of the activation is computed throughout the distributed network, the information that originated the particular marking is carried. This has two major advantages: It allows several sets of activation to proceed through the network at the same time, and it allows the marker-passer to continue as activation at some node is collected (i.e., activation strength is computed without disturbing the symbolic information on the mark).

Allowing the distributed memory to be computing several activations simultaneously is an important feature of our mechanism. One of the goals of our formulation of marker-passing is to let each set of marking happen in parallel. In the example we've been discussing, three sets of marks are started in parallel: the marking from *buying*, from *letter-opener*, and from *me* (the agent of the action). As these marks flow through the network, they may encounter the microfeature network at any time. If, for example, we marked *gun* (perhaps I'm holding one) near in time to when we marked *letter-opener*, some entities will gain activation from one, some from the other, and some from both. We must be able to sort out which is which, and our system uses the symbolic information as tags for this purpose.

When a node is activated from several microfeatures its activation for each separate tag is computed. When one of these gets above a threshold marking continues using that tag. Thus, if *letter-opener* activates *knife* (via *sharp*, *pointed* and *metallic*) while *gun* activates it also (via *metallic*), the node representing *knife* will only start passing marks from the *letter-opener* frame unless the activation from *gun* also reached the threshold. In our present model this threshold is a function of activation energy divided by the number of symbolic branches leaving the node. Thus, the more facts we know about the entity represented by a node, the more activation energy is needed to allow markers to continue flowing.

There are still two features which differentiate our defining-characteristic-link network from the standard local connectionist networks. Connectionist networks generally allow weighted links and inhibition links. Both of these appear desirable in the design of such networks.

As the zorch is turned into activation strength and passed through the network the computations can be as arbitrarily complex as desired. If weighted links were used, the zorch could be divided based on the weights. If inhibition is desired, it can be implemented by adding inhibition links between separate network nodes, or it can be simulated by using asymmetric weights and a competition-based activation scheme (Reggia, 1986). We have, however, avoided adding these to date, since the simple sorts of examples we've presented here do need such complex mechanisms. As more is learned about the formulation and performance of local, distributed mechanisms, we envision adding such information.

5. IN CONCLUSION

Having demonstrated that we can create systems that gracefully integrate symbolic operations and connectionist models (or at least a limited class thereof), we now return to the issue of why to do so. In particular, we want to briefly address the issue of why we advocate a hybrid model rather than a wholesale embracing of one or the other technique. Why not simply create either a connectionist planner or focus on a symbolic mechanism for recognizing the similarities between objects? The latter is a well-known, difficult problem—much work has been expanded along these lines with little success—but what about the former? Why not simply create connectionist models to replace the symbolic planning models we've been using?

Our answer to the above is that we are trying to exploit the known best features of each approach. That is, each approach has shown the ability to shine in certain ways while having trouble in others. Further, these strengths and weaknesses of the two approaches are complementary. Connectionist models are stronger at similarity-based reasoning (as in finding the commonalities between the knife and the letter-opener), but they are particularly weak when it comes to providing control and symbol processing (as, for example, in performing planning). Some models, for example the system of

Touretzky and Hinton (1985), have been proposed for performing symbolic manipulations on massively parallel, distributed memories, but even these models have still needed to use a traditional, VonNeuman-style architecture to control gating and control flow within the system.

Since each style of machine has proven suitable for certain problems, our approach, demonstrated herein, has been to look for a blending. Essentially, the technique described in this paper took a massively parallel algorithm, the most connectionist of many of the modern symbolic approaches, and combined it with a local connectionist model, the most symbolic of the connectionist approaches. The "buying the letter-opener" example, presented in this paper, is an example of a problem in which merging these approaches enables us to deal with a task which neither approach, alone, has yet been able to solve.

We also believe this approach, the hybridization of symbolic and subsymbolic systems, provides an exciting area of exploration. Evidence in the cognitive literature seems to argue for both connectionist and nonconnectionist systems (see, for example, Posner & Snyder, 1975). Integration of these approaches is an important step towards explaining these results. We believe that the work presented in this paper is a start in the correct direction and an indication that progress can be made.

REFERENCES

Brachman, R.J. (1978, May). *A structural paradigm for representing knowledge* (Tech. Rep. No. 3605). Boston, MA: Bolt, Baranek, and Newman.

Brachman, R.J., & Schmolze, J.G. (1984, September). *An overview of the KL-ONE knowledge representation system (Tech. Rep. No. 655)., CA: Fairchild Inc.*

Charniak, E.C. (1983). "Passing markers: A theory of contextual influence in language comprehension." *Cognitive Science*, 7(3), 171–190.

Charniak, E.C., (1986). A neat theory of marker-passing *Proceedings of the 5th National Conference on Artificial Intelligence* (pp. 584–588).

Charniak, E.C., Gavin, M.K., & Hendler, J.A. (1983, February). *The FRAIL/NASL reference manual.* (Tech. Rep. No. CS-83-06). Providence, RI: Brown University Dept. of Computer Science.

Cottrell, G.W. (1985a). *A connectionist approach to word sense disambiguation.* Unpublished doctoral dissertation, Computer Science Department, University of Rochester.

Cottrell, G.W. (1985b). Connectionist parsing. *Proceedings of the Seventh Annual Conference of the Cognitive Science Society* (pp. 201–211). Irvine, CA.

Fahlman, S.E. (1979). *NETL: A system for representing and using real-world knowledge.* Cambridge, MA: MIT Press.

Feldman, J.A., & Ballard, D.H. (1982). Connectionist models and their properties. *Cognitive Science*, 6, 205–254.

Hayes-Roth, B. (1977). Evolution of cognitive structure and processes. *Psychological Review, 84,* 260–278.

Hendler, J.A. (1985). Integrating marker-passing and problem solving. *Proceedings of the Seventh Annual Conference of the Cognitive Science Society,* (pp. 130–139). Irvine, CA.

Hendler, J.A. (1987). *Integrating marker-passing and problem-solving: A spreading activation approach to improved choice in planning.* Norwood, NJ: Ablex Publishing Corp.

Hinton, G.E., McClelland, J.L. & Rumelhart, D.E. (1986). Distributed representations, in D.E. Rumelhart & J.L. McClelland (Eds.), *Parallel distributed computing.* Cambridge, MA: MIT Press.

Hirst, G.J. (1983). *Semantic interpretation against ambiguity* (Tech. Rep. No. CS-83-25). Providence, RI: Brown University Computer Science.

McClelland, J.L., Rumelhart, D.E., & Hinton, G.E. (1986). The appeal of parallel distributed computing. In D.E. Rumelhart & J.L. McClelland (Eds.), *Parallel distributed computing.* Cambridge, MA: MIT Press.

McDermott, D.V. (1978). Planning and acting *Cognitive Science,* 2, 71–109.

McDermott, D.V. (1983). *The NISP manual.* (Tech. Rep. U/DCS/RR No. 274). New Haven, CT: Yale University Computer Science.

Norvig, P. (1986). *Recovering intended inferences from text.* Unpublished doctoral dissertation, Computer Science Department, University of California at Berkeley.

Posner, M.I., & Snyder, C.R.R. (1975). Attention and cognitive control. In R.L. Solso (Ed.), *Information processing and cognition.* Hillsdale, NJ: Erlbaum.

Quillian, M.R. (1966). *Semantic memory* (Scientific Rep. No. 2). Boston, MA: Bolt, Beranek, and Newman.

Reggia, J. (1986). Virtual inhibition in parallel models of associative memory. *Proceedings of the Ninth International Joint Conference on Artificial Intelligence,* (pp. 244–248). Los Angeles, CA.

Smolensky, P. (1986). Connectionist AI, symbolic AI, and the brain. *AI Review.*

Smolensky, P. (1988). Putting together Connectionism. *Behavior and Brain Sciences,* Princeton, NJ (to appear).

Touretsky, D.S., & Hinton, G.E. (1985). Symbols among the neurons: Details of a connectionist inference architecture. *Proceedings of the Ninth International Joint Conference on Artificial Intelligence* (pp. 238–243). Los Angeles, CA.

CHAPTER 7

SAPIENS: Spreading Activation Processor for Information Encoded in Network Structures*

ANDREW ORTONY
University of Illinois at Urbana-Champaign

DEAN I. RADIN
Princeton University

One of the hallmarks of intelligence is the ability to efficiently identify information relevant to the solution of a problem while largely discounting irrelevant information. In many areas (such as language comprehension or perception) this means that relevant information must be accessed from memory more or less immediately, thus precluding any kind of exhaustive, or near exhaustive, search. Consequently, the relevance problem—the problem of how to identify a potentially relevant subset of the totality of stored information—is an important theoretical question in psychology and an important practical question for artificial intelligence. The work we report in this chapter might today be described as a localist connectionist approach to the relevance problem in the context of natural language understanding. However, it was actually undertaken in the mid-1970s, at which time we simply viewed it as a cognitive science approach to the then-hot topic of frame selection.

Prior to the recent rise of connectionism, research in AI had devised various domain-specific mechanisms for dealing with the relevance problem. For example, in the domain of scene analysis Waltz (1975) solved the problem by taking advantage of the huge reduction in data storage resulting from distinguishing the physically possible from the logically possible line junc-

* The research was supported in part by the National Institute of Education under Contract No. US-NIE-C-400-76-0116. We wish to thank George Kiss for supplying a magnetic tape version of the Associated Thesaurus, and Harry Blanchard, Glenn Kleiman, Philip Johnson-Laird, and David Waltz for helpful comments on earlier versions of this manuscript.

tions in a two dimensional representation of a scene. In the area of problem solving proper, a general guiding principle was the careful choice of problem representation. It quickly became apparent that the choice of problem representation could have a dramatic influence on the ease of solution, the old problem of trying to cover a mutilated chess board with dominoes being a simple but convincing example (see, for example, Raphael, 1965).

During the early and mid seventies, an important development in AI and psychology was especially pertinent to the relevance problem. This was the emergence from earlier but vaguer accounts (e.g., Bartlett, 1932; Piaget, 1952) of increasingly detailed proposals about the nature of generalized knowledge representations, variously called *frames* (e.g., Minsky, 1975), *scripts* (e.g., Schank & Abelson, 1977), and *schemas* (e.g., Rumelhart & Ortony, 1977). Because the *utilization* of a frame (in an AI system or in human cognition) results in a great deal of potentially relevant information becoming available automatically, it offers a powerful way of dealing with part of the relevance problem, but it does only deal with part of it. The missing component is the frame *selection* mechanism which is responsible for bringing the appropriate candidate frames into play in the first place. The purpose of SAPIENS was to explore a simple mechanism (Ortony, 1978) for doing this.

Efficient access to the knowledge structures relevant on a particular occasion depends, not so much on the nature and contents of individual knowledge representations (which is the problem of chief concern to those working on scripts, schemas, etc.), as on the overall organization and interrelations of such representations. In other words, it depends on the overall structure of memory, rather than on the structure of the things represented in memory. Models of the mascrostructure of knowledge organization have tended to rely upon associative networks. However, preconnectionist network approaches generally viewed semantic networks merely as spaces in which to conduct a particular kind of search operation known as the *intersection search* (Quillian, 1968; Collins & Quillian, 1969, Collins & Loftus, 1975). The general mechanism employed was that of spreading activation or marker passing (e.g., Charniak, 1983; Fahlman, 1979; Quillian, 1968), and the search was considered to have succeeded when it discovered an intersecting node that was reached from the different source nodes. This limited use, however, fails to capitalize on the power both of network representations themselves and of the spreading activation mechanism. The principal purpose of SAPIENS was to harness the potential power of the spreading activation mechanism and the associative network representation to simulate frame *selection*. Furthermore, this was undertaken in the context of a data base of sufficient size that the principles of the system's design could be generally applicable rather than relegated to the category of ungeneralizable "toy" problems that resist "scaling up."

Given our goal of frame selection as opposed to frame utilization, it was possible to deemphasize the (presumed) internal structure of the nodes in the net. The nodes are simply English words treated as the names of frames. In a complete representation, there would have to be at least three different levels—a lexical level, a conceptual level, and an episodic level. The *lexical* level would represent connections between *words* without distinguishing between distinctly different meanings that a word might have. At the lexical level, a word like *bank* could have connections to other words, some of which (e.g., *money*) have to do with the "financial institution" sense of *bank*, some (e.g., *weeds*) with the "side of a river" sense, and some with other senses of the word such as those related to basketball, airplanes, etc. Thus, the lexical level represents associations between words, not between concepts. At the same time, elements at the lexical level would serve as labels for, and hence be connected to, elements at the next, *conceptual*, level, where there would be separate representations (frames) for the distinct meanings of a word like *bank*. Clearly, at this level of representation a great deal of semantic information is explicitly encoded, for example by means of labeled relations between constituents. Finally, we assume that representations involving some of the more noteworthy specific experiences centered around particular frames would be represented at an *episodic* level. At this level, particular events and individual exemplars of frames would be associated with particular frames, sometimes indexed in terms of notable deviations from the canonical representations (see Schank, 1982).

A major goal of SAPIENS was to explore the degree to which frame selection can be facilitated through processes that operate only at the lexical level. There are both practical and theoretical reasons why this is a worthwhile enterprise. The practical reason is that it is much easier to construct a data base of lexical associations than it is to construct a comparably sized data base of conceptual and episodic structures. The theoretical reason is that the frame-selection process is essentially presemantic, in the sense that it presupposes that a semantic representation of the input is yet to be achieved. Indeed, the goal of the process is to facilitate the construction of such a representation, which means that the frame-selection process has to operate at a relatively impoverished semantic level. Of the three levels that we have proposed, the lexical level is the least complex, both semantically and structurally.

A parallel spreading activation mechanism ought to have at least the following characteristics: (a) *context sensitivity*, which permits the production of different patterns of activation for the same input string under different context conditions; (b) *decreasing activation* over time, so as to prevent every input from uniformly activating the entire network forever (what Waltz & Pollack, 1985, refer to as "heat death"); and (c) *summation of activation* from different sources.

Based on these principles, a processor for operating on a network was developed. In it, spreading activation was used as a mechanism, not just for finding intersections, but for settling on stable configurations of nodes that could serve as a restricted set of potentially relevant, candidate, nodes for subsequent employment in the task at hand. The ability to select a small set of potentially relevant nodes for possible use in subsequent processing is, as we have already discussed, an important aspect of an intelligent system. While we acknowledge that this is not sufficient to endow a system with genuine wisdom, we were unable to resist the name SAPIENS—Spreading Activation Processor for Information Encoded in Network Structures. The program was written in MACLISP and implemented on a DEC-10 computer. The data base was an associative thesaurus consisting of over 16,000 words and all free-associative strengths between them (Kiss, 1968). Since the data base was empirically generated (by soliciting associative responses from students), and since it is quite large, the network can be regarded as a reasonable facsimile of (part of) some arbitrary individual's associative network.

It was considered important to use a large, empirically realistic, data base for two reasons. We wanted a large network in order to get sufficient interconnectivity and semantic diversity to avoid trivializing the problem. Most implemented models of the kind we are discussing contain only a few hundred nodes. Quillian (1968) employed about 850 nodes, and the Waltz and Pollack (1985) program used a network containing between 1000 and 2000 nodes (Waltz, personal communication). Clearly, the massively increased number of possible paths in a network of 16,000 nodes imposes severe constraints on the nature of the processor that works on it. The fact that many humans process language perfectly well with an active vocabulary of the same order of magnitude (e.g., Anderson & Freebody, 1981) suggests that scaling-up would not be a serious problem for us. In other words, if the average person has an active vocabulary of, say 40,000 words, only a three-fold increase in network size would be required. Such an increase in size would not constitute a formidable computational or representational obstacle.

We wanted an empirically realistic data base because we considered it essential for exploring performance on a broad range of tasks. In a network containing associative strengths between elements, one has to deal with the question "Where do the numbers come from?" Our solution to this question was to use numbers that came from data. In this respect, at least, the localist approach we used has an advantage over the parallel distributed approach using semantic microfeatures (e.g., McClelland & Kawamoto, 1986), for in that approach there is the problem of what the (input) microfeatures are, and where *they* come from.

The main result of SAPIENS is that, given several input words, it quickly identifies a restricted set of relevant words (typically, 10 to 20) without extensive searching. These words can be thought of as the names of the best candidate frames for subsequent use by other mechanisms. In this respect, SAPIENS is analogous to the filtering process in Waltz's (1975) program that generates semantic descriptions of scenes with shadows. This filtering process takes a scene and finds a small subset of the most likely line segments (out of thousands of possibilities) for further processing by a semantic description mechanism. SAPIENS takes an input string and finds a small subset of relevant labels (i.e., words) for concepts that are potentially usable for further processing by, for example, inference or problem-solving mechanisms.

Four tasks were used to examine the performance of SAPIENS. While some of these tasks were suggested by published experimental work, it is important to emphasize that they are not intended as simulations of experiments. Rather, they should be viewed as illustrations of the kind of problems that SAPIENS can handle and of the way in which it handles them. In particular, they demonstrate that a significant part of problems like frame selection and lexical disambiguation can be achieved through the use of a very simple mechanism operating on a realistically large number of nodes in a network.

The first task shows how context can be used to disambiguate ambiguous words. The second task seeks to show how standard typicality effects found in various laboratory tasks can be accommodated by SAPIENS. Third, SAPIEN's performance in a mock cued recall "experiment" is examined. Finally, we describe a simple examination of the effects of manipulating word order of an input string.

It should be emphasized that we do not claim that the mechanism we propose is sufficient to realize complete solutions to such problems. Clearly, frame utilization is equally important. However, we do claim that, properly conceived, a spreading activation mechanism may well be a fundamentally important ingredient insofar as it represents a viable solution to the frame selection problem. We consider it significant that so much can be achieved with so little. The only data that SAPIENS operates with are associative strengths between words, and the activation mechanism is trivially simple, performing perfectly well without even resorting to the expedient of lateral inhibition (Waltz & Pollack, 1985).

DESIGN FEATURES

Before discussing the design characteristics of the program, it will be helpful to define and clarify our use of a number of key terms. First, in the discussion that follows we do not distinguish between *nodes*, *words*, and *frames* or concepts. This is not because there is no difference, but because in the present context the differences do not matter. The network contains *nodes* which are representations of English *words*. However, our theoretical orientation views words as labels or names for *frames*. We can consistently maintain this position, even though we have made no attempt to represent the richer underlying frame structures that would be needed if the network were to be employed in a full-fledged natural language processor.

On average, each node in the network is connected to about 20 other nodes. The connections, which are directed, represent the empirical values of the *associative strengths* between connected nodes. Associative strength is an integer measure (1–100) of how many times a response (R) was given to a specific stimulus word (S) by 100 people (Kiss, 1968). When SAPIENS is running, only a proportion of the associative strength is transmitted from S to R. This is the *activation* of R by S. The *activation level* of a node is the sum of activations received by it from all directly connected nodes. The set of nodes that can be reached from (i.e., as responses to) a particular node we call the *forward environment*.

The words that are used as input strings to SAPIENS we refer to as *seeds*. Each seed is *weighted* so that the original associative strengths can effectively be altered to simulate context effects and decreasing activation over time. The single list that results from appending the lists of neighbors of each seed, we call an *expansion*. Any item occurring more than once in this expanded list is an intersection, and this sublist of intersections comprises what we call the *relevant forward environment* or *RFE*. The expansion of all the input seeds and the identification of intersections (and therefore the RFE) occur within one time-slice or *cycle*. At each cycle the original seeds plus the intersections are re-input as new seeds. This severely restricts the kinds of intersections that result. As more and more intersections are input as seeds, it becomes increasingly probable that most of the activation will circle in on itself, with the result that a relatively stable configuration of highly active nodes quickly emerges. This breadth-first expansion in one time-slice simulates a parallel process, so that spreading activation in SAPIENS can be regarded as a parallel process which simultaneously spreads activation out from several input seeds.

The operation of SAPIENS is analogous to growing a crystal in a saturated solution. Input seeds used to start spreading activation are similar to chemical seeds used to start a crystal-growing process, and the densely interconnected associative network is similar to a densely saturated solution. The seeds constitute a core around which layer upon layer of molecules (or nodes) cluster. This clustering produces an onion-like series of shells around the seed core. The first layer is formed of molecules that are most strongly attracted to the seeds. After several layers have formed, the tendency for the crystal to continue growing diminishes. At some further point, an attraction threshold fails to be reached and the growing process stops. In a similar fashion, the first SAPIENS cycle creates a cluster of strongly connected nodes around the input seed core. As each new layer is formed, the size of the cluster increases, and more nodes are pulled into the RFE. Each new layer, however, is formed of successively more remotely related nodes.

The first cycle creates a tight cluster around the seed core. The second cycle wraps another layer about the first, but is associatively less strongly related to the core. Normally, SAPIENS stops after the second cycle, partly because successive clusters are mildly related at best, but mainly because the goal of quickly finding a small stable subset of relevant concepts for further processing is usually achieved after about two cycles. Although the resulting set usually contains a few false positives—not all of the words found represent concepts that are likely to be useful for further processing—the ones that are useful are found without searching the entire network.

Design Example

We shall now describe a fairly detailed example of the operation of SAPIENS using, for simplicity's sake, fictitious but realistically representative data. Suppose the input seeds are the two words, *apple* and *fruit*, both with input weights of 1. First, *apple* and *fruit* are expanded to create a list containing all neighboring nodes and their associative strengths. It is important to note that the total activation output from each seed equals 100, as shown in Figure 1. The value 51, for example, from *fruit* to *banana*, refers to the fact that the proportion of subjects producing *banana* as a response to the

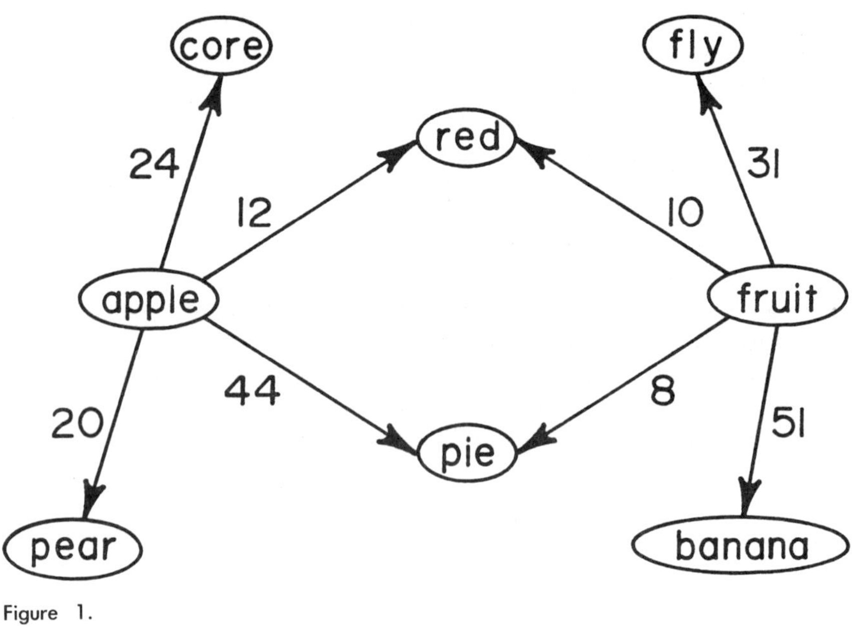

Figure 1.

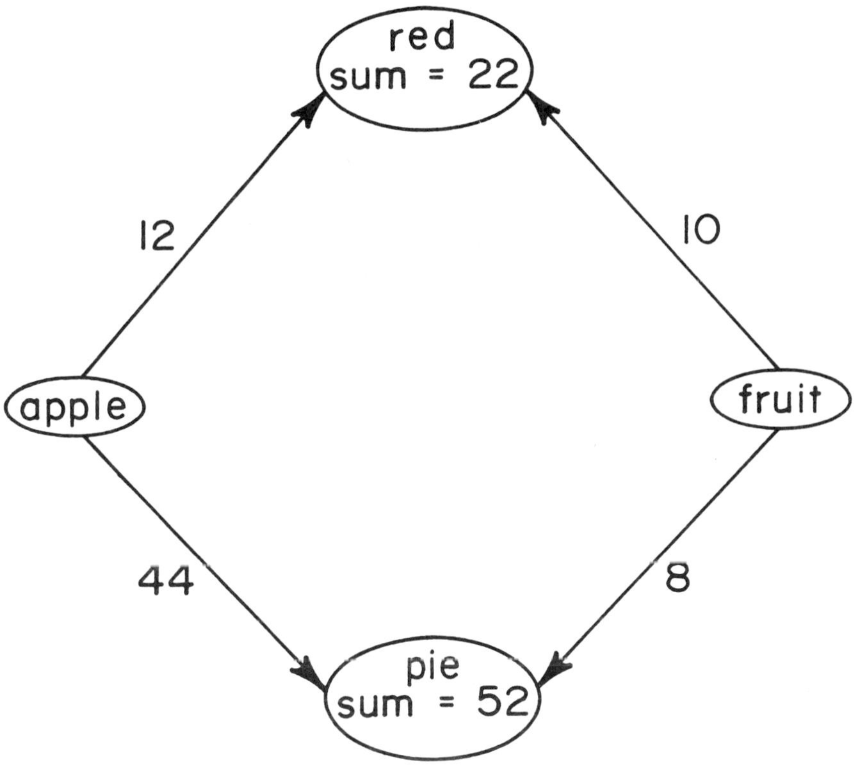

Figure 2.

stimulus *fruit* was 0.51. In other words, they are empirically based transition probabilities.

The next step is to find intersections in the expanded space and to create a new list of the intersections and summed strengths for use in the next cycle (Figure 2). The new input list contains the original seeds with input weights 1, together with the intersection words at weights which reflect the proportion of total activation on all words in the list (see Figure 3). The purpose of constraining the new list by adding in only newly found intersections is to control the spread of activation. In other words, it is functionally equivalent to employing thresholds, but it has the advantage that it obviates the need to preassign a threshold to every node in the network. The prevention of an activation explosion on the next cycle is also facilitated by the use of proportional weights on the intersected nodes. The weights on a seed are used to multiply the associative strengths of each response to that seed. For example, since the weight on the node *red* is 0.3, the activation sent to each response from *red* will be multiplied by 0.3. This means that the total activation sent from *red* will be 30, because each node originally outputs a total of

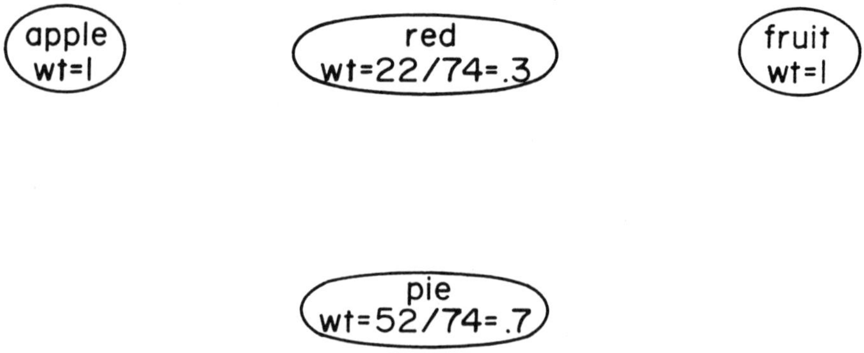

Figure 3.

100. The original seeds are always re-input at weight 1, and successive intersection nodes are always input at proportional weights which sum to 1, to reflect the attention being placed on the input seeds (high activation) and the process of spreading activation (low activation) operating on intersected nodes.

Because the total input weight of all intersected nodes is constrained to sum to 1, the activation on intersections decreases rapidly. As the new seed list increases in length on each successive cycle, the proportional activation on each word in the relevant forward environment (RFE) necessarily decreases. It is, of course, possible, and desirable, that the proportional activations on relevant RFE words will change as the spread continues because some nodes will receive additional activation from newly accessed nodes.

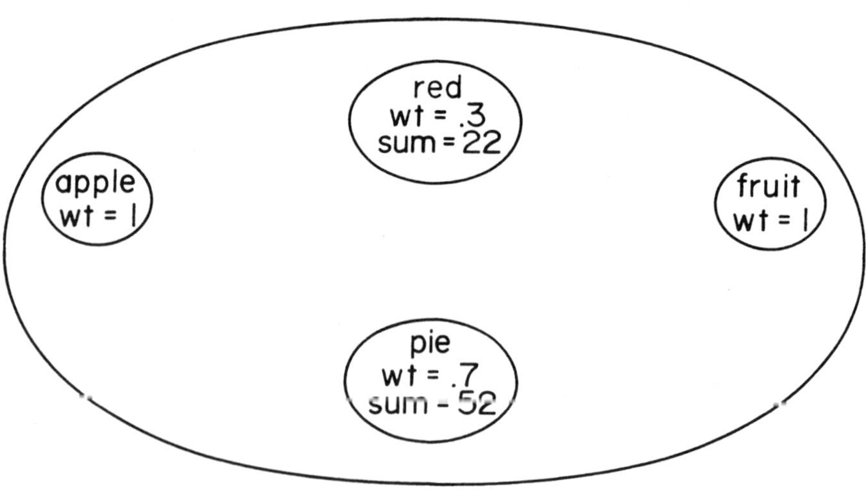

Figure 4.

Continuing the example, with the creation of the new input list, the cycle (one time-slice) is complete. At the end of each cycle the result is a new RFE. Notice, as can be seen in Figure 4, that the original input seeds *apple* and *fruit* do not have strength sums. Normally all nodes that appear in the RFE must have received activation from at least two sources. The exceptions are the original input seeds, which are always included in the RFE regardless of whether or not they received activation from other sources. Each word in the RFE has an activation strength associated with it which is the total amount of activation received by that word on the last cycle. The *activation strength sum* is the sum of all activation levels of nodes currently in the RFE. In the present example, the activation strength sum is 74.

The maximum strength sum attainable after any cycle is $100 \times N + 100$, where N equals the number of input seeds. This is because each input seed always outputs a strength of 100 (weight = 1), and the sum of strengths from all intersected nodes is forced to equal 100 (sum of weights = 1). In our example, if a total activation of 300 is injected into the network at the beginning of the second cycle, then 300 is the maximum attainable strength sum. This can occur only if all of the responses (from the input seeds and the intersections) happen to be intersections too, for these particular responses are the only nodes receiving activation in the network.

Because there is an upper bound on the strength sum, it is possible to make meaningful comparisons among different strength sums. For example, if the maximum strength is 300, and one pair of seeds produces a sum of 60 after two cycles, and another pair produces a sum of 30 after two cycles, we can say that the first pair represents 20% of the available activation and the second pair represents 10%. Thus, the first pair of seeds is better "integrated" (in the sense of "interrelated") than the second pair by 10% of the available activation.

Input seeds that are strongly associated with one another will produce an RFE containing more intersections and, hence, a higher activation strength sum. Weakly associated seeds may produce an RFE with only a few intersections, and a correspondingly smaller strength sum. *Association* in this sense is thus better thought of as a measure of "interrelatedness" between words.

PROGRAM PERFORMANCE

Lexical Disambiguation

The first task to be described is concerned with the problem of the selection of appropriate word meanings in the face of more than one alternative. The most typical example of this problem is the need to select the appropriate meaning of an ambiguous word on the basis of contextual information. The way in which SAPIENS deals with disambiguation was investigated by injecting a context-setting word and the target (ambiguous) word into the network as input seeds. The intersections found after only one or two cycles were usually sufficient to produce clusters of words that were clearly associated with the contextually appropriate meaning of the target word. For example, consider the words *mint*, *bank*, *bar*, *fruit*, and *game*. The first three are ambiguous in the usual sense: They each have more than one distinct meaning. In particular, we were interested in distinguishing between the sense of *mint* as candy, and as a place for manufacturing coins. Similarly, we were interested in distinguishing the sense of *bank* as a financial institution from the sense in which rivers have banks. And we wanted to distinguish the sense of *bar* as a place to have a drink from the sense in which a rod is a bar. The remaining cases, *fruit* and *game*, are not ambiguous words in the usual sense; rather, they are vague, being consistent with a wide range of very different kinds of referents.

If SAPIENS is appropriately sensitive to context, if ought to provide evidence of greater availability of words related to the appropriate meaning of an ambiguous word. In the case of the vague terms, context ought to influence the relative accessibility of different instances of the concept. Thus, for example, if *red* is part of the context for *fruit*, *apple* ought to be more available than *banana*, or *lemon*, whereas, if *yellow* were part of the context, one might expect the reverse to be true. The results of the simulation are presented in Table 1.

Recall that the activation strength sum can be taken as a measure of the degree of interconnectedness or integration of the word clusters associated with an input. Thus, for example, the fist two rows in Table 1 suggest that

Table 1. Results of Disambiguation Simulation

Context Word	Target Word	First Three Rank-Ordered Nodes in the RFE[a]	Number of Nodes in RFE	Strength Sum	Proportion of Total Activation
chocolate	mint	sweet, milk, taffee	3	22	.07
coin	mint	money, cash, gold	11	77	.26
river	bank	river, water, stream	9	87	.29
money	bank	money, silver, book	23	123	.41
metal	bar	iron, door, rod	3	23	.07
drink	bar	beer, drink, pint	7	83	.28
yellow	fruit	orange, green, banana	4	25	.08
red	fruit	apple, green, hot	3	31	.10
card	game	play, poker, board	3	28	.09
ball	game	play, football, tennis	3	33	.11
chess	game	board, play	2	40	.13

[a]Relevant Forward Environment

the pecuniary sense of *mint* is better integrated than the gastronomic sense. Another interesting property of SAPIENS is that the original seeds only appear in the RFE if they themselves receive activation from their associates. So, *river* appears high in the RFE (see Row 3 of Table 1) because it received activation from *water*, *stream*, and/or other activated elements in the RFE.

A general observation that can be made about these results is that the most highly ranked nodes in the RFE are, as a rule, semantically highly related to the appropriate sense of the target word. In cases where the RFE contains a relatively large number of nodes, for example *bank* in the context of *money*, the least highly ranked concepts are very weakly related, often representing syntagmatically, rather than paradigmatically, related concepts. For example, although not shown in the table, one of the least activated nodes in the RFE for *bank* is the word *make*. Perhaps this can be explained, at least in part, by the fact that one puts money that one *makes* into the bank. One way to interpret the varying numbers of nodes in the RFE is to suppose that the more nodes there are in it, the better integrated the RFE is, and the more knowledge there is associated with the particular use of the term (compared only, of course, to alternative uses of the same term). In other words, uses giving rise to larger RFEs might be regarded as the higher frequency, or more typical ones. However, caution should be exercised in this respect: It is not being claimed that, for example, the most probable use of the word *mint* is the pecuniary sense, but only that the sense is more probable than the particular alternatives that were investigated in the context of the present data base. It is perfectly possible that *mint*, as herb, would be the preferred sense in some other data base. Furthermore, a realistic test of this would need to control for the frequency of the context-setting word, which was not done in the present case.

If the program is viewed as a model of frame selection (presumably for later use in top-down processing), it can be concluded that SAPIENS is reasonably successful. The clustering process was supposed to quickly find a small subset of relevant nodes without resorting to extensive searching. This relevant subset, the RFE, ought to and did include nodes that were related to the input seeds *taken together*. Note that SAPIENS does not know which seed is the context-setting word and which is the target word. If both are somewhat ambiguous, a complementary disambiguation can occur. Consider, for example, *drink* and *bar*. both words are compatible with several different kinds of referents. *The bartender made a drink* and *The mother made a drink* suggest different kinds of drinks, while *The bar was crowded* and *The bar was rusted* imply different kinds of bars. Since SAPIENS finds items relevant to all of the input words, the RFE for *bar* and *drink* will reflect both a *drink* sense of *bar* and a *bar* sense of *drink*. Similarly, finding a relation between *red* and *fruit*, and *fruit* and *red*, entails a comparable process, except that the seeds are (technically) unambiguous.

Typicality and Verification Tasks

It is a well-established finding in cognitive psychology that the speed with which true sentences of the form "An x is a y" can be verified depends on how typical the example, x, is of the category, y, (see e.g., Rosch, 1973; Smith, Shoben, & Rips, 1974). If *apple* is a more typical fruit than *strawberry*, then the sentence *An apple is a fruit* will take less time to verify than the sentence *A strawberry is a fruit*. This fact is, of course, consistent with the view that more typical exemplars are more available than less typical ones. The second set of simulations was conducted to see whether or not SAPIENS would demonstrate such differential availability.

Injecting *apple* and *fruit* into the network results in an RFE containing the words *pie, pear, orange juice, tree, juicy, tart, banana, green, sauce,* and *summer* after just one cycle. These words are listed in order of activation level; the proportion of available activation was 0.32 (strength sum for this RFE was 97). By comparison, if the seed nodes are *strawberry* and *fruit*, the RFE comprises *cream, juice, jam, raspberry, pie, tart, summer,* and *food* after one cycle, with an activation proportion of 0.21 (strength sum equals 64). Thus, as expected, the more typical exemplar, *apple*, produces an RFE with greater overall strength than the less typical one. Similarly, if the seeds are *robin* and *bird*, the RFE consists of *bird, song, sparrow, swallow, thrush,* and *starling* after two cycles, with an activation proportion of 0.38 (strength sum of 114), whereas with *penguin* and *bird* the RFE contains *bird, fly, black, feathers, feather, flight, sky,* and *wing* after two cycles, with a smaller activation proportion of 0.29 (strength sum of 87) than in the *robin* case.

At this point it is worth making a couple of observations about the contents of RFEs. First, it should be remembered that the data that form the basis of the network were collected some 20 years ago and represent some amalgam of a large number of undergraduate students at Edinburgh, Scotland. Assuming the general tendency of the dialect to be closer to that of British English than to American English, it should be recognized that robins (being relatively uncommon in Britain) are not, in fact, the best exemplars of birds—sparrows, thrushes, and starling are all the better. Notice that these other good instances in fact find their way into the RFE for *robin* and *bird*. There are a number of other peculiarities of this kind. For example, it is the experience of most English people that apples are typically just as green as they are red (perhaps even more so) because there is a subcategory of apples known as *cooking apples* which are always green (and sour) and are used in pies and tarts. Another peculiarity by U. S. standards might be the occurrence of *pint* in the RFE, resulting from the seeds *drink* and *bar*. In Britain one of the most typical things to drink in a pub (especially at a bar) is a pint (of beer).

Notice also, that the RFE often contains more than one cognate as a word (e.g., *feather* and *feathers*). Sometimes this is just a plural form, sometimes a verb form, and so on. Ideally, these would have been removed, but their presence causes no real problems. Furthermore, it is interesting to note that the second highest RFE word for the pair *penguin/bird* was *fly*. Presumably, this is because words like *feathers* and *wing* contributed enough activation to *fly* for it to become highly activated. It does not seem likely that the high activation on *fly* can be explained by the fact that penguins are *unable* to fly because the data base does not incorporate the denial of properties. Finally, it should be mentioned that experiments on sentence verification routinely reveal not only that subjects are faster at verifying good examples of category members than poor ones, but also that they are very fast at rejecting non-examples like *A pencil is a bird*. The RFEs resulting from the corresponding inputs can be interpreted in a manner consistent with this finding, for, in the case where the term refers to a noninstance of the category, the RFE tends to be empty so that the strength sum is equal to or close to zero.

Cued Recall

In an experiment reported by Anderson and Ortony (1975), subjects studied a number of simple sentences such as 1 and 2.

(1) Nurses are often beautiful.
(2) Nurses have to be licensed.

Later, subjects were given either a "close" or a "remote" recall cue. In the present example *actress* would be the close cue for sentence 1 and the remote cue for sentence 2, while *doctor* would be the close cue for sentence 2 and the remote cue for sentence 1. The experiment showed that what was semantically close or remote depended on the entire sentence rather than on part of it, since the cues were not sufficient to permit the recall of control sentences which included key words like *beautiful* and *licensed*. For example, the cues were not differentially effective—indeed, they were not effective at all—for control sentences like 1a and 2a.

(1a) Landscapes are often beautiful.
(2a) Taverns have to be licensed.

SAPIENS appears able to accommodate these results rather easily. Assuming that different RFEs are created for the different sentences, then close recall cues can be expected to integrate more strongly with their respective RFEs, and distant recall cues less strongly. These integration strengths would reflect the probability of a cue eliciting recall of a sentence.

The task in the Anderson and Ortony (1975) experiment was simulated by first creating RFEs from input seeds corresponding to the substantive words in the to-be-learned sentences. Then the expanded recall cues were intersected with the input seed RFEs to find the amount of integration between them. The sequence of operations used by SAPIENS is illustrated in Figure 5.

The process begins when seeds A and B are injected into the network (Figure 5a). Intersections are then found between expanded A and expanded B. This area is called RFE1 (Figure 5b). The seeds and RFE1 are re-input to the network (Figure 5c) as explained in the basic design section. This gives rise to a second RFE (RFE2) resulting from the intersections of seeds A and B, with RFE1 (Figure 5d). As discussed earlier, any intersecting node, that is, any node receiving activation from at least two sources is part of this (and subsequent) RFEs. This is because SAPIENS assumes that all and only the nodes receiving activation from more than one source have their thresholds exceeded. A node receiving activation from just one source will not exceed

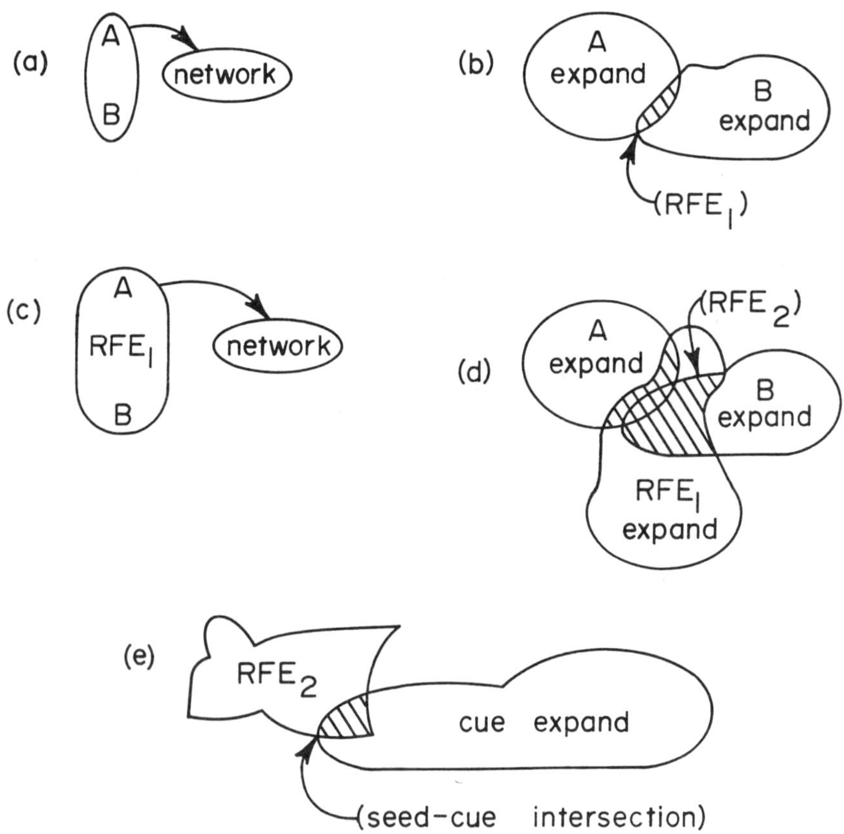

Figure 5.

its threshold because insufficient activation will be received (Figure 5d). The intersection between the RFE2 and the expanded cue is an example of a *seed-cue intersection* (Figure 5e). The activation strength sum of the seed-cue intersection can be viewed as being directly related to the probability of recalling the input given a cue because the activation strength sum is a measure of the overlap between the input seed RFE and the cue RFE. It is reasonable to assume that, if a cue RFE completely overlaps (and restimulates) the input seed RFE, the probability of recalling the inputs will be very high. On the other hand, if a cue RFE does not overlap any portion of the input seed RFE, it is unlikely that the input will be recalled because no close associates of the inputs were reactivated. Therefore, if one activation strength sum is greater than another, we can assume that the former sum will reflect a greater probability of input sentence recall than the latter sum. The results are summarized in Table 2.

In almost all cases the trends are in the direction found by Anderson and Ortony (1975). Not all the words from the original experiment were available in the network: consequently, some minor changes were needed. None of

Table 2. Results of Cued Recall Simulation

Target Sentence Pair[a]	Proportion of Total Activation With: Close Cue	Remote Cue[b]
	shampoo	*detergent*
The *nurse washed* her *hair*	.36	.18
The *nurse washed* her *clothes*	.18	.22
	sexy	*doctor*
Nurses are often *beautiful*	.23	.17
Nurses give health *care*	.09	.51
	saw	*scissors*
The *farmer cut* the *wood*	.25	.18
The *farmer cut* the *fabric*	.08	.29
	climbing	*hanging*
The *ivy covered* the *walls*	.20	.16
The *picture covered* the *walls*	.11	.14
	hammer	*fist*
The *man hit* the *nail*	.20	.11
The *man hit* the *jaw*	.07	.19
	tv	*radio*
The *man watched* the *show*	.11	.01
The *man listened* to the *show*	.13	.11

[a]Words in italics represent input seeds.

[b]As presented, the close cue usually produces a higher proportion of total activation for the first member of the pair and a lesser proportion for the second. Similarly, the remote cue produces a lesser proportion for the first member and a higher proportion for the second.

these wording changes in any way affected the validity of the test. For example, the use of the cue *sexy* for the sentences about nurses were necessitated by the fact that the original cue used in Anderson and Ortony (1975), *actress*, was not in the network. As an aside, it should perhaps be mentioned that this forced us to revert (for the purposes of science, only) to the standards of sexism that were pervasive at the time the data base was assembled 20 years ago.

It is important to emphasize that no claims are being made here about specific comprehension and memory processes. All that is being supposed is that, whatever the comprehension process is, it does involve the establishment of an RFE to limit the frames that are brought into focus for that process. It is also assumed, as discussed in Ortony (1978), that that part of the RFE could be used as (or reproduced from) the representation of the remembered sentence so as to facilitate subsequent retrieval.

Word Order Effects

Compare sentences 3a and 3b.

(3a) The boy smashed the bus.
(3b) The bus smashed the boy.

In sentence 3a, images of vandalism and boy-related items are likely to come to mind, whereas in sentence 3b, accidents and bus-related items come to mind. Because SAPIENS is unable to use (or even recognize) syntactic information, the normal input would treat the three input seeds *boy*, *smash*, and *bus* as one simultaneous parallel input string with no regard to case role assignments. Consequently, *boy, smash, bus* and *bus, smash, boy* produce exactly the same RFE. But, of course, sentences 3a and 3b have very different meanings, and ideally one would like the fact that they are different to be reflected in the resulting RFE.

One way in which this can be done is to assign different initial weights to input seeds as a function of their position. For example, one might assign an initial weight of 10 to the first seed, which in the case of examples 3a and 3b would be the agent, and weights of 1 to the second and third words (the verb and patient). Such weights could be regarded as reflecting the salience of each case role. Figure 6 shows the effects on the RFE that such a manipulation of weights has. As one might expect, boy-related items are ranked high when the input seed *boy* has a high weight, and bus-related items are ranked high when the input seed *bus* has a high weight.

This result suggests that the salience of the input words can play an important role in determining the activation levels of nodes in an RFE. The activation levels are important, because the nodes with the most activation tend to be more relevant than lower activated nodes.

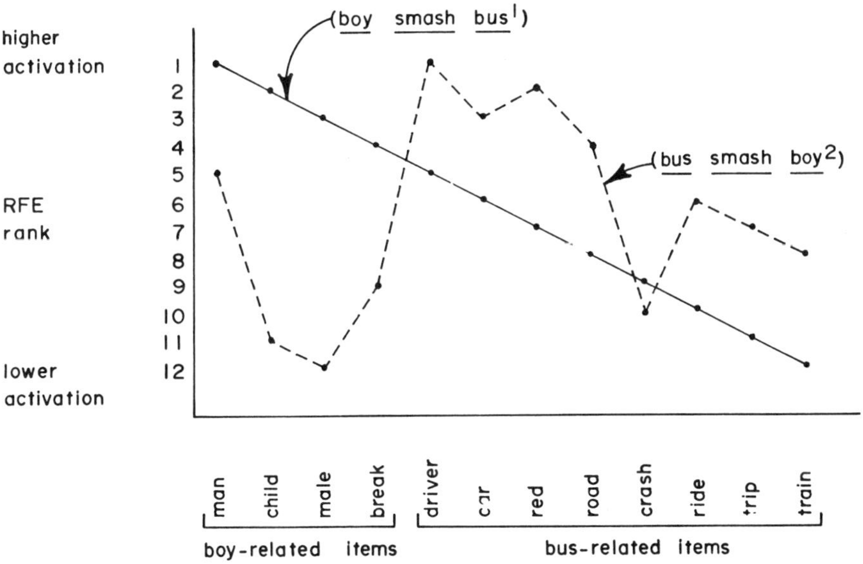

1 This straight line shows the ranking of nodes in the RFE produced by boy smash bus.

2 This curve shows how the nodes are ranked lower for boy-related items and higher for bus-related items when the input seeds are bus smash boy.

Figure 6.

Obviously, assigning weights of 10 to the agent, and 1 to the verb and the patient, is a completely arbitrary way of handling the problem. A more sophisticated approach would be to select the weights on the basis of some kind of optimization procedure. Since SAPIENS has no parsing ability, the user has to apply the weights to each case role. For handling natural language it would be necessary to recognize, for example, that the input was in passive form so as to permit appropriate adjustments to the weights to be made. The purpose of the present demonstration, however, is only to show that, in principle, SAPIENS has the flexibility to be sensitive to simple relational constraints ordinarily marked by syntax.

CONCLUSION

Natural language processing, like other areas of AI, has to face the problem of how to reduce the search set of candidate representations in order to quickly access appropriate ones to facilitate performance on the task at hand. This has proved to be a somewhat stubborn problem in the past, and one which becomes increasingly difficult as the size of the data base increases. SAPIENS appears to be a viable general solution to this problem, because it quickly produces a small set of candidates for a range of important natural language processing tasks. The principles behind the design of SAPIENS are essentially independent of the particular mechanisms that would be employed in a complete natural language processing system, and, in fact, there is no reason why SAPIENS could not be utilized in other AI domains as well.

Of course, there is nothing unique about the idea of using associative networks as the basis for dealing with natural language processing tasks such as lexical disambiguation (Small, Cottrell, & Tanenhaus, 1987). However, SAPIENS differs from previous proposals (e.g., Charniak, 1983; Fahlman, 1979; Hayes, 1977; Small, 1980; Waltz & Pollack, 1985) in a number of important respects. First, it is an implemented system rather than a theoretical proposal (e.g., Charniak, 1983). Second, it does not depend on building individual frame structures to superimpose on the network in order to reduce search (Hayes, 1977). It engages in minimal search without the need for such structures. Third, as already discussed, the network is sufficiently large to permit SAPIENS to be more than a mere feasibility demonstration. Fourth, the data base is simple in the sense that it contains only one kind of connection, thus avoiding the need to label hundreds of thousands of connections as *isa*, or *part-of* links, and so on (e.g., Fahlman, 1979). Finally, the mechanism and representation is very simple. Neither inhibitory connections nor explicit thresholds are used, yet there appears to be no serious problem with false positives, and no problem of heat death.

How can so much be achieved with so little? The answer lies in the "realistic" nature of the connections. People's free associations between words embody a great deal of tacit semantic information—information which is implicitly represented in the data base, both qualitatively and quantitatively. This information is virtually impossible to generate through introspectively based manual coding: One could not expect to build a realistically comprehensive set of connections and associative strengths in a nontrivial associative network on the basis of a priori reasoning. The success of SAPIENS lies in the extensiveness of the connections in the network (about a third of a million) and in their empirical validity as simple associative relations. Fortunately, we were spared the task of collecting these data ourselves.

REFERENCES

Anderson, R.C., & Freebody, P. (1981). Vocabulary knowledge. In J.T. Guthrie (Ed.), *Comprehension and teaching: Research reviews* (pp. 77–117). Newark, DE: International Reading Association.

Anderson, R.C., & Ortony, A. (1975). On putting apples into bottles: A problem of polysemy. *Cognitive Psychology, 7*, 167–180.

Bartlett, F.C. (1932). *Remembering*. Cambridge, England: Cambridge University Press.

Charniak, E. (1983). Passing markers: A theory of contextual influence in language comprehension. *Cognitive Science*, 7, 171–190.

Collins, A.M., & Loftus, E.F. (1975). A spreading activation theory of semantic processing. *Psychological Review*, 82, 407–428.

Collins, A.M., & Quillian, M.R. (1969). Retrieval time from semantic memory. *Journal of Verbal Learning and Verbal Behavior*, 8, 240 247.

Fahlman, S.E. (1979). *NETL: A system for representing and using real-world knowledge*. Cambridge, MA: MIT Press.

Hayes, P.J. (1977). On semantic nets, frames and associations. *Proceedings of the Fifth International Joint Conference on Artificial Intelligence*.

Kiss, G.R. (1968). Word associations and networks. *Journal of Verbal Learning and Verbal Behavior*, 7, 707–713.

McClelland, J.L., & Kawamoto, A.H. (1986). Mechanisms of sentence processing: Assigning roles to constituents of sentences. In J.L. McClelland & D.E. Rumelhart (Eds.), *Parallel distributed processing: Explorations in the microstructure of cognition* (Vol. 2, pp. 273–325). Cambridge, MA: Bradford Books.

Minsky, M. (1975). A framework for representing knowledge. In P.H. Winston (Ed.), *The psychology of computer vision*. New York: McGraw-Hill.

Ortony, A. (1978). Remembering, understanding, and representation. *Cognitive Science*, 2, 53 69.

Piaget, J. (1952). *The origins of intelligence in children*. New York: International Universities Press.

Quillian, M.R. (1968). Semantic memory. In M. Minsky (Ed.), *Semantic information processing*. Cambridge, MA: M.I.T. Press.

Raphael, B. (1965). *The thinking computer*. San Francisco, CA: Freeman.

Rosch, E.H. (1973). On the internal structure of perceptual and semantic categories. In T.F. Moore (Ed.), *Cognitive development and the acquisition of language*. New York: Academic Press.

Rumelhart, D.E., & Ortony, A. (1977). The representation of knowledge in memory. In R.C. Anderson, R.J. Spiro, & W.E. Montague (Eds.), *Schooling and the acquisition of knowledge*. Hillsdale, NJ: Erlbaum.

Schank, R. (1982). *Dynamic memory*. New York: Cambridge University Press.

Schank, R.C., & Abelson, R.P. (1977). *Scripts, plans, goals and understanding*. Hillsdale, NJ: Erlbaum.

Small, S. (1980). *Word expert parsing: A theory of distributed word-based natural language understanding* (Tech. Rep. No. 954). Baltimore, MD: Department of Computer Science, University of Maryland.

Small, S., Cottrell, G., & Tanenhaus, M. (Eds.) (1987). *Lexical ambiguity resolution*. Los Altos, CA: Morgan Kaufmann.

Smith, E.E., Shoben, E.J., & Rips. (1974) Structure and process in semantic memory: A featural model for semantic decisions. *Psychological Review*, 81, 214–241.

Waltz, D. (1975). Understanding line drawings of scenes with shadows. In P.H. Winston (Ed.), *The psychology of computer vision*. New York: McGraw-Hill.

Waltz, D.L., & Pollack, J.B. (1985). Massively parallel parsing: A strongly interactive model of natural language interpretation. *Cognitive Science*, 9, 51–74.

CHAPTER 8

Simulated Annealing as a Parsing Technique*

GEOFFREY SAMPSON
Centre for Computer Analysis of Language and Speech
University of Leeds

1. OPTIMIZATION BY SIMULATED ANNEALING

Writing in 1951, John von Neumann argued that problems including what we would nowadays identify as the combinatorial-explosion phenomenon, and the quasi-impossibility of total debugging, implied that computing in the future would need to rely less on the long chains of methodical, all-or-nothing reasoning characteristic of orthodox formal logic, and more on stochastic, probabilistic analytic methods characteristic of certain branches of physics:

> there are numerous indications to make us believe that this new system of formal logic will move closer to another discipline which has been little linked in the past with logic. This is thermodynamics, . . . [which] is that part of theoretical physics which comes nearest in some of its aspects to manipulating and measuring information. (von Neumann, 1963, p.304)

The discipline of information technology has been very slow to respond to this insight. Now, however, the technique of simulated annealing (see, e.g., Kirkpatrick, Gelatt, & Vecchi, 1983; van Laarhoven & Aarts, 1987) is emerging as a highly fertile computational technique which is achieving considerable success through applying thermodynamic concepts in domains

* I am grateful to my departmental colleague Peter Roach and to John Bridle of the Royal Signals and Radar Establishment, Malvern, for introducing me to simulated annealing. Parts of this paper appeared in *Proceedings of the 11th International Conference on Computational Linguistics (COLING '86)*, and this material is reprinted with the permission of the International Committee on Computational Linguistics.

which have proved resistant to more orthodox approaches for the kinds of reasons alluded to by von Neumann over 30 years ago.

Simulated annealing is currently being applied to the practical problem of optimizing the physical design of computer circuitry, and to the theoretical problems of resolving patterns of auditory and visual stimulation into meaningful arrangements of phonemes and three-dimensional objects. Grammatical parsing—resolving unanalysed linear sequences of words into meaningful grammatical structures—can be regarded as a perception problem logically analogous to those just cited, and simulated annealing holds great promise as a parsing technique. My aim in this paper is to explain the simulated annealing concept, and to describe a small-scale experiment already carried out to explore its practicality in the parsing domain. A larger-scale project to create a simulated-annealing parser for English is now being initiated at Leeds.

Simulated annealing can most directly be explained via a physical analogy. Consider the logical space of alternatives in some complex combinatorial problem as a chunk of mountainous terrain, in which the altitude of any point corresponds to the relative "goodness" of that particular solution (the lower, the better). We want to find the lowest point, but there are far too many points for each to be considered separately. We might try to locate the low point by dropping a ball onto the territory at random and hoping it will roll down to the low point. This corresponds to randomly choosing a particular overall solution to the combinatorial problem and then considering a series of modifications to individual components of the solution, adopting the modifications whenever they improve the overall solution and rejecting them otherwise. But the ball is very unlikely to reach the low point. Much more probably, it will roll a short way downhill and come to rest in a local minimum—a place where all *immediate* moves are uphill, even though, some distance away, there are places much lower than the spot where the ball has halted.

In this situation, a good way of improving the search technique would be to pick up the landscape and shake it, so that the ball does not invariably roll downhill but sometimes bounces over obstructions. One would begin by shaking hard, so that even the highest peaks can be cleared, and then gradually reduce the amplitude of shaking so that the ball searches for lowness in terms of progressively finer detail. In computational terms, rather then deciding whether to adopt each of a series of modifications to the randomly chosen initial position simply by reference to whether it yields a gain or a loss, one decides by reference to whether it yields a loss that is greater than a positive number whose magnitude tends towards zero as the process continues. This is simulated annealing. The metaphor comes from metallurgy: the desirable molecular structure of steel is obtained by gradual cooling from an initial molten state in which molecules rearrange their positions randomly, and "quenching"—cooling abruptly—leads to a flawed structure, just as using a simple gain/loss computation in modifying a random initial solution leads to an undesirable local minimum.

Not all combinatorial problems are amenable to the annealing technique. If the desired low point in the mountainous terrain of the analogy happened to be at the bottom of a deep, narrow mineshaft sunk from a high place, annealing would not help to find it. But the logical geometry of many real-life combinatorial phenomena is more like the geometry of natural mountains, where there is a strong tendency for relatively low-lying points to be adjacent to many other relatively low-lying points. For such phenomena, simulated annealing can be a good practical way of arriving at optimal solutions. It exhibits the same capacity to build on unforeseen success which, writing in a different intellectual connection, Friedrich Hayek (e.g., 1955) attributes to blind evolution with natural selection as opposed to conscious planning. (Cf. Kirkpatrick et al., 1983, p. 679.) In the case of auditory and visual perception, simulated annealing is arguably also a psychologically plausible model of human cognitive processing. Quoting Geoffrey Hinton, Bridle and Moore (1984, p. 315) advocate a concept of perception as a " 'a parallel, distributed computation in which a large network settles into a particular state' under the influence of the sensory input."

2. STATISTICAL METHODS IN LANGUAGE PROCESSING

The applicability of annealing as a parsing technique presupposes a statistical approach to natural language (NL) analysis which will itself be unfamiliar to many readers. At this point I must therefore digress from the annealing concept in order briefly to describe the statistics-based NL research paradigm within which I have been working for some time, and to which simulated annealing appears to offer an important contribution.

Much current work in parsing, as represented in books such as King (1983) or Jones and Wilks (1983), analyses input by systematically checking its properties against the various possibilities allowed by a grammar which specifies the language in question as a well-defined class of sentences. The grammar may be in the form of an ATN, a GPSG, or in some other form, and the checking process may operate in very diverse ways; but all these approaches have in common the notion of a clear-cut boundary between a (probably very rich and complex) class of well-formed inputs, and other inputs which simply do not belong to the language under analysis. (This generalization is not invalidated by a parser such as that of Charniak (1983), which is capable of recognizing various ill-formed sentences; Charniak's parser, too, implies a rigorous distinction between the class of well-formed and the class of ill-formed sentences, although, unlike most parsers, it also implies a further, equally rigorous classification of ill-formed inputs into those which are nevertheless recognizable and those which are not.)

I do not mean to undervalue the remarkable results achieved in practice by some of these parsers, from W.A. Woods's LUNAR system onwards. Nevertheless, there are major difficulties in making such parsers work adequately with authentic, unedited input drawn from unrestricted domains. NLs are so endlessly diverse and unpredictable in the turns of phrase they display that many people find it extremely difficult to believe that any sets of rules, no matter how complex, can fully define them. It remains as true as it was 60 years ago that "All grammars leak" (Sapir, 1921, p. 38). Because of this problem, the strategy adopted by many computational linguists is to concentrate on developing systems which work for artificially delimited "toy" subsets of language, in the hope that such systems, if successful in their limited domains, can be enlarged and linked up until they eventually cover entire languages. Occasional remarks suggest that these researchers do not always appreciate how very wide the gap is between their toy languagettes and real-life language (cf. Sampson, 1984).

An alternative strategy insists on working only with authentic language, recognizing that this makes any attempt to incorporate well-formedness definitions into automatic processing systems at best impractical and perhaps impossible in principle. The Unit for Computer Research on English language (UCREL) of the University of Lancaster, led by Geoffrey Leech and Roger Garside, has made considerable progress in recent years in developing automatic routines which succeed in analysing authentic text. The work of this group is surveyed in Garside, Leech, & Sampson (1987). Most UCREL work uses the Lancaster-Oslo/Bergen Corpus, a million-word machine-readable collection of text extracts designed to form a representative sample of printed British English in its diverse genres from government reports to thrillers.

The first major UCREL achievement was the CLAWS word-tagging system. Since 1982 CLAWS has been assigning part-of-speech tags, drawn from a finely-differentiated, 133-member tagset, to words of authentic running English text (which are often grammatically many ways ambiguous in isolation), with a consistent accuracy level of 96%–97% of words correctly tagged—a figure which we believe can be further improved by tuning the system in various ways. This is an achievement to which we have been unable, despite extensive inquiries, to discover near rivals.

The significant point about CLAWS is that it embodies no knowledge whatever of the overall grammatical architecture of English sentences. Instead, it uses an empirically derived matrix of relative transition-probabilities between pairs of adjacent tags (together with information enabling a set of candidate tags to be identified for any given word taken in isolation). Simplifying greatly, CLAWS works by forming all possible paths through sequences of tags which are candidates for the words of a sentence, evaluating each path by multiplying together the probabilities of the transitions between successive pairs of tags contained in the path, and assigning to each word that tag for which the summed values of all paths passing through the tag is highest. As a matter of policy, no entries in the matrix of tag-transition probabilities are zero; we know, as well as other linguists, that failure to observe a particular transition in our data does not imply that the transition is "ungrammatical", and therefore even unobserved transitions are assigned a small positive probability. Thus it is true to say that the system "knows" nothing about English in the sense of drawing sharp distinctions between grammatical and ungrammatical sequences; it deals exclusively in *relative* likelihoods. Yet this wholly "unintelligent" system works extremely well. It is easy to make CLAWS fail by inputting trick sentences of the kind often encountered in linguistics textbooks, but the lesson of CLAWS is that such sentences are notably rare in real life.

The performance of CLAWS goes some way towards vindicating a point of view about language processing which, among researchers on grammar, currently ranks as somewhat heretical, but which has counted as received wisdom among workers on speech processing for several years. In the 1960s it was widely believed by phoneticians, on grounds of *a priori* linguistic and psychological theorizing, that successful automatic speech-analysis systems would require subtle artificial-intelligence techniques simulating general human intelligence and native-speaker-like linguistic competence. This view was put to empirical test over the years 1971–76 by the U.S. Department of Defence's Advanced Research Projects Agency, which funded rival research groups to develop speech-understanding systems capable of achieving set performance criteria (see, e.g., Lea 1979; Cole, 1980). The clear winner in this contest was Carnegie-Mellon University's HARPY, which differed from its rivals in making little use of AI techniques in the ordinary sense and instead relying heavily on statistical information about simple transition probabilities between adjacent linguistic units, and on heuristic methods for eliminating low-probability analyses. Since the ARPA Speech Understanding Project, it has been widely accepted by speech researchers that the earlier enthusiasm for "intelligent", top-down approaches was misguided, and that successful systems are likely to be statistics based rather than logic based and to use conceptually simple (e.g., Markovian) models of language structure. (Recent representatives of this trend include Bahl, Jelinek, & Mercer, 1983; de Mori, 1983; Cravero, Fissore, Pieraccini, & Scagliola, 1984; Derouault & Mérialdo, 1984; Smith, Denenberg, Slack, Tan, & Wohlford, 1985.) Newell (1980) has argued that the point is valid, not merely for practical automatic language-processing systems, but also for models of human psychological machinery. Mainstream computational linguistics has been very slow to take on board the lessons of HARPY, but I take these lessons to be relevant for all levels of language processing.

3. PROBABILISTIC PARSING

Having largely solved the problem of grammatically disambiguating words, our group are currently attempting to develop a CLAWS-like solution to the larger problem of grammatical parsing of authentic language. We have built up a data base of material parsed manually according to a detailed standard parsing scheme which allows labelled parse-trees to be drawn for any written English "sentence" (including such phenomena, frequent in practice though often overlooked by mainstream linguistics, as chapter headings, captions, addresses and dates, disjointed speech quoted in fiction, academic prose including formulae, bibliographical citations, etc.); from this we extract statistics on the relative frequency of various local properties of parse-trees. Our overall assumption is that the correct parsing of an input sentence will in the great majority of cases coincide with that labelled tree-structure over the words of the sentence which maximizes some function (probably a function making heavy use of Markovian transition-probabilities) representing degree of conformity to the statistical norms of English grammar. (A range of such functions are being explored; I discuss an example below.) It is important to understand that, in our approach, the concept of an "ill-formed" tree plays no role. The domain of a tree-evaluation function is the class of *all logically possible labelled trees*, that is, all structures conforming to the mathematical definition of *ordered tree* and in which each node has a label drawn from a recognized vocabulary of grammatical category-symbols. Many logically possible trees, in this sense, will be quite absurd as linguistic structures (e.g., a complex tree in which each node without exception is labelled "prepositional phrase"); in our terms, a tree which is "absurd" is simply one whose value is relatively low.

The most direct way to imitate the CLAWS technique in the parsing domain, having settled on a suitable tree-evaluation function, would be to evaluate all possible labelled tree-structures having terminal nodes corresponding to the words of an input sentence and to choose the tree whose overall value is highest. Unlike in the case of word-tagging, however, for parsing this approach is wholly impractical. Our statistics suggest that an average sentence is about 23 words long (including punctuation marks, which in our work count as words and have their own place in parse trees). Our parsing scheme yields parse-trees with about 0.6 nonterminal nodes per terminal, so the correct parse-tree for a 23-word sentence is likely to have about 13 nonterminal nodes, apart from the root (whose label is known in advance, since it is always S). The number of alternative choices of label for a nonterminal node in our scheme is 26 or thereabouts.[1] There will thus be on the order of 26^{13} alternative labellings of 13-node tree-structure; if a million alternatives were considered per second, it would take on the order of 30,000 years to examine the alternative labellings for a single structure, and comparable periods for each of the many other ways of linking the given sequence of words into a tree-structure. The quantity of processing needed might be reduced through dynamic-programming techniques (see, e.g., Baker, 1979), but it does not seem plausible that the savings achievable would be enough to make the technique of exhaustively checking alternatives a practical one, even if—which is far from clear—the appropriate tree-evaluation function permits parse-trees to be decomposed into independently assessable subtrees as required for dynamic programming.

4. THE PILOT PROJECT

The Lancaster group have been pursuing their own, nonstochastic approach to solving this problem. However, the situation described seems to be precisely the kind to which simulated annealing is applicable. The annealing technique is regarded as particularly useful in domains characterized by a high degree of frustration, in the sense of Toulouse (1977)—that is, where an analytic decision which makes an overall solution better in one respect will commonly at the same time worsen it in another respect; and the parsing domain is full of frustration, since a decision to group a sequence of nodes together as a constituent of a given category will often yield a good way of accounting for the daughter nodes at the cost of creating a problem about fitting the mother into the surrounding tree-structure. I therefore conducted a small experiment in an initial attempt to see whether simulated annealing might be used for probabilistic parsing.

The task I set myself in this pilot project was to design a system which yields parsings of sentences that are input to it in the form of sequences of wordtag-symbols. (Since the CLAWS system is already capable of assigning wordtags to words of running text with high accuracy, it was reasonable to eliminate this step in my initial annealing experiment.) The desired output

[1] The number of distinct labels is actually far higher, on the order of 60,000, but this is because the labelling system provides many "subcategory symbols", combinations of which are added automatically to a main category by reference to the labels of its daughters: e.g., the verb phrase *has not been* would be labelled not simply V but Vzeb, to show that it is inflected for third person singular, is negated, and has *be* as main verb. Since these subcategory symbols are assigned automatically, they are not relevant to the combinatorial complexity of the problem of locating optimal parse-trees. When automatically-assigned subcategory symbols are eliminated, there remain 29 distinct labels, but a few of these are in perfect complementary distribution with others with respect to their daughters (e.g., X always dominates the wordtag XNOT, representing *not*, and R never dominates XNOT), so that, with respect to combinatorial complexity, they could be treated as allographs of a single label.

parse-trees were those assigned to sentences by our existing parsing scheme; however, for my pilot project the refined range of grammatical categories in this scheme were collapsed into a very crude system which distinguished just 30 rather than 133 wordtags (including six tags for classes of punctuation-mark), and 14 nonterminal (i.e., phrase and clause) categories. My chief reason for using such crude categories was the practical one that I already had access to relevant statistics for these categories but not for the full set of grammatical distinctions in our scheme. As the crude category-set is no more than a temporary expedient, there seems little to be gained by describing it in detail here, and I shall simply explain those categories which occur in cited examples as they arise.

Since the system is given wordtags as its input, the small size of the wordtag-set used was a conservative decision, making the task of the parser harder rather than easier. On the other hand, because nonterminal node-labels are part of what a parser is required to output, the fewness of these categories does in one way make the task of my pilot system unrealistically easy: the space of possible solutions to be searched is considerably smaller than it would be if more nonterminals were distinguished. But 14, although a low number, is not absurdly low by comparison with the roughly 26 distinctions that I quoted above as being made by our full parsing scheme (the details of which, incidentally, were decided without any consideration of the special problems posed by the annealing technique). Furthermore, in another way, the crudity of the nonterminal categories severely hinders the success of the parser: because the subcategory information referred to in Footnote 1 is entirely ignored by my statistics, the best available tree-evaluation function based on these figures must inevitably be far less efficient at discriminating between plausible and implausible analyses than a similar function based on more refined statistics would be. On balance, while my pilot system is extremely simple by comparison with any system likely to be used for serious parsing, I believe it may offer a useful indication of whether the annealing technique is worth pursuing

The system was given statistics derived from manual parsings of sentences totalling about 40,000 words taken from various parts of the LOB Corpus. (This database of manually parsed material has since been enlarged.) Following a suggestion of my colleague Eric Atwell, I extracted, from this set of parse-trees, figures on the frequency of each type of transition between daughter-labels below a given mother-label. For instance, the occurrence in the database of a node labelled J (adjective phrase) and having daughters labelled n j P (singular noun, adjective, prepositional phrase) would add one to each of the entries J /[n,J/nj,J/jP, and J/P] in the statistical table. (I deliberately illustrate using a node-type which may strike the reader as ill-formed but which occurs in real-life material: an authentic example is the phrase italicized in "the value obtained must be *assignment compatible with the type of the variable* . . .".) The number of logically-possible transition types is 28,336, since each type is defined by a mother label drawn from 14 nonterminal categories and two daughter labels each drawn from 45 possibilities (the set of 44 terminal and nonterminal categories together with left- or right-hand boundary symbol), except that a transition from boundary symbol to boundary symbol is disallowed: $14 \times (45^2 - 1) = 28{,}336$. Of these logical possibilities, 1251 had nonzero entries. It is a measure of the difficulty of capturing the diversity of authentic NL constructions by clear-cut well-formedness definitions that about one in three of the 1251 transition-types found in 40,000 words each occurred once only.[2] At the other extreme, the commonest transition-type was P / [i (prepositional phrase beginning with preposition), which occurred 4381 times.

I defined a function which uses these figures to compute a value for any logically-possible labelled tree (whether or not it is linguistically "sensible") as follows. Write *f(x/yz)* for the observed frequency of transition-type *x/yz*. Now, consider any non-terminal node a: call its label m_a, and the labels of its daughters $d_{a,1}, d_{a,2}, \ldots, d_{a,n_a}$ (where $d_{a,1}$ and d_{a,n_a} will always be left- and right-hand boundary markers, respectively). Then, if a tree contains nonterminal nodes $a_1, a_2, \ldots, a_k$, its value is defined as:

$$(1/k) \sum_{i=1}^{k} \ln(1/\sqrt{}((1/(n_i - 1))(\sum_{j=1}^{n_i-1} 1/(f(m_i/d_{i,j}d_{i,j}+1)+0.1)^2)))$$

The rationale of this formula is as follows. In evaluating each individual node, I take the reciprocal root mean square reciprocal of the frequencies of its daughter-transitions, rather than simply taking their mean, in order to ensure that one or two "bad" transitions drag down the overall value of a node with many daughters, most transitions between which are "good"; this prevents the system being unduly tolerant of non-standard structures. The small quantity 0.1 is added before taking reciprocals in order to guard against division by zero when a tree to be evaluated contains one or more daughter-transitions not attested in the database. The part of the formula which derives a value for the tree from the values of individual nodes averages the logarithms of the latter, rather than averaging raw node-values, again as a device to ensure that a minority of "bad" nodes drag down the overall value of a tree containing many "good" nodes.

I do not pretend that this formula will ultimately prove to offer the best possible estimator for the plausibility of alternative candidate parses for a string. But it has emerged from a certain amount of experimentation, and in conjunction with the statistical data discussed it does commonly seem to succeed in preferring correct to incorrect parses.

---

[2] On this point, cf. Sampson (1986).

5. MOVE SET AND ANNEALING SCHEDULE

In order to apply the annealing technique to the task of maximizing this function, it is necessary: (a) to define a class of local changes to trees which is ergodic, in the sense that any logically-possible tree can be converted to any other by applying a series of changes drawn from the class (we cannot allow the initial choice of tree to eliminate the possibility of ever reaching some other tree, which might be the correct one); and (b) to define an annealing schedule.

For (a), I define a possible move as follows. Given a parse-tree, select a node other than the root at random. Disconnect it from its mother. It will then be located within an "arch" of nodes whose left and right bases are respectively the last word before and the first word after the disconnected constituent, and whose "keystone" is the lowest node dominating both of those words. Choose at random *either* any node located on the arch other than the two bases, *or* any link between two nodes in the arch. In the former case, attach the disconnected node to the chosen node as an extra daughter, and relabel the new mother of the disconnected node with a randomly-chosen label. In the latter case, create a new node on the chosen link between existing nodes, label it with a randomly chosen label, and attach the disconnected node to it as a sister of the node at the lower end of the chosen link. In either case, if the ex-mother of the disconnected node is left with only one daughter, then delete the ex-mother by merging its upward and downward links (if the ex-mother is the root node, its remaining daughter becomes the new root and is accordingly relabelled S).

It is easy to show that this class of moves is ergodic. There is a straightforward algorithm for converting any tree by means of such moves into the simple tree in which each terminal node is a daughter of the root (keep detaching the leftmost terminal node which is not a daughter of the root and reattaching it to the root, deleting its ex-mother as required); and (except with respect to relabelling) the reverse of each of these moves is also a legal move. These two facts establish that any structure can be transformed into any other structure; and a given structure can acquire any legitimate labelling through successive detachment of each nonroot node and reattachment with re-labelling to the node that was previously its mother.[3]

---

[3] Ideally the move-generating function should be defined in such a way that, in the long term, each point in the possibility-space is visited with equal frequency (in the terms of Cox and Miller (1965, p. 9), the "equilibrium probability distribution" should be flat). That is not true for the function described, though empirical tests suggest that the deviation from flatness is not very great. Since carrying out the pilot project reported here, I have replaced the move-generating function with one which is much more efficient computationally and which yields a flatter equilibrium probability distribution, but the function described above is the one which produced the results reported.

There is no magic formula to determine the ideal annealing schedule for a given class of combinatorial optimization problems: this depends on the geometry of the solution-space, and has to be discovered by experimenting. (Geman & Geman, 1984, discuss the possibility of determining annealing schedules theoretically, but as I understand it their work applies only to problems of special types.) The requirements on an annealing schedule are that annealing must begin at a high enough "temperature" for the system to be thoroughly "melted" (that is, the factor by which the negativity of locally negative moves is discounted must be sufficiently large for moves to occur at random with no significant bias towards locally positive moves), and "cooling" must take place slowly enough for adequate searching of the possibility-space to occur. (If the initial temperature is unnecessarily high, or cooling unnecessarily protracted, a heavy penalty will be paid in extra processing for little or no gain in ultimate accuracy of search.) "Temperature" might be treated as a constant figure which is added to the result of subtracting previous likelihood-value from subsequent likelihood-value in determining whether a move under consideration yields a net gain and is therefore adopted. Commonly, however, users of the simulated annealing technique strengthen the thermodynamic analogy by adding, not a constant figure, but a figure drawn randomly from the positive half of a Gaussian distribution with zero mean, with "temperature" standing for the standard deviation of the Gaussian. Thus, even at a high temperature it will sometimes happen that a slightly locally negative move is rejected, and even at a low temperature it will occasionally happen that a strongly locally negative move is accepted. (Locally positive and neutral moves are always accepted at any temperature.)

6. AN EXAMPLE OF PERFORMANCE

Let me illustrate by reference to one of the first annealing runs carried out by my pilot system, on a short sentence input as:

d j j n o v i d j n .

Brief glosses for these symbols are: d, determiner; j, adjective; n, singular noun; o, modal verb or *do*; v, main verb; i, preposition; ., sentence-final punctuation mark. Thus the sequence stands for a sentence such as *The quick brown fox will jump over the lazy dog.* This is of course an artificially simple example, and if authentic language were commonly as well behaved as this example, then the case for using probabilistic parsing techniques would be weak. However, notice that my technique embodies no concept of a contrast between well formed and deviant strings, so that in principle it should be as easy to set an annealing parser to seek the "least implausible" analysis of a highly deviant input as to seek the correct analysis of a thoroughly well-formed input. The reason for beginning with a simple example is that I anticipate that the performance of the system will become more sensitive to details of the evaluation function and annealing schedule as inputs become more complex and less well formed, and at present I have only begun to explore alternative evaluation functions and annealing schedules.

The schedule used for the run to be illustrated was as follows. The structure initially assigned to the string was the flat structure in which each word is an immediate constituent of the root:

[Sdjjnovidjn.]

(In displaying trees as labelled bracketings, I place the capital letter labelling any constituent immediately inside the left bracket which opens that constituent; closing brackets are unlabelled.) This tree has the value −2.26. The initial temperature (standard deviation of the Gaussian) was 1. The temperature was reduced by 3% after every fiftieth successive attempt to change the tree. The system was deemed to have "frozen" at the first temperature-drop at which each of the 100 preceding attempts to change the tree either had been rejected or left the value of the tree unchanged. (A small proportion of randomly generated tree "changes" yield a new tree which is in fact identical to the old tree: it would complicate the programming to exclude these cases from the class of legal tree-changes, and there is no practical advantage in doing so. Since such moves do not affect tree-value, when generated they are always counted as accepted.)

To give the reader a feeling for the way that an annealing run proceeds, I display in Figure 1 the situation reached after every hundredth attempted tree-change. On each line I show the temperature reached immediately before the drop which occurs at that point, the proportion of the last hundred attempted changes which were accepted, the value of the current tree, and that tree itself. Decimals are rounded to three significant figures. The system froze at temperature 0.287, after 2100 attempted changes, of which 1173 were accepted. (The five "changes" recorded as having occurred between attempts 2000 and 2100 were null changes, as discussed at the end of the previous paragraph.) The structure attained at freezing is the correct structure for the input sequence, according to our parsing scheme. (Categories N, P, and V stand for noun phrase, prepositional phrase, and verb phrase—the latter in our scheme refers to a sequence of auxiliary and main verbs, not including object, complement, etc.; we treat sentence-final punctuation marks as immediate constituents of the sentences they conclude. We recognize no internal structure in a noun phrase such as *the quick brown fox*.)

I should immediately add that the success of this particular annealing run was not attained consistently. This run was one of a series of three over the same input and using the same schedule, the other two were less successful.

Attempts	Temp.	Changes	Value	Current Tree
100:	0.970	93	−1.31	[Sdjj[Rn[Lo[G[Fvi][J[N[Gdj]n].]]]]]
200:	0.913	93	−1.52	[S[Wd[D[Jjj]no]][Fv[Vidj]][Pn.]]
300:	0.859	92	−1.02	[S[Pd[Dj[Jjn]][Nov]i[R[Adj]n]].]
400:	0.808	88	−2.30	[Sd[Dj[V[Fjn][P[V[Gov][Ji[Ldj]]n].]]]]
500:	0.760	82	0.00976	[S[Vdj][Njn[W[Tovidj]n]].]
600:	0.715	90	−0.536	[S[N[Jd[T[Pjj][S[N[Tn[Jov]][Tid]]j]]]n].]
700:	0.673	66	1.64	[S[Sd[Njj[Nn[No[Vvi]]]d[Njn]]].]
800:	0.633	69	−1.51	[S[N[Nd[Sjj]]no][J[A[Tvi][Tdjn]].]]
900:	0.596	73	−0.0562	[S[N[D[F[Ad[S[Fjj][Nno]]]v][Tid]j]n].]
1000:	0.561	75	−0.984	[Sd[D[L[Gj[Njn]][Wov]][F[F[Lid]jn].]]]
1100:	0.527	58	1.50	[S[P[Mdj][Njn][Jovi][Ndj]n].]
1200:	0.496	66	−0.184	[S[Ndjjn][Novid]j[Mn.]]
1300:	0.467	63	0.668	[S[N[Dd[Njj]n]ov[Lidj]n].]
1400:	0.439	54	0.325	[S[Nd[Njj]n][P[Jo[Rvi[Sdj]]n].]]
1500:	0.413	57	0.760	[S[Nd[V[Njj]n]o][S[T[Tvi][Dd[Njn]]].]]
1600:	0.389	18	4.93	[S[Ndjjn[F[Vov][Pid]]][Njn].]
1700:	0.366	8	5.21	[S[Ndjjn][F[Vov][Pi[Nd[Njn]]]].]
1800:	0.344	11	5.46	[S[N[Ndjjn][F[Vov][Pi[Ndjn]]]].]
1900:	0.324	7	5.70	[S[Ndjjn][F[Vov][Pi[Ndjn]]].]
2000:	0.305	5	6.63	[S[Ndjjn][Vov][Pi[Ndjn]].]
2100:	0.287	5	6.63	[S[Ndjjn][Vov][Pi[Ndjn]].]

Figure 1.

One of the other runs yielded the structure [Sdjjnov[Pi[Ndjn]].] (value 4.43) at temperature 0.270, after 2200 attempted changes of which 1328 were accepted; here part of the correct structure has been discovered but no structure has been built in the first half of the sentence. The third run froze on the structure [Sdjjn[Vov]id[S[Njn].]] (value 4.18) at temperature 0.166, after 3000 attempted changes of which 1378 were accepted. In this case parts of the structure are downright erroneous: for instance, the closing noun phrase and full stop have been analysed as an embedded sentence, such as a direct quotation.

However, although this experiment yielded failure as well as success, on balance I believe it gives considerable grounds for optimism. Notice the scale of the system's achievement when it does arrive at the correct parsing. It has genuinely winnowed this structure out from among the total range of logically possible labelled tree-structures, without using any knowledge of grammatical rules in the ordinary generative-grammar sense: in the early stages of the run it considers trees containing quite crazy structures, such as a genitive phrase (G) consisting of a finite clause (F) followed by an adjective phrase (J). In reality, a genitive phrase must by definition have as its last IC a genitive-inflected word. (In a more sophisticated version of the annealing parser, rules of this kind which are totally reliable might be used to reduce considerably the size of the solution-space to be searched, though it will be important to err on the side of caution in shrinking the solution-space, since it involves risks of two types: structures might inadvertently be eliminated although they would be regarded as the preferred analyses of deviant but authentic inputs, and elimination of individual points in the solution-space could make it difficult to ensure that a given move-generating function remained ergodic.) Considering how many alternative trees are available to the system, a few thousand steps seems a small number by which to reach the correct solution or at least (in the other two cases) its vicinity.

When the system has erred, this may well be due in part to the crudity of the statistics used. Thus, the fact that, on one run, the last three input elements were wrongly parsed as a subordinate S may have to do with the fact that my statistics do not distinguish between direct quotations within sentences, and independent sentences: in reality, independent sentences virtually always include sentence-final punctuation, while direct quotations commonly do not. But note that, despite this limitation, the evaluation function used has in fact assigned considerably lower values to the two mistaken outputs than to the correct parse. This suggests that even without refining the statistics it should be possible, by modifying the annealing schedule (e.g., raising the freezing threshold, or lowering the cooling rate), and perhaps by adjusting the definition of the class of legal tree-changes (in order to modify the adjacency-relationships between various points in the solution space), to ensure that future runs would not freeze on these erroneous structures.

Parenthetically, I should point out that the fact that the system has produced different outputs on separate annealing runs over the same input is not in itself objectionable, though the particular outputs produced for this particular input on two of the cited runs are indeed unacceptable. When an input sentence is genuinely ambiguous, this will correspond in probabilistic-parsing terms to a solution-space in which two or more solutions are higher-valued than all other solutions but do not differ significantly in value among themselves. In this situation an annealing parser will find just one analysis on any run, and the analysis found will not be the same from run to run. Psychologically, this behavior seems natural (we notice one interpretation of an ambiguous sentence, unless some special consideration causes us to become conscious of the ambiguity)—though arguably this feature of the annealing technique might be disadvantageous for practical automatic language-processing purposes.

Results like those described here led to the initiation in December 1986 of a much more sophisticated research project using a richer statistical database, sponsored by the Royal Signals & Radar Establishment under contract MOD2062/0128 (RSRE), and directed by myself and Eric Atwell. For an up-to-date account of this project see Haigh, Sampson, and Atwell (1988).

7. LEFT-TO-RIGHT PROCESSING

If the annealing technique can be made to yield good results with a wide range of authentic language material, its chief advantages will be robustness and computational simplicity. The annealing algorithm is highly robust, in that it cannot fail to yield *some* output analysis for any input, no matter how deviant. In this respect its performance seems quite human-like; normally we appear to impose an interpretation of some kind on even bizarre word-sequences, unlike conventional automatic parsers which often "fall over" and yield no output at all for deviant inputs. Simulated annealing might be described as a "relaxed" computational technique, by comparison with more usual methods of computing which depend on long chains of logical deduction in which a single false step is commonly fatal; programming in the conventional style tends to involve obsessive and often unsuccessful searches for the "last bug", whereas an annealing system is expected to make all sorts of mistakes, which are gradually eliminated by an "unintelligent" but nevertheless effective technique. This system of extracting good results from fallible mechanisms may again seem psychologically more realistic than techniques which require perfection at every step. From the practical programmer's point of view the annealing technique is highly attractive; the annealing algorithm is conceptually simple and easy to program, and an annealing parser has the computationally desirable feature of high modularity. Modifications to the evaluation function, or to the statistics it refers to, do not affect the annealing routines at all; they merely lead to different decisions with respect to individual moves that are considered as the routine proceeds (and, again, it seems psychologically realistic that a parser might use statistical information which is modified as a person's language experience grows, without the general parsing mechanism being affected). Thus the annealing technique seems potentially very interesting both as a practical method of automatic parsing and as a model of psychological machinery.

There is one respect, however, in which the model presented so far is quite implausible psychologically: it ignores the left-to-right sequential manner in which humans process written as well as spoken language.

Introducing the time dimension is sometimes seen as problematic by writers on simulated annealing (e.g., Bridle & Moore, 1984, p. 318). Fortunately, however, there is a natural way to incorporate time into an annealing parser which not only is psychologically plausible but promises greatly to increase its efficiency as a practical technique.

The "timeless" version of the annealing parser discussed so far uses statistics on constituent-likelihood derived from a database of manually parsed complete sentences. For a "dynamic" annealing parser, the statistics should be derived from a database augmented as follows. For each sentence of length w words, w-1 partial-sentence trees would be derived from the full-sentence tree by eliminating from the full tree the last n terminal nodes

($1 \leq n < w$), together with all nodes dominating the eliminated leaves (thus leaving a sequence of independent trees representing the constituents which are complete before the cutoff point), and linking these constituents into a single tree by making their respective roots into successive daughters of a new root node. Parsing, using these augmented statistics, would proceed in a series of rounds. The output of the (n-l)th round would be an annealed parsing of the sequence consisting of the first n-1 words of the target sentence; the nth round would begin by adding the nth word of the sentence as an extra daughter of the root, and annealing would begin anew at melting temperature on the resulting tree.

The opportunity for efficiency would arise from the fact that NL grammar only rarely forces the reader to backtrack—the insight on which Mitchell Marcus's PARSIFAL system was founded (Marcus, 1980). Marcus's strategy involved a total exclusion of backtracking from his central parsing system, with "garden path" sentences being handed over to a quite separate "higher level problem solver" for processing. However, Marcus's predictions about a sharp categorization of NL sentences into garden-paths and non-garden-paths have provoked considerable criticism. In a dynamic annealing parser, all parts of the current tree would at all stages be available to revision, but the relative rarity of the need for backtracking could be exploited by adding a bias to the function which randomly selects nodes for reconsideration, so that nodes are reconsidered less frequently as they become "older". Since the bulk of processing time in an annealing parser would undoubtedly be consumed in calculating gains and losses for candidate tree-changes, this system of concentrating the search for profitable tree-changes on the areas of the tree where such changes are most likely to be found could be a good means of saving processing time by reducing the total number of moves considered.[4]

(Another, simpler device which could improve the performance of the system, suggested by several writers on simulated annealing, would be to use annealing in order to locate the vicinity of the optimum and then switch to systematic search in order to explore that small area of the solution-space; apart from saving processing time, this might eliminate the danger that an annealing parser would freeze on an analysis which was high-valued but not the highest-valued of all possibilities. One difference between the problem of parsing natural language and some of the problems to which simulated annealing has previously been applied is that, in parsing, arguably, a miss is as good as a mile.)

---

[4] This approach is rather different from Bridle and Moore's concept of *continuous flow annealing* (1984, p. 321), which reduces the *termperature* in older parts of a network rather than the frequency with which its nodes are reconsidered; Bridle and Moore's approach would presumably not reduce processing time per sentence in a dynamic annealing parser

8. PROCESSING TIME AS A LIMITING FACTOR

Even with these modifications, speed of processing is likely to be a serious practical drawback of annealing as a parsing technique. A process which chooses moves at random and weeds out those which prove undesirable is intrinsically much slower than a process which uses systematic rules to lead it to desirable moves. Just how slow an annealing parser runs will depend on how much processing is needed to evaluate an individual move. Clearly, this will be far less if the value of a parse-tree is a simple function of the values of all constituents in the tree (in which case only the few nodes to be changed need be compared for any move under consideration) than if the overall value depends on relationships between nodes at different places in the tree. But, even for the dynamic parser, it is probably unrealistic in the present state of hardware technology to think that parsing of word-processor input might occur in less than a large multiple of real time. However, as hardware speeds increase, this is a diminishing problem. (One particularly interesting feature of the annealing technique is that the geometry of the solution-space seems likely to grow increasingly favorable as input sentences grow longer, since the nodes affected by any tree-change will be a smaller proportion of total nodes in a tree for a longer sentence, so that value-differences between adjacent solutions will tend to be smaller. For conventional parsing techniques, on the other hand, increasing the length of sentences tends greatly to increase processing difficulties.)

9. DEEP PARSING

Another problem that will not have escaped the reader's attention is that I have discussed parsing purely in terms of finding surface parse-trees, which is the task on which our group are currently engaged. Many of the detailed phenomena which distinguish real-life texts from the kinds of example sentence found in mainstream linguistics textbooks are phenomena for which no theories of "underlying structure" or "logical form" exist; accordingly, it seems unavoidable that our initial approach to parsing this sort of material must deal with "surfacy" analyses, which are the only kind that can be defined fairly controversially. However, there is nothing about simulated annealing that makes it intrinsically inapplicable to the task of deep parsing. What will need to be done in order to achieve that further goal is to define a class of logically possible deep structural analyses and a class of moves between them, and to find an evaluation function which takes any pairing of a deep structure with a surface word-sequence into a plausibility value. This task is very different in kind from the work currently done by theoretical linguists and AI researchers interested in underlying or logical grammar, who tend to have little time for statistical thinking, but that does not mean that the task is senseless or impossible. Deep parsing will presumably need to exploit semantic/"inferencing" considerations as well as information about grammar in the narrow sense, but nothing says that these matters cannot be built into an evaluation function.[5]

10. THE PDP ALTERNATIVE

Finally, it will be appropriate to compare the parsing technique I have outlined here with various computational methods of a broadly similar type which have emerged in the last few years. My parsing technique is in one sense a rather nonstandard application of the simulated annealing concept. Annealing is commonly associated with the research paradigm known as *parallel distributed processing* (e.g., Rumelhart, McClelland, & the PDP Research Group, 1986) or *connectionism* (see Feldman & Ballard, 1982, or the special issue of *Cognitive Science*, vol. 9 nos. 1–2, 1985).

Within that paradigm, what changes as computation proceeds are the properties (degree of activation, "strength") of individual links in a network whose structure remains constant. Although I find the PDP paradigm intuitively attractive, I believe that it could not easily be extended to the problem of grammatical parsing. The obvious analogy, in the grammatical domain, to the approach discussed by PDP researchers to tasks such as word recognition would presumably involve, for a given input sentence, varying the pattern of activation in a network of labelled nodes which was such that any individual logically possible parsing of the sentence would form a subset of the network. A network of this kind, embracing within itself all possible analyses of a sentence, would admittedly not need to contain as many connections as the total of all links in all the distinct possible trees: for instance, there would be many trees which share the feature that, say, the first two words form an A constituent, and the general network would need only to contain one instance of this piece of structure. But the general network would nevertheless need to be colossally complex: so much so that I am sceptical whether this kind of computation with inputs of realistic length would be a practical proposition in the near future. (Notice that, not only would every link in any individual tree have to be reflected by some link in the general tree, but there would also have to be very many inhibitory links in the general tree, having nothing corresponding to them in the individual trees, in order to ensure that high-valued solutions contain no upward

[5] I have recently been awarded a grant by the Economic and Social Research Council to convert the Gothenburg Corpus (Ellegård 1978) into a database of deep grammatical and semantic statistics usable for research of this kind.

branching—a lower-level unit must always belong to just one unit at the next higher level.)

Furthermore, it is not at all clear even in principle how such a general network should be constructed. There will be very many logically-possible analyses of a reasonably long input in which *at some level* between the root S node and the leaf nodes the first constituent is an A: how many A nodes should there be in the general network to allow for these possibilities? (We cannot say that *all* the possible analyses will be covered by a single A node in the general network with diverse links to higher and lower nodes: some logically-possible trees will have an initial A constituent at one level which dominates a smaller A constituent at an initial position within it at a lower level.) The problem stems, I think, from the fact that NL parse-trees are disorderly with respect to number of links on the paths between various leaves and the root. It will often happen that, say, the correct parsing of a 20-word sentence has one word separated from the root by 18 intermediate nodes, while the next word is immediately dominated by the root, the next word again is three nodes from the root, and so forth: there is very little predictability about how these figures are distributed over the words of an entire sentence. It seems to me that the domains in which the PDP paradigm is commonly applied tend not to be disorderly in this sense.

If it were possible to construct and compute with a general network of the kind outlined, one advantage might appear to be that the computation could proceed in parallel so that various aspects of the analysis would be optimized simultaneously. My own approach buys computational simplicity (at any particular time, the system has to deal only with a single tree rather than with a data-structure incorporating all possible trees) at the cost of requiring the tree-modifying routine to be sequential: only one local aspect of the analysis is varied at a time. Connectionists often point out that the known low speed of information transmission in neural networks casts doubt on the psychological plausibility of any sequential approach. However, the distinction between sequential and parallel processing is subtler than is sometimes supposed. Rumelhart, Hinton, and McClelland (1986, p. 61) point out that, unless a parallel process is *synchronous* (that is, it is controlled by a clock issuing pulses which cause the separate units of a network to update simultaneously), then with sufficient magnification of the time-scale it can be seen as a sequence of independent events. The low speed of neural information transmission is evidence, not against sequential processing in this sense, but only against sequential processes in which events depend on their immediate predecessors. Synchronicity seems prima facie implausible and, as Rumelhart, Hinton et al. (1986, p. 61) point out, theoretically problematic as a property of a mental processing model. Conversely, my sequential approach to parsing is one in which successive tree-changes are chosen randomly and essentially independently of one another. I conclude that my model of parsing is not hit by standard connectionist arguments against serial processing.[6]

11. SELMAN & HIRST'S APPROACH

Although the concept of a general network incorporating every possible tree-structure seems to me quite intractable, a connectionist approach to parsing which is somewhat akin to this has been developed by Bart Selman and Graeme Hirst (e.g., Selman, 1985), in work which I encountered only after the body of the present paper was already complete. Selman and Hirst use annealing to locate optimal patterns of activation in a network representing the various possibilities allowed by a generative grammar: that is, their network is intended to include, not every topologically possible way of drawing a tree and assigning labels to nodes, but only those labelled trees which are grammatically well formed. My central problem with Selman and Hirst's approach is that, if it is indeed possible to design generative grammars which are adequate to cover the messy complexities of authentic natural languages, I am not clear what advantage there would be in using a stochastic technique such as simulated annealing for parsing—since, at least for well-behaved types of grammar such as the context-free example Selman uses, very efficient deterministic parsing algorithms are available. This issue is not addressed in the work I have seen.

12. THE HOPFIELD-TANK ALGORITHM

Not all connectionist models are stochastic. J.J. Hopfield and D.W. Tank (1985; see also Bounds, 1986) describe a deterministic connectionist algorithm for combinatorial optimization. Their approach involves analysing the class of logically possible solutions to a problem as different assignments of the values True and False to a fixed set of elementary solution-components. Thus, in their example of the travelling salesman problem (in which the goal is to find the shortest path passing through each of a given set of towns once), solution-components are of the form "town *X* is the *i*th visited," where *X* ranges over the set of towns and *i* over the integers from 1 to the number of towns in the set. The alternative truth-values of any such solution-compo-

[6] Since writing the above paragraph, I have worked out a method of parallelizing my own approach to tree-optimization, and I expect to begin implementing this technique shortly on a transputer array currently being installed at Leeds.

nent are modelled by the maximum and minimum outputs of an amplifier having a sigmoid response-function (conceptually, a *neuron*). Hopfield and Tank are able to show that, if a set of these neurons, representing solution-components, are linked via conductances (conceptually, *synapses*) and supplied with external inputs to form a network meeting certain general constraints, then, from any initial pattern of voltages, the circuit will settle into a state in which each neuronal output is at either its maximum or its minimum. The fact that the problem to be solved is the travelling salesman problem, and the individual town-locations which define a particular instance of the travelling salesman problem, can be represented by properties of the synapses and external inputs, in which case the circuit will settle fairly rapidly from an arbitrary initial state into a state whose pattern of neuronal outputs represents a good solution to the problem. Choice of initial state is random, but, thereafter, the behavior of the system is deterministic. An interesting conceptual feature of this computational technique is that, at any instant before a stable state is attained, the output voltage of a neuron will commonly be intermediate between maximum and minimum—in other words, the system works by treating True and False as opposite poles of a continuum of truth-values.

It might well appear that (if adequate hardware were available) this fascinating optimization technique would be an attractive way to attack the parsing problem. I believe that it is not, for reasons having to do with the need to analyse the solution-space in terms of a fixed set of elementary propositions. Consider the issue of defining alternative tree-shapes in such terms, leaving aside the question of node-labels. The solution-components could hardly consist of statements such as "Node i immediately dominates node-sequence $j\ k\ l$" (as would seem natural if we were considering an individual tree in isolation), since we have no basis for identifying a node in one tree with a node in another tree of different structure, apart from the special cases of root and leaves. (This point relates to the comment made earlier about the disorderliness of NL parse-trees.) The obvious way of using a fixed set of propositions to define a variety of tree-structures over a sequence of n leaf nodes would be to let the propositions be of the form "Segment $[a, b]$ of the leaf-sequence is a constituent," where $1 \leqslant a < b$ and $b \leqslant n$; provided legal trees have no nonterminal nodes with fewer than two daughters, any tree can unambiguously be defined as an assignment of truth-values to the set of elementary propositions of this form. But to define tree-structures in this way would be to make the relationship between a node and its immediate constituents logically derivative from the relationship between the node and its ultimate descendants. Linguistically, this is so grossly counterintuitive that I feel quite sceptical about the applicability of the Hopfield-Tank algorithm in this domain.

13. CONCLUSION

The last sections have given my reasons for preferring the specific parsing technique that I described earlier to various related techniques. But there can of course be no guarantee in advance that my own approach will prove successful. We have seen that the applicability of simulated annealing depends on the logical geometry of the solution-space, and, although my approach allows some degree of control over this geometry (different move-generating functions determine different adjacency-relations between points in the solution-space, and different evaluation functions determine different "vertical" relationships between adjacent points), there are no doubt intellectual domains whose logical geometry cannot be rendered suitable for annealing by any choice of adjacency function and evaluation function.

Perhaps the space of English parse-trees—whether surface or deep—necessarily resembles the Witwatersrand rather than the Cotswolds, being an upland plateau riddled with deep goldmines rather than a rolling landscape whose treasures lie exposed in valley bottoms. I conjecture that NLs are Cotswold-like rather than Rand-like, and that, if they were not, humans could not understand them. Only detailed empirical research will settle the question.

REFERENCES

Bahl, L.R., Jelinek, F. & Mercer, R. (1983). A maximum likelihood approach to continuous speech recognition. *IEEE Transactions on Pattern Analysis and Machine Intelligence, PAMI-5*, 179–190.

Baker, J.K. (1979). Trainable grammars for speech recognition. In D.H. Klatt & J.J. Wolf, (Eds.), *Speech Communication Papers for the 97th Meeting of the Acoustical Society of America*, pp. 547–550.

Bounds, D.G. (1986). *Physics for travelling salesmen: some new approaches to combinatorial optimisation*. Report of Royal Signals & Radar Establishment, Malvern.

Bridle, J.S., & Moore, R.K. (1984). Boltzmann machines for speech pattern processing. *Proceedings of the Institute of Acoustics*, 6 (4) 315–322.

Charniak, E. (1983). A parser with something for everyone. In M. King (Ed.), *Parsing natural language*. New York: Academic Press.

Cole, R.A. (Ed.) (1980). *Perception and production of fluent speech*. Hillsdale, NJ: Erlbaum.

Cox, D.R., & Miller, H.D. (1965). *The theory of stochastic processes*. London: Methuen.

Cravero, M., Fissore, L., Pieraccini, R., & Scagliola, C. (1984). Syntax driven recognition of connected words by Markov models. In *Proceedings, 1984 IEEE International Conference on Acoustics, Speech and Signal Processing*.

Derouault, A.-M., & Mérialdo, B. (1984). Language modeling at the syntactic level. In *Proceedings of the Seventh International Conference on Pattern Recognition.*

Ellegård, A. (1978). *The syntactic structure of English texts.* (Gothenburg Studies in English, 43.) University of Gothenburg.

Feldman, J.A., & Ballard, D.H. (1982). Connectionist models and their properties. *Cognitive Science, 6.* 205–254.

Garside, R.G., Leech, G.N., & Sampson, G.R. (Eds.), (1987). *The Computational Analysis of English: A Corpus-Based Approach.* Longman.

Geman, S., & Geman, D. (1984). Stochastic relaxation, Gibbs distributions, and the Bayesian restoration of images. *IEEE Transactions on Pattern Analysis and Machine Intelligence, PAMI-6* (6) 721–741.

Haigh, R., Sampson, G.R., & Atwell, E.S. (1988). Project APRIL: a progress report. In *Proceedings of the Annual Meeting of the Association for Computational Linguistics.* Buffalo, N.Y.

Hayek, F.A. (1955). *The counter-revolution of science.* New York: Free Press.

Hopfield, J.J., & Tank, D.W. (1985). 'Neural' computation of decisions in optimization problems. *Biological Cybernetics, 52,* 141–152.

Jones, K.S., & Wilks, Y.A. (Eds.), (1983). *Automatic natural language parsing.* Chichester, England: Ellis Horwood.

King, M. (Ed.) (1983). *Parsing natural language.* New York: Academic Press.

Kirkpatrick, S., Gelatt, C.D., & Vecchi, M.P. (1983). Optimization by simulated annealing. *Science, 220,* 671–680.

van Laarhoven, P.J.M., & Aarts, E.H.L. (1987). *Simulated annealing: theory and applications.* Dordrecht: D. Reidel.

Lea, W.A. (Ed.) (1979). *Trends in speech recognition.* Englewood Cliffs, NJ: Prentice-Hall.

Marcus, M.P. (1980). *A theory of syntactic recognition for natural language.* Cambridge, MA: MIT Press.

de Mori, R. (1983). *Computer models of speech using fuzzy algorithms.* New York: Plenum.

von Neumann, J. (1963). *Collected works* (v). Oxford, England: Pergamon.

Newell, A.N. (1980). Harpy, production systems, and human cognition. In R.A. Cole (Ed.), *Perception and production of fluid speech.* Hillsdale, NJ: Erlbaum.

Rumelhart, D.E., Hinton, G.E., & McClelland, J.L. (1986). A general framework for parallel distributed processing. In D.E. Rumelhart, J.E. McClelland & the PDP Research Group, *Parallel distributed processing: Explorations in the microstructures of cognition.* (vol.1). Cambridge, MA: MIT Press.

Rumelhart, D.E., McClelland, J.L. & the PDP Research Group (Eds.), (1986). *Parallel distributed processing: Explorations in the microstructures of cognition.* Cambridge MA: MIT Press.

Sampson, G.R. (1984). Review of P. Jacobson & G.K. Pullum (Eds.), *The Nature of syntactic representation, Lingua, 64,* 419–25.

Sampson, G.R. (1986). Evidence against the 'grammatical'/'ungrammatical' distinction. In W. Meijs (Ed.), *Corpus linguistics and beyond,* Amsterdam, Netherlands: Rodopi.

Sapir, E. (1921). *Language.* New York: Harcourt, Brace & World.

Selman, B. (1985). *Rule-based processing in a connectionist system for natural language understanding.* (Tech. Rep. No. CSRI-168). Toronto, Canada: Computer Systems Research Institute, University of Toronto.

Smith, A.R., Denenberg, J.N., Slack, T.B., Tan, C.C., & Wohlford, R.E. (1985). Application of a sequential pattern learning system to connected speech recognition. In *Proceedings, 1985 IEEE International Conference on Acoustics, Speech and Signal Processing.*

Toulouse, G. (1977). Theory of the frustration effect in spin glasses. I. *Communications on Physics, 2,* 115–119.

CHAPTER 9

Image Compression by Back Propagation: An Example of Extensional Programming*

GARRISON W. COTTRELL
Computer Science and Engineering
University of California, San Diego
La Jolla, California

PAUL MUNRO
Department of Information Science
University of Pittsburgh
Pittsburgh, Pennsylvania

DAVID ZIPSER
Institute for Cognitive Science
University of California, San Diego
La Jolla, California 92093

INTRODUCTION

The recent discovery of powerful learning algorithms for parallel distributed networks has made it possible to program computation in a new way. These new techniques allow us to program massively parallel networks by example rather than by algorithm. This kind of *extensional programming* is especially

* This research was supported by Training Grant MH 14268-11 from the National Institute for Mental Health, the System Development Foundation, and the Air Force Office of Scientific Research, Air Force Systems Command, USAF, under Grant Number AFOSR-86-0062. Requests for reprints should be sent to Institute for Cognitive Science, C-015; University of California, San Diego; La Jolla, CA 92093.

useful when there are no known techniques for solving a problem. This is often the case with the computations associated with basic cognitive processes such as vision and audition. In this paper we compare extensional programming to the traditional approach using a well-understood problem in image analysis, i.e., bandwidth compression. Our results demonstrate that a very simple learning method is able to program networks to do image compression about as well as the best known techniques.

The technique we employ is known as *back propagation*, developed by Rumelhart, Hinton, and Williams (1986a). It can be considered a generalization of the perceptron learning procedure for multilayer nonlinear networks of neuronlike computing elements and answers the basic objections of Minsky and Papert (1969) to such learning mechanisms. Training the network consists of repeated presentations of input–output pairs representing the function to be learned. The learning algorithm operates by adjusting the weights between the elements of the network in such a way as to reduce the overall error in the output. In many cases, the network finds a solution to the problem that was unknown in advance to the user. In doing so, it develops its own *internal representation* of the input that is useful for solving the problem. It is often difficult to analyze this representation, because many units are involved and the representations are highly *distributed* over the set of internal units. A subgoal of the present research is to make a first step towards unraveling the nature of these representations by applying the learning mechanism to a domain where the types of useful representations have been well studied.

Another aspect of this work is that the representation of images in an efficient format by neuron-like computing elements may give us clues to the way such information is represented in actual neural tissue. The learning procedure itself is not particularly biologically plausible (Zipser, this volume, Chapter 10), but the mechanisms it discovers for solving problems are. Whether or not there is anything like back propagation in the brain, we learn something about how the brain *could* solve problems from the solutions it discovers. Such information could be useful in guiding neurobiologists in their observations of cell firings during cognitive tasks.

The rest of the paper is organized as follows. First, we provide an intuitive introduction to the back-propagation algorithm; its application to *encoder problems*, of which image compression is an example; and an abbreviated look at current techniques in image compression. Details of back propagation and its derivation can be found in Rumelhart, Hinton, and Williams (1986b). Some important mathematical properties of the image compression technique we consider are provided in the Appendix. Readers who are familiar with these areas are encouraged to pass on to the results section.

It should be noted that we knew none of the techniques for image compression until after we completed the main part of the project. They are presented for comparison purposes and because they are useful in analyzing the solution discovered by the network. Following the background, we present the results of our simulations. Several simple network architectures are shown to demonstrate near state-of-the-art compression of image information, and the nature of the internal representation developed by the networks is explored. This is followed by a discussion of the significance of the results. Finally, we suggest problems left unsolved.

BACKGROUND

The Back-Propagation Algorithm for PDP Networks

Parallel distributed processing (PDP) models[1] consist of networks of simple neuron-like processing units connected by weighted links. The units communicate by passing real-valued activation across the links. In the kind of PDP networks we will be considering, the networks are arranged in layers, with the outer layers corresponding to input and output (see Figure 1), and activation only passing one way through the network. A model of a particular function is built by picking a representation of the domain and range of the function as patterns of activation across input and output units. Units in the internal layers are referred to as *hidden units*, and correspond to an *internal representation* of the input environment that is useful in producing the desired output. *Supervised learning schemes* are ones in which the network is given repeated examples of the desired input–output mapping, and the network adjusts the weights between units to produce the desired function. The network thus *programs itself* to solve the task.

Back propagation is a supervised learning scheme for PDP networks that is currently accepted in the PDP community as the best known technique. The basic idea is to change the weights between the units in such a way as to reduce the error in the output. An input pattern is presented to the network, and activation is propagated forward through the network to the output units. The correct output pattern is provided to the output units in the form of a *teaching signal*. It is clear how output units should change their weights to reduce the error. If their value is too low, they should raise the weights on active input lines. If their value is too high, they should lower the weights. This will tend to make them attain the proper value in the same situation in the future. What is difficult to determine is how the hidden units should change their weights—there is no explicit teacher for them. Back

[1] These are also known as *connectionist* models (Feldman & Ballard, 1982).

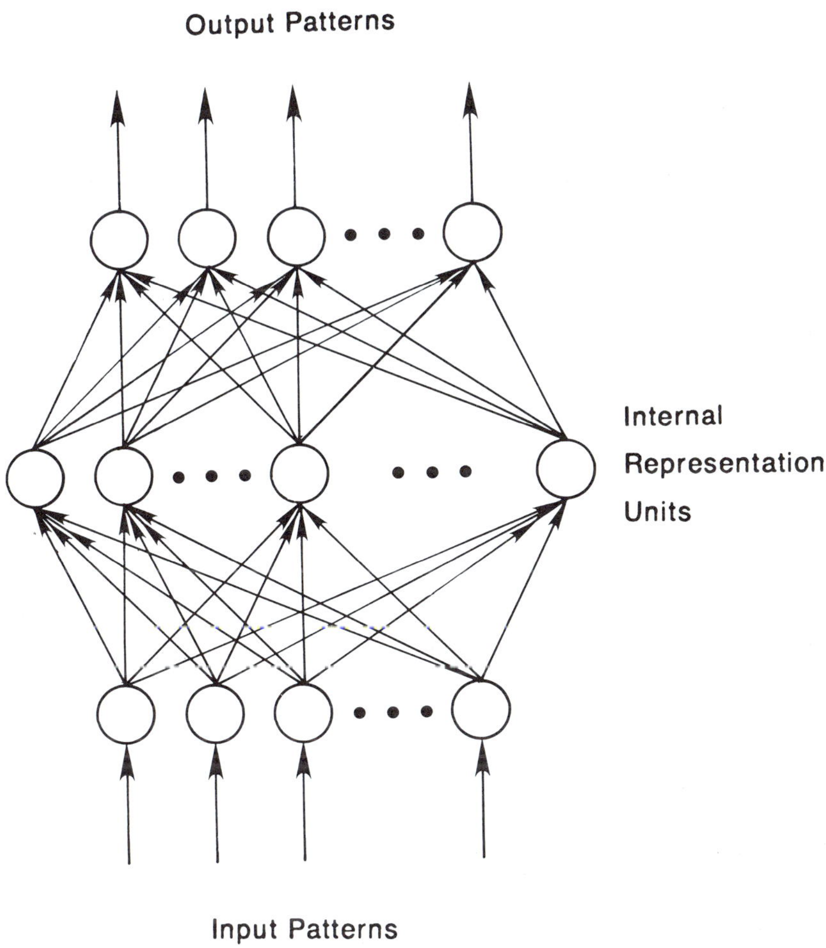

Figure 1. A feed-forward PDP network.

propagation provides a rule for propagating an error signal back through the network from the output units, which tells the hidden units which way to change their weights. The surprising thing is that the rule for changing the weights is a purely *local* one, in the sense that every unit can find out its error through the connections that already exist in the network.

The weight-changing rule is derived by defining an error measure—the mean square error of the output—and taking the partial derivative of this with respect to the weights. Intuitively, we can think of the error as a surface over the space of possible weights. Taking the partial derivative tells us which way to move in weight space in order to go downhill in the error. Because this is a gradient descent algorithm for reducing the error, it is subject to the major problem associated with gradient descent procedures, i.e., *local minima* in the error surface. However, in the problem examined here, we have never run into this situation.

The Encoder Problem

Image compression is a type of *encoder problem.*[2] That is, a network is given the problem of performing an identity mapping over some set of inputs. The network is constrained to perform this mapping through a narrow channel of the network, forcing it to develop an efficient encoding in that channel. There are two interesting aspects to this: (a) the network is developing a compact representation of its "environment"; and (b) although the algorithms used were developed as supervised learning schemes, this problem can be regarded as unsupervised since the output is the same as the input—the system self-organizes to encode the environment.

An example is shown in Figure 2, from Rumelhart et al. (1986a). The net-

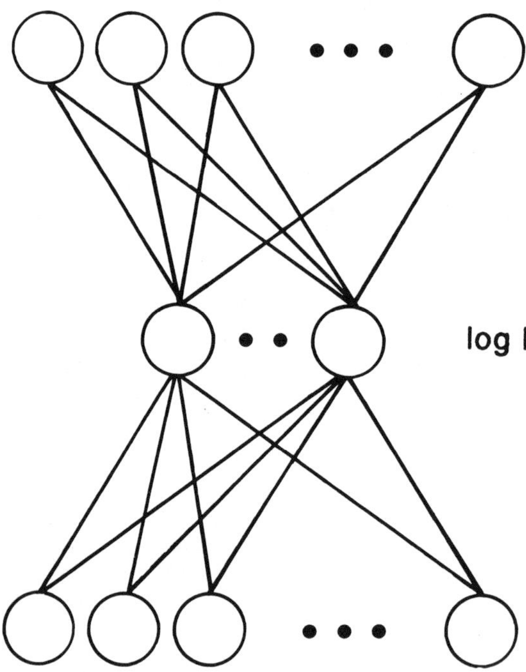

Figure 2. An *N-K-N* encoder (from Rumelhart et al., 1986a). One unit at a time in the input vector is "on." The network learns the binary encoding of the position of the "on" unit in order to turn on the same unit in the output.

---

[2] Ackley, Hinton, and Sejnowski (1985) were the first to demonstrate a learning algorithm for PDP networks that could solve the encoder problem.

work shown here is an *N-K-N* encoder—a three-layer network in which the input and output layers have *N* units each, and an intermediate layer with *K* units ($K < N$) that is the channel between the other two. The problem presented to the network is to associate one unit being on in the input layer with a corresponding one being on in the output layer, with no a priori mapping existing between the two. In their simulations, they found that, with $K = \log_2(N)$, the network could learn a binary encoding of the environment.

Image Compression Techniques

Introduction

The efficient encoding of the information contained in a digitized image is of considerable practical interest due to the sheer number of bits in a typical image. Applications include reducing storage and transmission requirements, and using the encoding as a feature analysis of the image. The idea is to take a section of the image and *transform* it into a domain where the information can be encoded more efficiently. If we are willing to allow some degradation of the image, it is possible to take an image which is represented by eight bits per picture element (or *pixel*) and transmit it using less than 1 bit/pixel.

Basic Concepts

For our purposes, we define an *image* as a discretized $M \times N$ light intensity function $f(x,y)$, where x and y correspond to the spatial coordinates within the image, and $f(x,y)$ is a light intensity value from 0 to $2^K - 1$, where K will typically be 8 for us. One element of $f(x,y)$ is often referred to as a *pixel*, for *picture element*. *Image compression* then refers to the process of converting these $M \times N \times K$ bits to another representation where fewer bits can be used, and a new version of the image, $g(x,y)$ can be recovered which is a reasonable reconstruction of the original. By *reasonable*, we mean within some *fideltiy criterion* of the original image.

The choice of a fidelity criterion is problematic, since most analytic measures that have been used do not correspond in a neat way to people's subjective ratings of images. However, for objective comparison, usually some form of mean-square error measure is used. If $g(x,y)$ is the reproduced image, then the error is given by

$$e(x,y) = g(x,y) - f(x,y),$$

the mean-square error is given by

$$MSE = \frac{1}{M \times N} \sum_{x=0}^{M-1} \sum_{y=0}^{N-1} e^2(x,y),$$

and the normalized MSE with respect to the average squared intensity of the image is given by

$$NMSE = \frac{MSE}{\frac{1}{M \times N} \sum_{x=0}^{M-1} \sum_{y=0}^{N-1} f^2(x,y)}.$$

This is what we will use, expressed in percent.

Image compression is measured in terms of the bits per pixel of the output image actually transmitted over the channel. Although one might think it should be the ratio of bits in the original image to bits in the encoded image, this can be a misleading measure. If the original image uses 15 bits/pixel, the final image could use significantly less without any apparent degradation, due to limitations in the resolving power of the human eye. Hence the number of bits used in the encoding per pixel of the restored image is a typical measure used. This is independent of the number of bits used for the original pixels. This is then plotted relative to some fidelity criterion, for example, the NMSE defined above.

Current Techniques

The following discussion has been adapted from Gonzales and Wintz (1977). Depending on the requirements of the application, techniques for image compression fall into two classes: error-free encoding, and encoding relative to a fidelity criterion. We will be concerned with the second class, since we wish to find something comparable to our networks' behavior. In particular, we will consider *transform encoding* techniques.

A typical encoding system consists of three processes, as shown in Figure 3. The first stage maps the input data from the pixel domain to another domain that is useful for reducing the number bits that must be used to encode the data. The quantizer rounds off the output of the mapper, and the coder assigns a code word (such as a Huffman code) to the quantizer output. In our work, we have simply used the output of the quantizer as the final result to be transmitted, so we will not consider the coder further.

The Mapper

The kinds of mapping we will be considering are *linear transformations*.[3] First, the $M \times N$ image is divided into $n \times n$ subimages and the subimages are encoded separately. In *nonadaptive* transform encoding, the same coder

[3] Other types of mappings are also used, such as run-length and DPCM encoding, but are not comparable to the network model, so we will not consider them further.

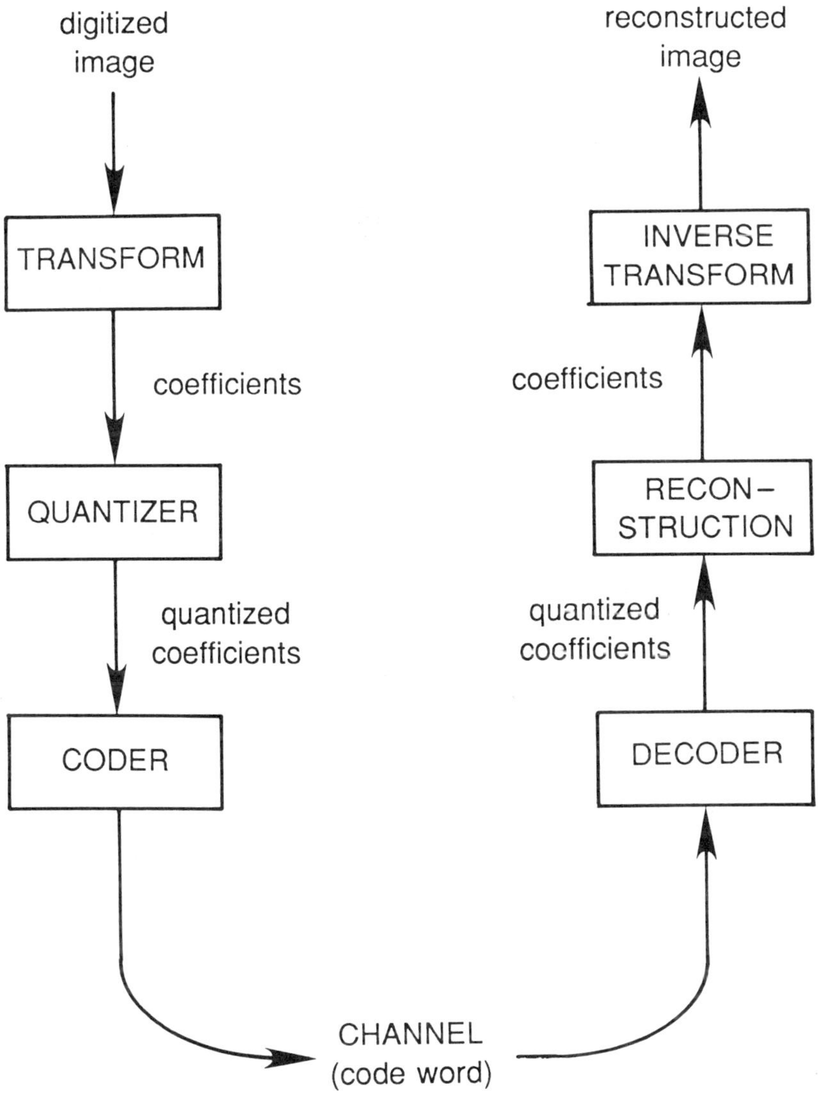

Figure 3. A typical image compression system.

is used for all subimages. In *adaptive* transform encoding, the subimage is classified in some way first by a pattern-matching algorithm, and the best available encoder is used for each type of subimage. Some researchers advocate a different type of adaptive coding in which the same transform is used for each, but the quantizer is adaptive for each subimage's coefficients (Tescher, 1979). We will consider this possibility later.

Considering each $n \times n$ subimage as a vector **x**, a linear transformation is of the form

$$\mathbf{y} = \mathbf{A}\mathbf{x},$$

where **A** is an $n^2 \times n^2$ array, giving a new n^2-vector **y**. If **A** is invertible, then so is the encoding. The reconstructed image is the inverse transform

$$\hat{\mathbf{x}} = \mathbf{B}\mathbf{y}$$

where **B** is usually $\mathbf{A}^{-1}$. The point of the transformation is to take the pixels, which are usually highly correlated, and produce numbers which are less correlated. These can often be encoded more efficiently than the original pixels.

The idea is illustrated in Figure 4. This is a scatter-plot of height vs. weight for some hypothetical set of people. Notice that height and weight are highly correlated and thus, to some degree, redundant information. If we rotate the axes of the coordinate system along the direction of maximum variance, with the restriction that the axes be orthogonal, we get a new set of axes as shown. We now have two sets of numbers that are less correlated than the original pair of coordinates—the distance along the y_1 axis, which might be called *size*, and the distance from that axis, y_2. Notice that the magnitude of the y_1 coordinate is about the same as either of the original x coordinates, and the y_2 coordinate is considerably smaller. Thus we should

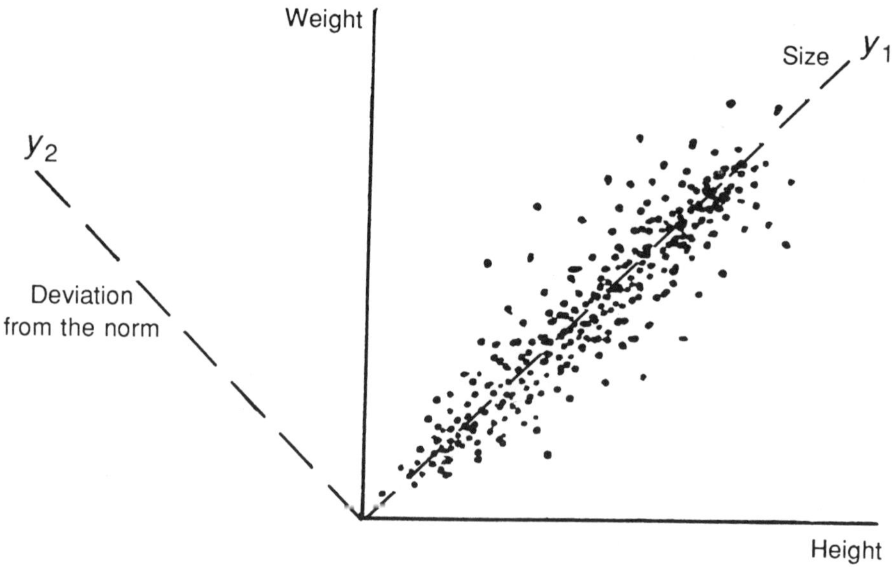

Figure 4. Height vs. weight scatter plot for a large population. (Adapted from Mozer, 1983.)

be able to use a smaller number of bits to encode them. Also, if we were willing to tolerate some error in the reproduction, we could simply transmit the y_1 coordinate and get most of the information, since the y_1 coordinate captures most of the variance of the plot. This scatter-plot could just as well be a plot of neighboring pixel values, which are typically highly correlated.

The same procedure can be used for an $n \times n$ array of pixels. If each pixel is represented by an intensity level between 0 and $2^K - 1$, then we can represent any such image in a n^2-dimensional coordinate system with each axis labeled 0, . . . , $2^K - 1$. Each image corresponds to a point in this space. In this case, the desired rotation of the coordinate system involves n^2 axes. The same intuitions apply: Neighboring pixels are going to have similar values; hence, they will be highly correlated, just as height and weight are. The difference is that we have, say, 64 axes instead of two, if we encode 64 pixels at a time. Regularities in the data are going to cause them to "clump" in parts of that space. If we can find a set of axes that go through those clumps, we can get away with sending the coordinates of the points on those axes (e.g., y_1 in Figure 4) instead of the whole 64-element vector.

There is a procedure that finds just this set of axes, called the *Hotelling* or *principal components* transform. Its derivation and the computational properties discussed below are given in more detail in the Appendix. Briefly, the rows of the transform matrix **A** are formed from the eigenvectors of the covariance matrix of the **x**-vectors. This could be done for each patch independently (adaptive encoding), but typically the patches to be reconstructed (the **x**-vectors) are treated as multiple instances of the same vector variable, and the same transform is used for each patch.

For example, the axes shown in Figure 4 are the eigenvectors of the covariance matrix of the height and weight variables. They are arranged in **A** in order of amount of variance accounted for, so row 1 of **A** is y_1 and row 2 is y_2. Suppose this is a set of data for a population of Martians discovered by a probe, and we want to send this data back to Earth using the smallest number of bits. Each point is represented as a (height, weight) pair. These are the **x**-vectors. We multiply **A** times each pair. What this does is form the inner product of the pair with each of the new axes. This is just the coordinate of the point on that axis (since the eigenvectors are of unit length). So, the **y**-vectors we send back to Earth are these coordinates. Notice that the y_1 coordinate (size) is going to vary the most as each point is sent. This corresponds to the fact that that eigenvector accounts for most of the variance of the data.

There are several things that are of note. First, it can be shown that the coefficients y are *uncorrelated*, as desired. Second, if one uses the first K rows of the matrix **A** formed in this way, rather than the whole thing, and thus send only K coefficients to the quantizer, then this is provably optimal in the mean-square error sense for K coefficients (see the Appendix). The problem with the Hotelling transform is the computational expense involved in computing the eigenvectors anew for each image. There are deterministic transforms (that is, that don't depend on the image), such as the discrete cosine/Fourier and Walsh-Hadamard transforms, which approach the Hotelling transform in variance accounted for in the first K coefficients. Hence the Hotelling transform is more a standard for comparison than a useful algorithm. However, we considered it because it will be a useful solution to compare and contrast to the solution discovered by back propagation.

Basis Images

Each pixel of the reconstructed image patch is a linear combination of the transmitted y_i coefficients. But taken as a whole, the image patch can be regarded as a linear combination of *basis images*, one for each coefficient.[4] The set of basis images for the Hotelling and Hadamard transforms of a picture of a cameraman (from Gonzales & Wintz, 1977) are shown in Figure 5. The Hadamard transform is a deterministic one, so these basis images are the

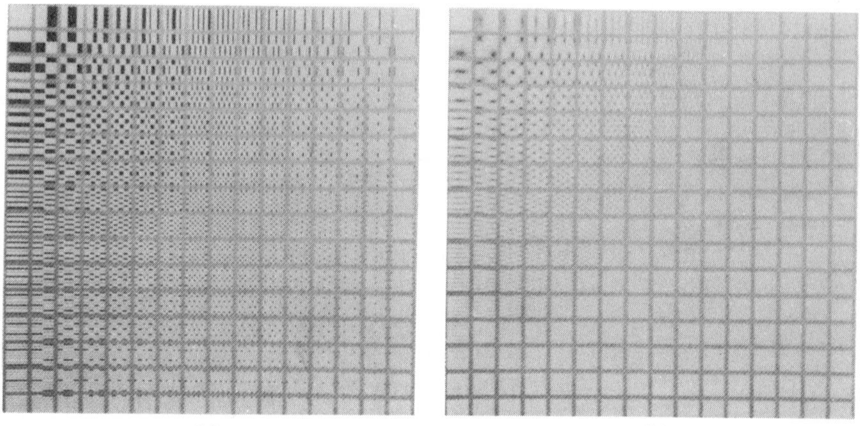

(a) (b)

Figure 5. The set of basis images from two types of transforms. A: The set of Walsh-Hadamard basis images. B: The set of Hotelling basis images for a particular image. (From Gonzales & Wintz, 1977. Reprinted by permission.)

[4] This really requires viewing the transform as a two-dimensional one, but the idea is clear enough without getting into a jumble of subscripts. For each coefficient, the basis image is the corresponding column of the reconstruction matrix **B**.

same for all images it is applied to. The Hotelling basis images depend on the transformation matrix **A**, which in turn depends on the statistics of the particular image. These actually correspond to a visual representation of the eigenvectors. The picture is formed by adding these images together, weighted by the coordinate along that eigenvector. The ones shown here are for a 256×256 image, with 16×16 subimages. For image compression, only a subset of the coefficients are actually transmitted—those that account for most of the variance. These correspond to some square of basis images in the upper left-hand corner of the total array of basis images. Typically, at least half of the coefficients can be discarded this way without visible degradation of the image.

The Quantizer

The number of possible values of the coefficients transmitted, the y_i's, can be considered infinite. Hence we have to have a strategy for assigning them to some finite number of values if we wish to achieve true data compression. This is called *quantization*. There are two aspects to quantization. First, for a particular range of values, one could use a *uniform* quantizer, i.e., equally spaced values, or, depending on where the coefficients "spend the most time," one could use a *nonuniform* quantizer, with more values in the "busy" regions (Max, 1960).

Second, there is the issue of whether to use the same quantizer across all coefficients or not. Since the variance of the coefficients can be considerably different from one to the other, this is usually unwise. One strategy, then, is to normalize the coefficients by their standard deviation, resulting in unit variance for all the coefficients, so that one quantizer (and hence the same number of values) may be used for all.

However, the coefficients with the largest variance contribute the most information; thus, it is better to assign them more quantization levels. This is called *block quantization*. Huang and Schultheiss (1963) found that the number of bits used to code a particular coefficient should be proportional to the log of the variance of the coefficient. This minimizes the mean square error in quantized coefficients. In a completely linear system, this also minimizes the mean square error in the output due to quantization.

RESULTS

The problem to be solved is the compression of the information in a single image. The network used here does this by learning an identify mapping through a narrow channel. The result is that the image can be represented with very little loss of information with 1 bit/pixel, representing an eight-fold compression of the information in the image. The network also does a good job of compressing several images it was not trained on.

We have conducted several trials with three-layer networks, generally training them on a single image. The network we have used most often consists of 16 hidden units each of which receive input from an 8×8 patch of the image. These then are completely connected to an 8×8 unit output patch. This is diagrammed in Figure 6.

The function computed by the units is the *logistic* function:

$$output = \frac{2}{1 + e^{-x}} - 1$$

where the x is the weighted sum of the inputs and a bias term. This is an S-shaped function that ranges from -1 to 1 (see Figure 7). It is sometimes

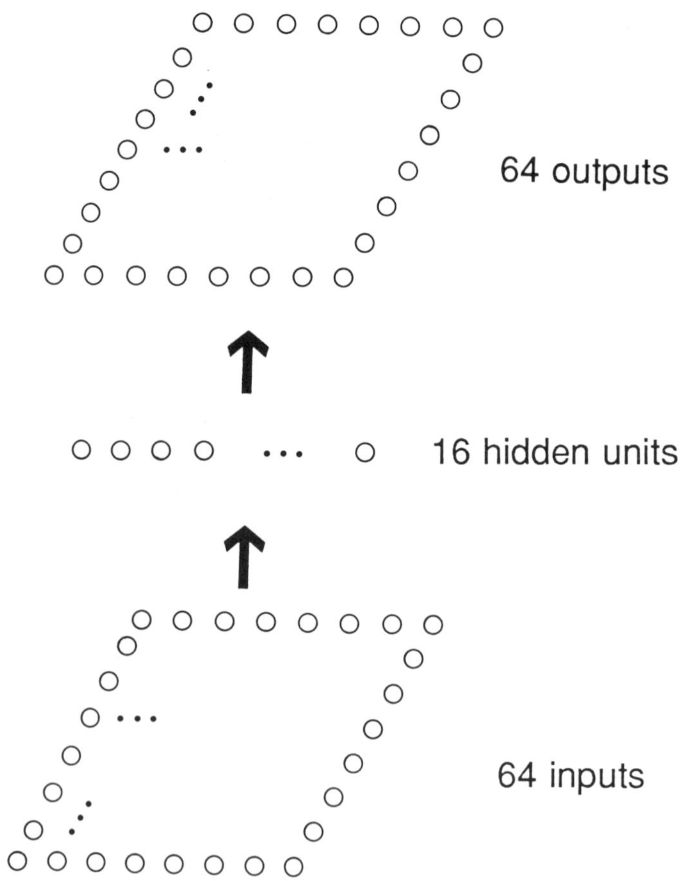

Figure 6. The network used in most of our examples.

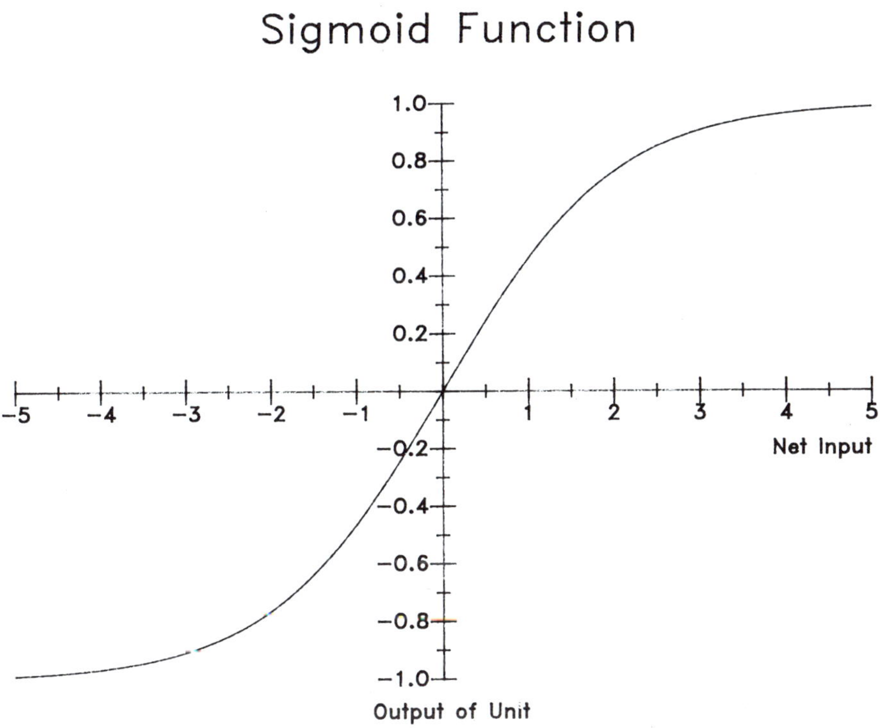

Figure 7. The logistic activation function.

referred to as the *squashing function* because it takes the linear input and "squashes" it into a small range.

In order to have a representation of the gray levels of the image that could be produced by this function, the gray levels of the image were converted linearly from the 0 to 255 range to the range 0 to 1, or 0 to .85. Although the latter representation appears to work the best, the results presented here derive from the former for historical reasons. Other conversions, such as to the range -1 to 1, didn't work as well because they force the output units into ranges hard to get at by the logistic function.

The training regime consisted of randomly sampling patches of the image, allowing activation to propagate through the network, and then back-propagating the error on each trial, adjusting the weights. It was found that a learning rate of .25 worked well initially, and better results were obtained by then lowering the learning rate to .01. This was a hard problem, and typically we would let the network run for 100,000 iterations at .25 and another 50,000 at .01. The use of momentum (see Rumelhart et al., 1986a) did not seem to help.

After training, the resulting "patch compressor" was tested by using it to compress and reconstruct the whole image by dividing the image into nonoverlapping patches and reconstructing one at a time. In order to achieve true compression, the outputs of the hidden units must be quantized. Two things made this easy for us. First, the squashing function forces all outputs into the range -1 to 1, so no scaling is necessary. Second, back propagation tended to distribute the error relatively evenly over the hidden units in this situation. Thus the hidden unit output variances were about equal, so block quantization was not used. We simply used a uniformly spaced quantizer that retained the endpoints (-1 to 1) of the output values (see Figure 8). Quantization then consists of linearly transforming the outputs of the hidden units (which were in the interval $[-1,1]$) to some integer range corresponding to the number of bits required, and then back again, effectively restricting their information to the number of bits used. Better results could be obtained by taking into account the output values most often used and using nonuniform bucket spacing. It turns out that, in this problem, the hidden units rarely reach the outer limits of the squashing function, so the endpoints of the quantizer were wasted.

Compression Results

The original image and the reconstructed image are shown in Figure 9. In this reconstruction, the outputs of the hidden units were restricted to eight

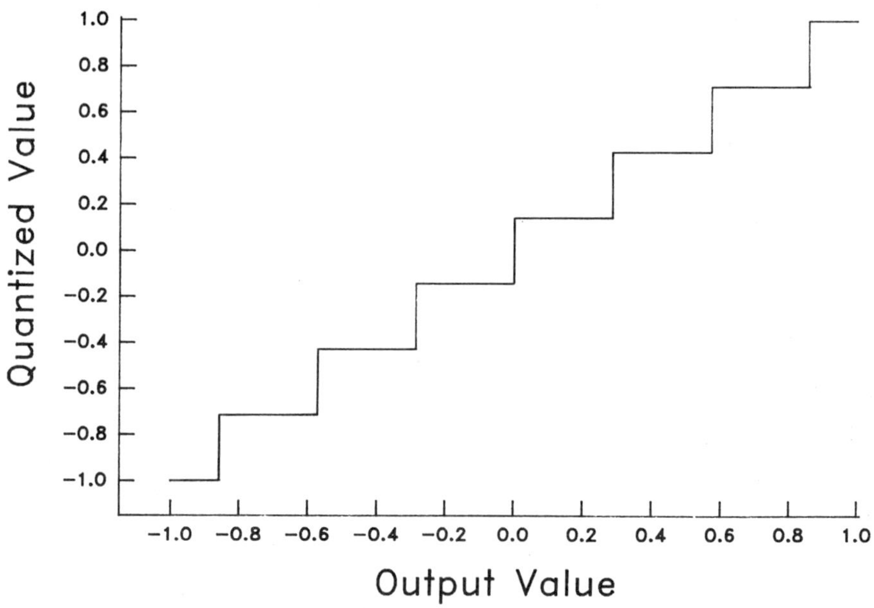

Figure 8. The quantization buckets used in reconstructing the image with three bits of output values (four or five bits are usually necessary).

Figure 9. *A*: The original image of the Intelligent Systems Group (ISG) at UCSD. *B*: The reconstructed image using eight bits (or 256 quantization values), representing 2 bits/pixel, resulting in NMSE of 0.413%.

bits (or 256 quantized values). This corresponds to a compression rate of 2 bits/pixel, with a corresponding error of 0.431%. It turns out that this is as good as this network does without any quantization.[5] One thing to notice here is that the network has done a good job of avoiding edge effects around the patches from the reconstruction process. Most of the errors are at the edges of objects in the image and in the stripes on the seated man's shirt.

By reducing the quantization of the hidden unit outputs, better compression results can be obtained without sacrificing much in mean square error. Figure 10 shows the effects of quantization. Notice that apparently most of the information is captured by restricting the hidden units to 32 output values (Figure 10A), with minor mean square error cost. This corresponds to an encoding rate of 1.25 bits/pixel. When quantization is reduced to four bits, the effects become pronounced.

Another approach to getting better compression results is to enlarge the input patch or reduce the number of hidden units. Figures 11 and 12 show the results of two experiments along these lines. Figure 11 shows the result

Figure 10. Quantization effects. *A*: 5 bits, 1.25 bits/pixel, NMSE 0.474%. *B*: 4 bits, 1 bit/pixel, NMSE 0.676%. C: 3 bits, .75 bits/pixel, NMSE 1.796%.

[5] As expected, throwing away half the coefficients (i.e., using 32 hidden units) results in no visible degradation. With six bits of quantized values, this corresponds to 3 bits/pixel.

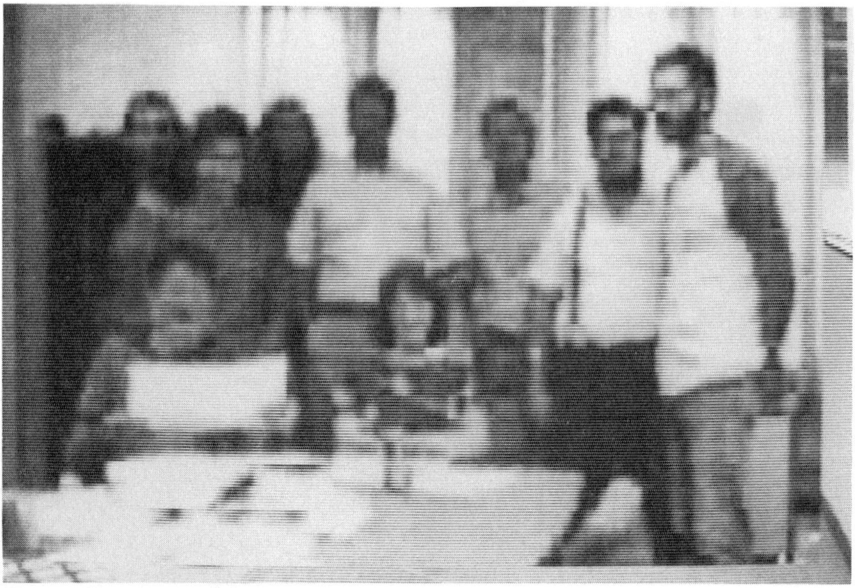

Figure 11. A reproduction using a 16 × 16 input patch and 16 hidden units. Compression rate is .312 bits/pixel, NMSE 2.670%.

Figure 12. The reproduced image using eight hidden units, 32 quantizer levels, resulting in .625 bits/pixel, NMSE 1.182%.

of using a 16×16 input patch with 16 hidden units, and five bits of quantization on the hidden unit outputs—a pretty lousy picture, but it is based on only .312 bits/pixel. Figure 12 is more successful. This is the result of using the original 8×8 input patch, but only eight hidden units, with 32 quantization levels. Here we have a relatively decent picture with a compression rate of .625 bits/pixel.

There is a possible trade-off between the number of hidden units and the number of bits used in quantization. It seems reasonable to expect that, if there are more hidden units, we can get away with fewer output values. This is partially true. The image in Figure 13 was obtained using an 8×8 input patch through 32 hidden units and four bits of output. This is the only network we have trained that reproduces the shirt stripes, so in that sense it is a "good" image. However, the NMSE is comparable to the case of using four bits with 16 hidden units, and there is a patchiness effect—the patches used to reconstruct the image become visible. Also, due the increase in the number of hidden units, this is an unimpressive compression rate of 2 bits/pixel, and reducing the number of quantization levels any more makes the picture unacceptable. (On the other hand, raising the rate to 3 bits/pixel produces a near-perfect image). The reason is that we are not working the network very hard. Taken to its extreme, suppose we had 64 hidden units. Then each hidden unit would just have to duplicate the output of the corresponding input

Figure 13. The reproduced image using 32 hidden units, four bits of quantized levels, 2 bits/pixel, NMSE 0.625%.

unit. They would require most of the output range to be faithful, and no compression would be attained because restricting the bits of the hidden units would incur too much error. Thus, it is important to force the network through a bottleneck, or no efficient encoding is achieved.

Another possible way to achieve better performance at 1 bit/pixel is to increase the patch size along with the number of hidden units. In Figure 14, we show the result of using a network with a 16×16 input patch and 64 hidden units. This is the same ratio of units as in the 64×16 case; hence, the same quantization level results in the same bit rate. Using four bits of quantization levels results in only marginally less mean square error cost than in the 64×16 case, but the image is qualitatively better than the one in Figure 10B using 1 bit/pixel. This beneficial effect is due to using a larger input patch. Figure 15 is a graph relating normalized mean-square error (NMSE) to the ratio of patch size to bit rate for two different patch sizes. It shows that the bigger patch attains a lower NMSE for the same bit rate.

At the outset of this research, we expected that perhaps a network could be trained that would work well for all images, justifying the expense of the initial training. This is a somewhat misplaced dream, given that our network learns, in some sense, the statistics of the image it is trained on, and different images have different statistics. However, we thought, perhaps it will

Figure 14. The reproduced image using a 16×16 input patch, 64 hidden units, and four bits of quantization levels. This is a compression rate of 1 bit/pixel, NMSE 0.590%.

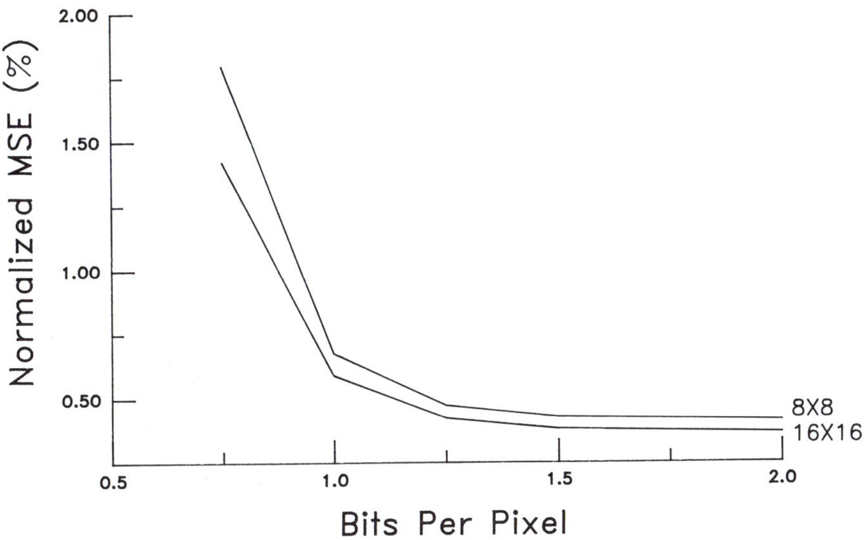

Figure 15. Graph showing effects of larger input patch.

work well for a class of images. It turns out that it does a creditable job of reproducing some images that it wasn't trained on. Two of the images we tested it on, and their reproductions, are shown in Figure 16.

An interesting side here is that a linear version of the network without quantization[6] does a much better job of generalizing to other images, in some cases resulting in less NMSE than on the training image. However, this does not necessarily mean compression, because, in the linear reconstructions, the outputs of the hidden units were not quantized. Since the units are linear, their activation is not restricted to a particular range, and quantization of the hidden unit outputs requires knowledge of their statistics to be done properly. Preliminary study shows that the linear version is comparable to the nonlinear version when quantized, but we have not looked at its generalization behavior under these circumstances. It is likely that the range of hidden unit values will vary more widely with images of different energy. The linear network seems a perfect candidate for adaptive quantization as proposed by Tescher (1979). This approach could make the linear version quite attractive by allowing use of the dynamic range of the linear version, which is probably how it achieves such better generalization results.

---

[6] In this case, all of the units in the network are linear, and back propagation is the same—the derivative of the activation function is simply a constant, and so can be ignored. This is a generalization of the Widrow-Hoff delta rule to layered networks.

Figure 16. Two images reproduced by the network trained on the image in Figure 9A. *A*: The Symbolics Graphics group. *B*: Reproduced image, using six bits of quantized values, 1.5 bits/pixel, NMSE 1.267%. C: Cadillac. *D*: Reproduced Cadillac, 1.5 bits/pixel, NMSE 0.764%.

The Network's Solution

Now the question is: What is the nature of the solution the network has discovered? What is the internal representation at the hidden unit layer? We don't have an analytic answer to this question.[7] However, we can compare the network's solution to the principal components transform to get an idea of what it does.

First we set up some correspondences between our network and the usual image compression system. Recall that the first step in transform encoding is to multiply the patch vector by a matrix to obtain less correlated coefficients:

$$\mathbf{y} = \mathbf{A}\mathbf{x}.$$

The analog in our network is that A is the weight matrix between the input and hidden unit layers, with each row of **A** corresponding to the input weights on one hidden unit, and each hidden unit output a semilinear version of y_i.

---

[7] Actually, as of this revision, we have an empirical result, to be reported in future work. The basic result is that the linear version finds a linear transformation of the principal components solution, and the nonlinear version approximates this. For a general result, see Baldi & Hornik, to appear.

Both back propagation and principal components analysis (PCA) minimize mean square error, but probably in different ways, since there are many ways to accomplish this. Principal components does it by having each row of **A** be the projection of the image onto axes corresponding to decreasing amounts of the variance. The first hidden unit responds to the direction of maximum variance, the next to an orthogonal axis that accounts for the next highest amount of variance, etc. Each hidden unit then has variance equal to the eigenvalues of the corresponding eigenvectors, which start large and quickly diminish. Thus, all hidden units, or rows of the matrix **A**, are not equal in accounting for the error of the output.

The variance of the coefficients is strongly correlated with the amount of image variance along the axis the hidden unit reflects. We ran an experiment where, during the reconstruction process, we recorded the mean square error of the resulting image if each hidden unit's output was set to 0.[8] The results of this experiment are given in Table 1. We can see from the table that, while the hidden units don't account for equal amounts of error, they are not spread out in the way that a principal component analysis would be. The variance of each unit is also shown in Table 1. The variances are about equal and are highly correlated with the output error without the corresponding hidden unit. If the network were doing a (localist) principal components analysis, the hidden units would be individually responding to the eigenvectors of the covariance matrix. Their variances would be equal to the

Table 1. Hidden Unit Statistics

Unit	Average	Variance	MSE Without the Unit
0	0.342	0.10347	286.828
1	0.344	0.11533	397.380
2	−0.364	0.12112	355.864
3	0.349	0.11144	347.517
4	−0.346	0.12042	434.365
5	−0.348	0.10504	319.268
6	0.355	0.11542	391.452
7	−0.346	0.12734	459.412
8	−0.341	0.11305	358.808
9	−0.361	0.11763	428.412
10	0.320	0.09896	295.973
11	−0.343	0.12459	414.386
12	−0.343	0.11999	414.046
13	−0.332	0.10390	312.565
14	−0.351	0.11661	417.271
15	0.359	0.10821	359.726

[8] This is analogous to pulling the plug wires on an engine one at a time and watching the RPM to see how much each cylinder is contributing to the engine speed.

eigenvalues of those eigenvectors, which typically differ by orders of magnitude from the first to the last. The network, on the other hand, finds axes where the variances are much closer together. The tendency of the network to produce variances within a similar range is probably due to the constraint that the hidden units operate in a fixed range. Figure 17 is a histogram of the outputs of all of the hidden units collapsed together during reconstruction of the training image. It is handy that the squashing function has this "built-in" range constraint, essentially solving some aspects of the quantization problem for us as part of its solution. This explains why it was easy to obtain good results with a simple-minded quantizer. On the other hand, the histogram shows that there are much better quantizers for this network—one that takes into account where the outputs are spending most of their time would have given better results.

Figure 18A shows the weight matrix for each of eight hidden units thresholded at various levels, one hidden unit per row. This shows what pattern in the input excites each hidden unit most. Figure 18B shows the output patch driven by each hidden unit alone. Again, each row corresponds to one hidden unit, and the columns correspond to different levels of activation from the hidden unit. The left-hand column corresponds to the hidden unit outputting a -1, the right-hand column a $+1$. One obvious thing to note here is that the hidden units try to reproduce what they "see." Figure 18C shows the same information as 18B, but on a color monitor to show the gray scale. (18B is a thresholded version of 18C.) These are the basis images for

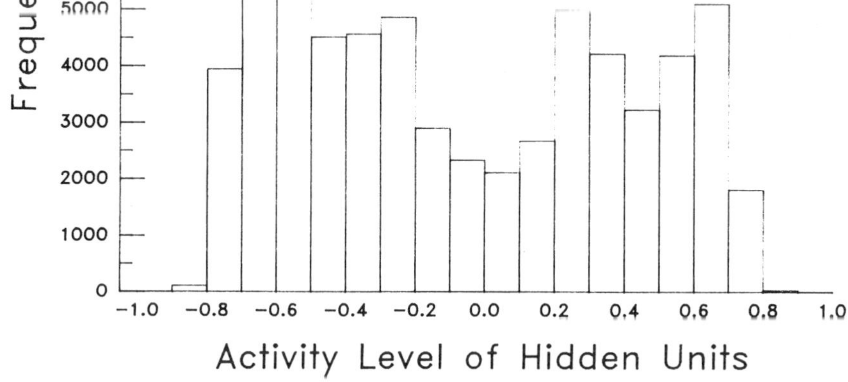

Figure 17. Histogram of output values of hidden units during reconstruction of ISG picture.

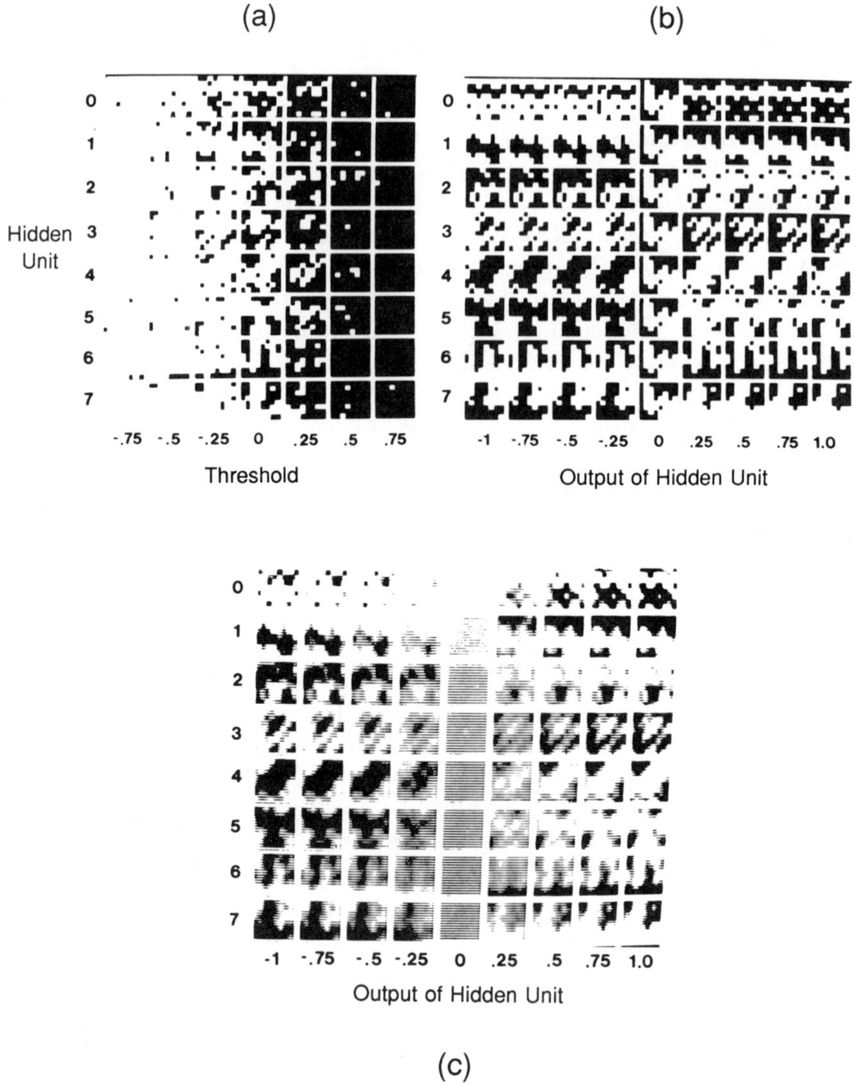

Figure 18. The internal representation. *A*: The weight matrices from the input patch to eight hidden units, thresholded from −.75 to +.75. The middle column (zero threshold) shows the "canonical" feature responded to by that hidden unit. *B*: The output patch driven by each hidden unit at different output values from −1 to 1. C: The same picture as *B* on a color monitor.

the network. An interesting point here is that they look more alike than different in their level of complexity. That is, there is not a clear ordering, as there is for the basis images of the Hotelling transform.

DISCUSSION

This study has produced results that have implications for both connectionist networks and image compression. These are discussed and summarized below.

Implications for Connectionist Networks

Extensional Programming

The major result of this study is that a relatively straightforward application of the back-propagation learning procedure to a problem that has been studied for many years results in near state-of-the-art performance. This performance was not obtained by programming a connectionist solution to the problem, but by the process of *extensional programming*. In this procedure, many examples of the desired behavior are presented and the network must program itself to achieve the behavior. This suggests that other problems for which solutions are not known in advance may be solved by back propagation.

A major problem with this technique is determining post hoc how the network solved the problem. In our case, we have some pieces of the answer, mainly because image compression is a well-studied problem. Hence we have some idea what to look for, if not an analytical solution. By comparison of our network to the techniques of image compression, we can gain insight into the solution found. However, this will not be the case in general. The importance of back propagation is that, whether or not we know of an algorithm for the solution of a problem, back propagation will in many cases find a solution to the mapping simply from examples of the input–output patterns.

Linear Networks

There is currently a bias in the connectionist community, shared by the authors of this article, against linear networks. This is partly due to the assumption that interesting problems must require nonlinearity for their solution. While the results were only alluded to, we found that a linear version of the network produced results compatible with the nonlinear version. It is thus useful to check whether nonlinearity is necessary for a particular problem. If not, the elimination of evaluating the logistic function can lead to more efficient solutions and a more tractable analysis of the solution discovered. If both approaches appear viable, comparison of the two can lead to a better understanding of nonlinear solutions. This approach needs to be carried further in future work. The linear network lends itself to analysis much more readily than the nonlinear one. Williams (1985) gives an empirical test that would give us an idea of how well the principal components are reflected by the linear network.

Internal Representations

One of the typical ways to speak of the solutions discovered by back-propagation learning is to say that the network discovers regularities in the input. This report adds at least a new vocabulary for discussing the kinds of regularities discovered in the case of autocoding. We can look at the *variance* of the hidden units as indicative of their usefulness in the resulting solution. It may not be the case that hidden units span the principal subspace of the covariance matrix, that is, the space spanned by a PCA solution, but it is possible that the hidden units are finding the best approximation to this within the constraint that the logistic function imposes of a limited range on the coefficients. If this turns out to be true, then we may speak of the hidden units as finding at least an analog of the principal subspace and as discovering useful *axes of covariance* of the input.

Implications for Image Compression

A major result of this work is the application of a new way of minimizing mean square error to a real-world problem that shows it is competitive with PCA. This new technique has several possible advantages over PCA and other current techniques. Some of these need confirmation by further investigation.

One major advantage of the nonlinear version is the relatively equal distribution of error among all of the coefficients. This should lead to a reduction in the effects of channel errors. In PCA and other techniques that approximate it, channel errors that affect the coefficients with high variance can result in a patch that is dominated by the corresponding basis image. The relatively equal contributions of each basis image in the network solution should mitigate these effects. In particular, we know in advance what range the value should be in, and, if a coefficient is suspected of being in error, an acceptable restoration of the patch can probably be effected by simply eliminating that coefficient or replacing it with its average value.

Second, because of the fixed range of the coefficients, problems with "tracking" the coefficients by an adaptive quantizer are mitigated. Adaptive quantizers try to follow coefficients as they change, changing the quantization as the coefficients shift. They can "lose track." In our system, we know in advance the range of the coefficients, which should make this less of a problem.

Third, the ability of our network to generalize to novel images is striking. The performance of the linear network is especially encouraging in this regard. This requires some qualification. First, it is likely that this generalization does not apply to images with very different statistics, such as text. Second, investigation is needed to compare this network to the results of current techniques on the same images. We do not know the results in this area, and we are not aware of the existence of a testbed of images for this purpose, although one probably exists.

OPEN QUESTIONS

There are more possibilities suggested by this work for future research than we could possibly pursue. We discuss a few of them here.

First, the parameter space of training image types, patch sizes, number of layers, and sizes of those layers is large. It is certain that there are better ones than the ones we considered. Some interesting possibilities suggest themselves. One is to use larger input patches than output patches, to give the hidden units some context for their target. This should lead to faster elimination during the training of edge effects. Another possibility, suggested by Geoff Hinton (personal communication, 1986), is to use larger output patches than the input patches. This forces the hidden units to learn to predict the context. This should also work to eliminate edge effects of reconstruction patches. Finally, it is possible that several hidden layers may be able to do what one cannot. Not enough is known about the ability of back propagation to scale up to many layers to make a sound judgment on this.

Another research direction is to try to improve the performance of the nonlinear network on novel images. One of the sources of error in the reconstructions of novel images is that the average intensity of these images differs from that of the training image. The average intensity of the training image appears to be reflected in the biases of the output units. It is a simple matter to estimate the average intensity of the novel image from a few samples. The question is whether appropriately modifying the biases of the output units will give a better result.

An alternative approach suggested earlier is to apply Tescher's adaptive quantizer to the linear network. If the linear network is capturing most of the useful information of the image, but the quantization based on the training image defeats it, this should give considerable improvement.

Initially, we had planned to compare the operation of the hidden units to what is known about neurons found in the visual system. Time constraints have made this another avenue for future investigation.

A final project is to use the outputs of the hidden units as inputs to a pattern classification device. The beauty of back propagation is that it can be used to train the pattern classifier as well, giving the procedure an elegance not obtained by heterogeneous techniques.

CONCLUSIONS

The major result of this work is in demonstrating the efficacy of current connectionist techniques for programming by example rather than algorithm. We have termed this *extensional programming*. The results here suggest that this technique is a powerful one. Its naive application to a problem of current interest among engineers resulted in respectable performance compared to current techniques.

However, back propagation is not a panacea—it brings new problems of its own. Designing a connectionist representation of the input (and output for non-autocoding problems) is itself an art. The representation must contain enough information to license solution of the problem, without providing so much that the solution is trivial. This is not always hard. Hinton (1986) has shown that, at least in some domains, back propagation can even design the input representation simply from the occurrence of a token in context.

The results of this study suggest that one useful approach to problems for which no algorithm is known, or for which no parallel algorithm is known, is to use connectionist representations of the problem and allow the network to discover the program itself. Analysis of the programs thus discovered may aid in our understanding of the problem and lead to methods for doing the programming ourselves. A variety of problems that cognitive science is concerned with are of this character—the input–output behavior is known, but the algorithm is not. With extensional programming we can begin to investigate algorithms that we did not invent ourselves.

APPENDIX: SOME DETAILS OF PRINCIPAL COMPONENT ANALYSIS

The transformation matrix for the Hotelling transform is formed by first computing (or estimating) the covariance matrix of the $\mathbf{x}$ (image) vectors

$$\mathbf{C}_{\mathbf{x}} = E\,\{(\mathbf{x}-\mathbf{m})(\mathbf{x}-\mathbf{m})'\}$$

where

$$\mathbf{m_x} = E\ \{\mathbf{x}\}$$

is the mean vector, E is the expected value operation, and prime (′) indicates transposition. Note $\mathbf{C_x}$ is an $n^2 \times n^2$ matrix if x is of length n^2. Then the eigenvectors of this matrix are computed.

Let $\mathbf{e}_i$ and $\lambda_i, i = 1,2, \ldots, n^2$ be the eigenvectors and eigenvalues of $\mathbf{C_x}$, such that $\lambda_i \geqslant \lambda_j$, for all $i<j$. Then the transform matrix $\mathbf{A}$ is formed by taking these eigenvectors in order as the rows. Then the Hotelling transform is

$$\mathbf{y} = \mathbf{A}(\mathbf{x} - \mathbf{m}).$$

The eigenvectors that form $\mathbf{A}$ correspond to the axes of Figure 4, and the y_i's to the coordinate along each axis.

It can be shown that the covariance matrix of the $\mathbf{y}$ vectors is a diagonal matrix consisting of the eigenvalues of $\mathbf{C_x}$, i.e., the elements of $\mathbf{y}$ are *uncorrelated*. Also, the variance of each y_i is λ_i.

To reconstruct $\mathbf{x}$ from $\mathbf{y}$, we need to invert the mapping. Since $\mathbf{C_x}$ is a real, symmetric matrix, it is always possible to find a set of orthonormal eigenvectors, hence $\mathbf{A}^{-1} = \mathbf{A}'$ and

$$\mathbf{x} = \mathbf{A}'\mathbf{y} + \mathbf{m_x}.$$

For image compression purposes, an important property of this mapping is that we can use the first K eigenvectors that account for the most variance. Then we will have K coefficients in $\mathbf{y}$, corresponding to K hidden units in the network. Now the reconstruction is an approximation, but it is an optimal one. Let

$$\hat{\mathbf{x}} = \mathbf{A}'_K \mathbf{y} + \mathbf{m_x}$$

represent the approximation to $\mathbf{x}$ obtained in this way. It can be shown that the mean square error between $\mathbf{x}$ and $\hat{\mathbf{x}}$ is given by

$$MSE = \sum_{j-1}^{N^2} \lambda_j - \sum_{j-1}^{K} \lambda_j$$

$$= \sum_{j=K+1}^{N^2} \lambda_j .$$

Since the λ_j decrease monotonically, the error is minimized by taking the eigenvectors associated with the largest eigenvalues. Hence, the Hotelling transform is optimal with respect to mean square error for a given number of coefficients retained.

REFERENCES

Ackley, D.H., Hinton, G.E., & Sejnowski, T.J. (1985). A learning algorithm for Boltzmann machines. *Cognitive Science, 9*, 147–169.

Baldi, P., & Hornik, K. *Neural networks and principal components analysis: Learning from examples without local minima.* Working Paper, Dept. of Mathematics, UCSD, La Jolla, Ca 92093.

Feldman, J.A., & Ballard, D.H. (1982). Connectionist models and their properties. *Cognitive Science, 6*, 205–254.

Gonzales, R.C., & Wintz, P. (1977). *Digital image processing*, Reading, MA: Addison-Wesley.

Hinton, G.E. (1986). Learning distributed representations of concepts. In *Proceedings of the Eighth Annual Conference of the Cognitive Science Society.* Hillsdale, NJ: Erlbaum.

Huang, J.J.Y., & Schultheiss, P.M. (1963). *IEEE Transactions on Communications Systems, 11*, 289–296.

Max, J. (1960), *IRE Transactions on Information Theory, 6*, 7–12.

Minsky, M., & Papert, S. (1972). *Perceptrons.* Cambridge, MA: MIT Press.

Mozer, M.C. (1983). *Principal component analysis using parallel computation* (ICS Progress Report, First Quarter). La Jolla, CA: University of California, San Diego, Institute for Cognitive Science.

Rumelhart, D.E., Hinton, G.E., & Williams, R.J. (1986a). Learning internal representations by error propagation. In D.E. Rumelhart, J.L. McClelland, & the PDP Research Group, *Parallel distributed processing: Explorations in the microstructure of cognition. Vol. 1. Foundations.* Cambridge, MA: MIT Press/ Bradford Books.

Rumelhart, D.E., Hinton, G.E., & Williams, R.J. (1986b). Learning representations by back-propagating errors. *Nature, 323*, 533–536.

Tescher, A.G. (1979). Transform image coding. In W.K. Pratt (Ed.), *Image transmission techniques.* New York: Academic Press.

Williams, R.J. (1985). *Feature discovery through error correction learning* (Tech. Rep. No. 8501). La Jolla, CA: University of California at San Diego, Institute for Cognitive Science.

CHAPTER 10

Programming Neural Nets To Do Spatial Computations*

DAVID ZIPSER
Institute for Cognitive Science
University of California, San Diego

INTRODUCTION

Cognition is now widely viewed as a computational process. Many of our ideas about these computations come from experience with digital computers. In the brain, however, computations are carried out by networks of neurons which compute quite differently. In the context of digital computers, computations are explicitly described using programs that state the operations to be performed. In the brain it is unlikely that all computation is specified by such explicit programs, which could come only from intelligent beings or the genetic code. Rather, it seems likely that much computation is specified by learning. Learning differs from traditional programming in that the computation is described implicitly, using examples of input and output. Learning can take place while experiencing normal behavior and may require little or no explicit instruction.

Very little work has been done using learning to program computations in neural-like networks because, until quite recently, there was no learning procedure powerful enough to deal with computations of real interest. This situation has changed with the discovery of a very powerful new learning procedure for neural-like networks. This new learning algorithm, discovered by Rumelhart, Hinton, and Williams (1986), is variously called the *general-*

* This research was supported by the Air Force Office of Scientific Research, Air Force Systems Command, USAF, under Grant Number AFOSR-86-0062. The U.S. Government is authorized to reproduce and distribute reprints for Governmental purposes notwithstanding any copyright notation thereon.

Requests for reprints should be sent to David Zipser, Institute for Cognitive Science, C-015; University of California, San Diego; La Jolla, CA 92093.

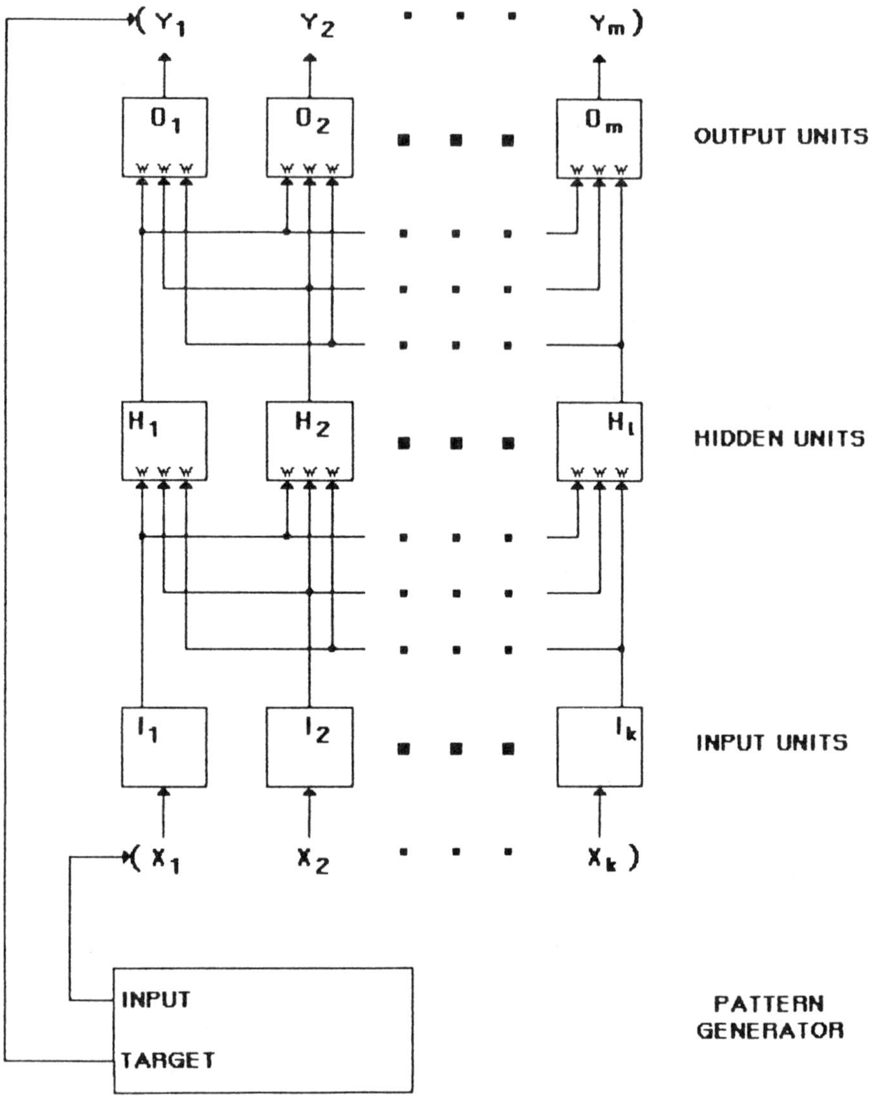

Figure 1. Strictly layered back propagation networks of the kind depicted in this figure were used for all the work described in this paper. Before training begins, all the weights and biases are set to random values between −1.0 and 1.0. At the beginning of each learning cycle, i.e., at time $t = 0$ for that cycle, the pattern generator provides an input pattern $(x_1\ x_2\ .\ .\ .\ x_k)$ and a target pattern $(y_1\ y_2\ .\ .\ .\ y_k)$. At time $t = 1$ the outputs of the input units are set equal the input pattern, that is: $I_1 = x_1, I_2 = x_2, \ldots, I_k = x_k$). The output activities of the hidden units are set at time $t = 2$ to

$$H_i\,(t+1) = squash\,(bias_{H_i} + \sum_{j=1}^{j=k} W_{H_i j}\, I_j\,(t))$$

ized delta rule or, more simply, *back propagation*. Back propagation learning is so powerful that it has radically effected much of the research being done on computation in neural-like systems. By providing a technique for programming networks, back propagation allows us, for the first time, to really see what these networks can do. What is being found is that neural-like networks are much better at computation than was previously demonstrable. It is now clear that neural networks are powerful computational elements capable of computing, with good accuracy, a wide range of useful functions. Many of these functions can, at present, only be defined extensionally. This is particularly true of those functions involved in perceptual processes,

---

where $k = n^0$ input units and where

$$squash\ (x) = \frac{1}{1 + e^{-x}}.$$

Similarly, the activity of the output units are set at $t = 3$ to

$$O_i\ (t+1) = squash\ (bias_{O_i} + \sum_{j=1}^{j=l} W_{O_{ij}} H_j(t))$$

where $l = n^0$ hidden units.

All these values are held constant until the end of the current learning cycle. The next step is to change all the weights and biases according to the back propagation rule. At $t = 4$ the output unit weights are changed by the rule

$$\Delta W_{O_{ij}} = \eta \delta_{O_i} H_j$$

where $\Delta bias_{O_i} = \eta \delta_{O_i}$. The term η is the learning rate constant and δ_{O_i} is defined by

$$\delta_{O_i} = (y_i - O_i)\, O_i\, (1 - O_i)$$

where $O_i\ (1 - O_i)$ is the derivative of $1/(1 + e^{-x})$ with respect to the total input to O_i and where y_i is the *i*th target unit.

At $t = 5$ the hidden unit weights are changed by

$$\Delta W_{H_{ij}} = \eta \delta_{H_i} I_j$$

where $\Delta bias_{H_i} = \eta \delta_{H_i}$ and where the term δ_{H_i} is defined as

$$\delta_{H_i} = \grave{e}_i\ (1 - H_i) \sum_{k=1}^{k=m} \delta_{O_k} W_{O_{ki}}.$$

where $H_i\ (1 - H_i)$ is the derivative of $1/(1 + e^{-x})$ with respect to the total input to H_i.

For the derivation of these rules see Rumelhart, Hinton, and Williams (1986).

which the brain does so well and which are so difficult to emulate with standard computer programming.

The task of learning schemes such as back propagation is to program networks to associate pairs of input and output activity patterns. The complete set of pattern pairs to be learned can be thought of as the extensional definition of a vector-valued function whose domain is the set of input patterns and whose range is the set of outputs. As learning schemes become more powerful, the set of functions that can be programmed gets larger. For example, the perceptron learning rule can program networks to compute linear Boolean functions such as "AND" and "OR" but not nonlinear ones such as "XOR." The early generalizations of the perceptron rule that extended the learning set to patterns with continuous rather than Boolean values are also limited to learning linear functions. Back propagation, on the other hand, can program multilayer networks to compute all the Boolean functions and, since it is applicable to patterns with continuous component values, it can deal with a much wider range of functions. It is not yet clear just what back propagation can do in this regard. One key issue is the degree to which continuous functions can be learned from a finite set of examples.

A good part of the significance of back propagation learning stems from the fact that it is defined on a neural-like network. An example of such a network is shown in Figure 1. Each of the units in the network can have any number of inputs. The effect of activity at these inputs is biased by a weight parameter in just the way it is in previous, well-known simple neural models such as the linear threshold unit. It is these weights that are changed as learning proceeds. Each layer in the back propagation network can have any number of units. Typically, strictly layered networks are used in which each unit in a layer receives inputs from every unit in the layer below. Training proceeds in cycles. In each cycle, a pattern pair is chosen from the function definition. The input pattern is applied to the first layer of the network, and the activity it generates is passed successively to the other layers until it produces a pattern of activity on the output layer. This output pattern vector is then subtracted from the correct output pattern to produce an error pattern. This error in turn is used to first adjust the weights in the output layer and then, by "back propagation," to adjust the weights in lower layers. The exact way the error is propagated back down through each layer constitutes the novel part of the learning rule. Discovering just how to do it correctly was the significant accomplishment of Rumelhart, Hinton, and Williams (1986). The details of what they have done are given in the caption of Figure 1.

When computations are programmed using learning, values from the range of the function must be available. Sometimes these values are supplied by an external source or "teacher." There is something unbiological about this, because teachers do not supply us with actual patterns of neural activity that, in effect, are needed by the back propagation learning procedure. This raises the question of how to configure a system so it can learn, either without an external teacher or with the kind of teaching we actually use. Several solutions have been proposed. One of the simplest and most elegant of these is to use teaching patterns that are the same as the input, or some fixed transformation of the input. While this limits us to learning the identity function or some fixed transformation, Hinton and Rumelhart (personal communication, June 6, 1986), have shown that, with this constraint, the hidden units learn to represent the input patterns in terms of salient features. When the number of hidden units is less than the number of input units, the information in the input is represented at a lower dimensionality. In many perceptual problems, this lower-dimensional feature representation is just what is wanted as a basis for further processing. In this paper I report studies that show what representations are formed when these feature learning strategies are applied to simplified models of the visual system.

FROM IMAGES TO COORDINATES

When vision is used to guide spatial behavior, an important early step in the process is to extract location information from the original retinal image and convert it to a form that can be used to guide effector movement and for further computation. It is known from neurophysiological studies that spatial location is represented in the parietal region (Andersen, Essick, & Siegel, 1985), but the way that this information is extracted from the retinal image and combined with eye position is still not understood. One possibility is that the hidden units in a network trained to identify map images of discrete objects will encode their location in the image.

To test this hypothesis, a network embodying a very simple model of a visual system was used. The input and output layers of this network consisted of identically configured image arrays as required for identity mapping. Rather than using complex images, the initial studies were done with pattern sets consisting of circular Gaussian spots appearing at random locations on the input image plane. These spots are simplified examples of the images of discrete objects as they would appear at low resolution after figure–ground separation. The validity of using such simple stimuli is analyzed in the discussion section.

A network typical of those used is shown in Figure 2. It consists of input and output layers, each with 100 units arranged as 10×10 arrays. The hidden layer has a small number of units ranging from one to nine in different runs. The locations to be coded are represented as spots of activity on the input layer. The position of the center of a spot is picked at random on each learning cycle. To reduce edge effects, the center of the spot is limited to a disk, 8 units in diameter and centered on the input array. The activity of

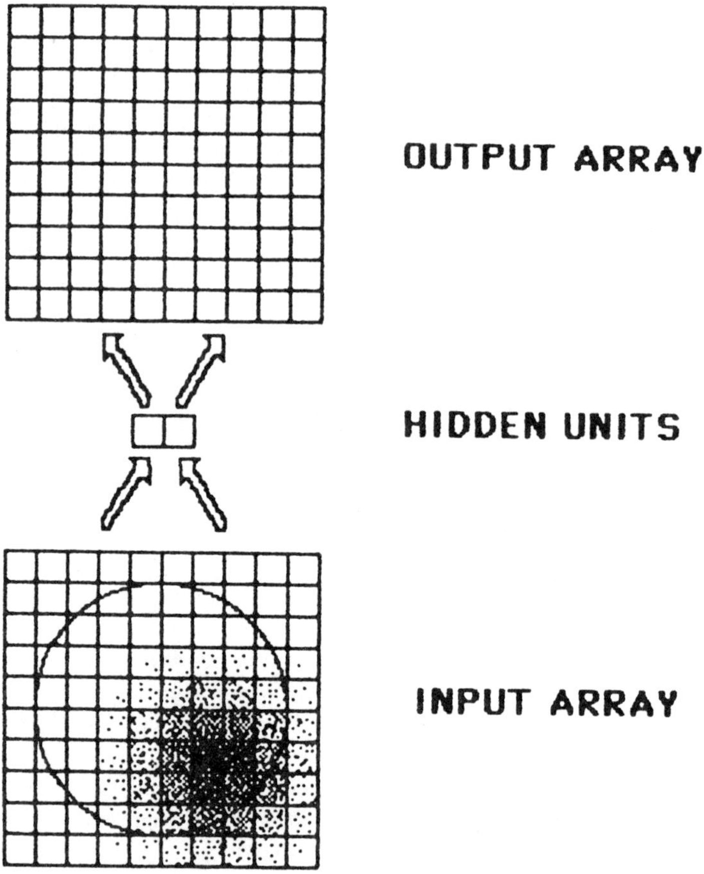

Figure 2. Identity mapping network with two hidden units and 10 × 10 unit input and output arrays. An example of a Gaussian spot, with 1/e radius = 0.2 times the array width, is shown superimposed on the input array. In identity mapping, the input and target patterns are the same. This means that the number of input and output units must be the same. It is important to note that the hidden units have no initial information about the spatial locations of the input units. The spatial arrangement of the input and output units is known externally to the pattern generator so that the target and input stimuli can be generated. Any spatial features that develop in the hidden units are a result of training.

each unit in the input array is determined as a Gaussian function of its distance from the center of the spot. Gaussian spots are used so that spot location can vary continuously, and also to allow a smooth, nondiscrete representation of the input stimuli. Training consists of applying a spot to the input and using this same spot as the target output. The job of the network is to learn to copy the input spot on the output layer. To accomplish this task, the small number of hidden units must somehow learn to encode the location of the spot. This network, as well as all others in this paper, was simulated using the P3 parallel system programming environment, implemented on a Symbolics 3600 LISP machine.

Figure 3 shows results obtained after extensive identity map training using a network, of the kind described above, with two hidden units. It is not completely trivial to determine what features the hidden units encode. A graphic display that shows the relationship between unit activity and spot position has proven very useful in this regard. To generate this type of display, the input spot is scanned over the disk of possible input positions. At every position of the spot, the activity of each hidden unit is plotted at a corresponding position on the computer display. A separate disk-shaped pattern is generated for each hidden unit. The pixels in the display used can be set only to black or white, so, to get a graded effect, the pixel closest to the spot position is set to black with a probability approximately proportional to the unit's activity. To clearly define the disk, the probability of setting a pixel to black is slightly greater than zero even when activity is zero. This produces a display in which the degree of darkening over an area is roughly proportional to unit activity. To more clearly show the activity pattern, a set of contour lines is superimposed on the display by plotting white pixels whenever the activity of a unit falls within certain evenly spaced narrow bands.

The two hidden units in Figure 3 can be seen to have formed an explicit encoding of spot location using an orthogonal coordinate system. Although

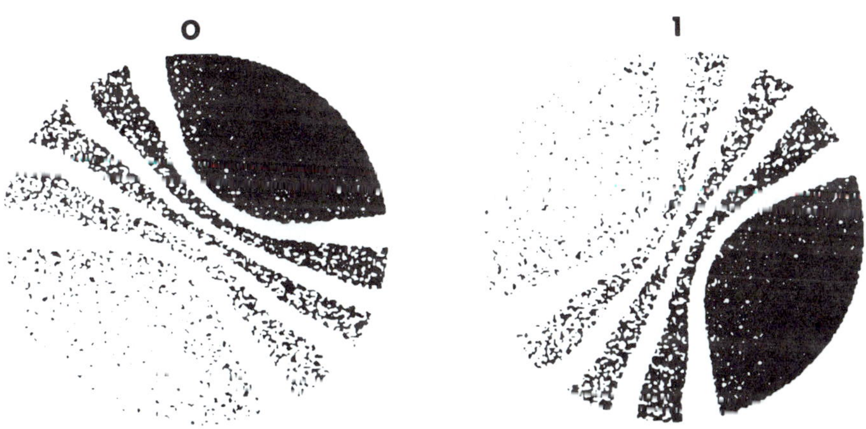

Figure 3. The activity pattern after identity mapping in a network with two hidden units. Each circular pattern represents the activity of a hidden unit as a function of the location of the center of the stimulus spot. Two ways of representing the value of activity are used. One of these represents activity as the density of black pixels, with all-black regions representing an activity of 1.0. In addition, a set of contour lines has been superimposed on the display by setting pixels to white if the value of activity lies within fixed bands. The bands used here are 0.16-0.24, 0.36-0.44, 0.56-0.64, and 0.76-0.84.

the activity profiles of the two units are everywhere orthogonal, they are not linear. This is presumably a consequence of constraints other than spot location, such as spot size and spot shape at the edges. When the training is repeated with the same network, using different sets of random starting weights and training sequences, the activity patterns learned by the hidden units are the same except for the angle of rotation of the coordinate axis. This angle is a free parameter, and its final value is a result of unpredictable details of random parameters in the training procedure. When a differently sized Gaussian spot is used for training, the pattern of hidden unit activity is similar, but there are some subtle differences in the curvature and separation of the contour lines. This indicates that spot size information is being coded by the hidden units. In some learning experiments the square input array was replaced by a toroidal one with the same number of units. In this case the spot did not have to be confined to a central disk, since the torus has no edges. In the toroidal case the contour lines defining the coordinate system were straight rather than curved when only two hidden units were used, suggesting that the curvature is, in part, an accommodation to the edges of the disk-shaped input space.

The activity pattern learned by the hidden units depends critically on the number of hidden units used. The result of using one, three, four, and nine hidden units is shown in Figure 4. The patterns become progressively more complex. The change in these patterns with increasing numbers of hidden units involves a change from a symmetrical subdivision of the input space to an unsymmetrical one. With one or two units, spot location is a simple function of unit activity. With nine units, location is represented in terms of complex receptive fields. With increasing numbers of hidden units, there seems to be a continuum between an explicit or "variable" and an implicit or "value" representation of location.

The accuracy with which the position of the spot is reproduced on the output increases with the number of hidden units, as shown in Figure 5. Since the spot locations used to measure error are, in general, not identical to those used in training, position error is a measure of the degree to which the network has learned to generalize and compute a continuous function. The error in the location of the spots on the output, as measured by the center of gravity of activity, is less than the distance between units in the input array. This degree of accuracy is reminiscent of our own visual hyperacuity which makes it possible to resolve features smaller than the distance between retinal receptors. These error measurements were carried out after 350,000 cycles, by which time there was no further improvement. The error after 20,000 cycles was 2 times the minimum value, while, after 50,000 cycles, it was 1.3 times the minimum.

The shape of the activity patterns that develop in the hidden units indicate that they might have an orientation tuning similar to neurons in the vis-

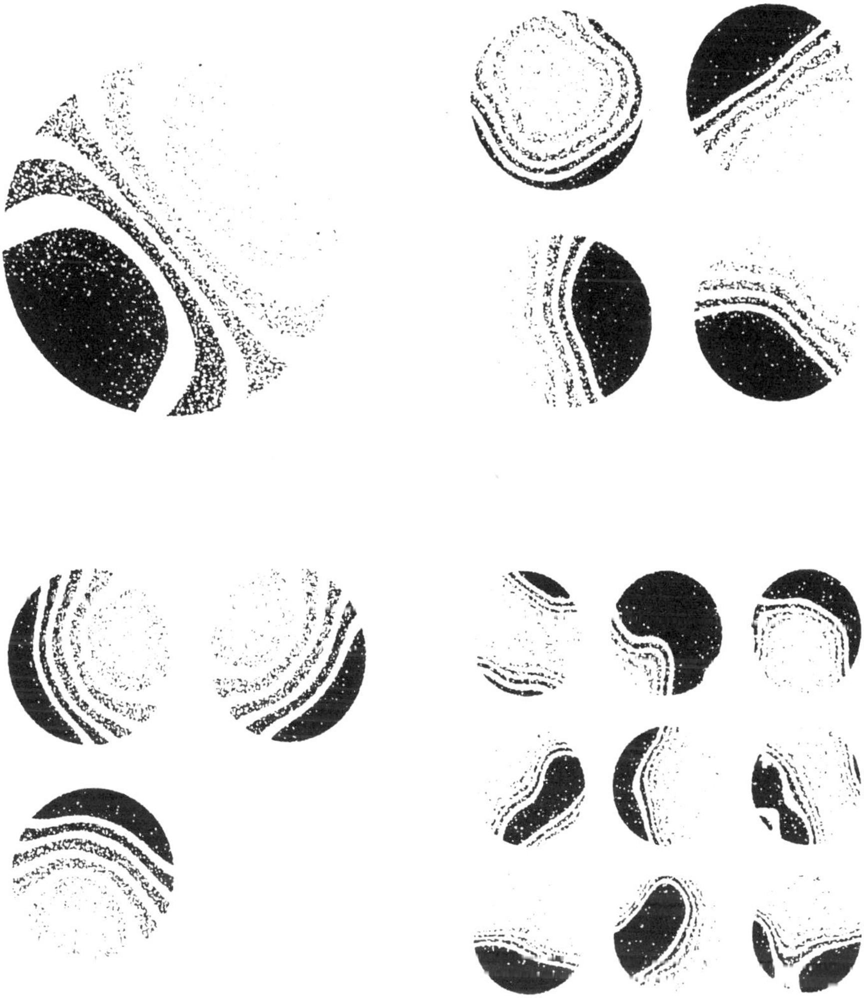

Figure 4. Identity mapping as in Figure 3 but with one, three, four, and nine hidden units.

ual cortex. This was tested by using a progressively rotated dark bar as input to the fully trained networks. The result is shown in Figure 6, which is a plot of hidden unit activity as a function of the orientation of a Gaussian bar having the same width as the spots used to train the network. The pair of hidden units have a strong orientation tuning. It is interesting that one unit has a decrease in activity at the preferred orientation, while the other has an increase. The preferred orientations of the two units are just 90° out of phase. At present this result is just a curiosity, but it does demonstrate how something akin to back propagation identity mapping could be used by the brain to generate feature detectors. If a process analogous to back propagation

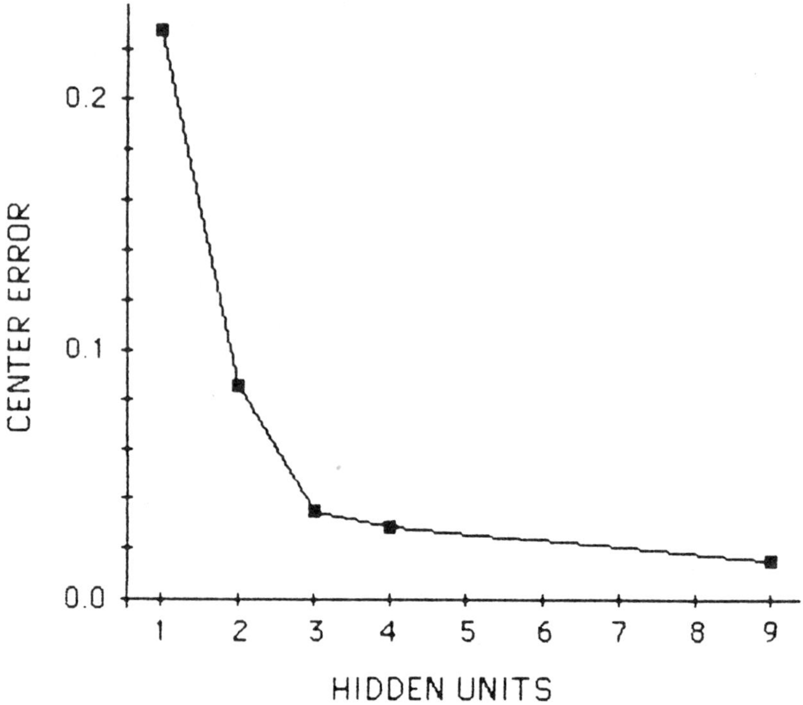

Figure 5. Center error as a function of the number of hidden units. The error measure used is the root mean square of the distance from the center of the target to the center of gravity of activity on the output array. Note that this is not the error that is being minimized by the learning procedure, but rather a measure of the accuracy with which spatial locations are being represented in the hidden units.

identity mapping does occur in the brain, then computer simulations could provide a way to predict the features to which real neurons respond.

It is the value of the connection weights that determines the particular computation carried out by a network. Examination of these weights should give some insight into how the computations are accomplished. In the case of the identity mapping computation, the product of the hidden unit weight matrix and the output unit weight matrix must equal a constant times the identity matrix. One way to accomplish this is for the hidden unit weight vectors to become orthogonal, and the output unit weight matrix to be the transpose of the hidden unit weight matrix. This is just what happens in the case of image compression by identity mapping (see Cottrell, Munro, & Zipser, this volume, Chapter 9). To see what is happening in the case being studied here, hidden unit vectors were compared with corresponding rows of the output weight matrix. Figure 7 shows the weights after training for the case of two hidden units. Weight vectors are rendered in a three-

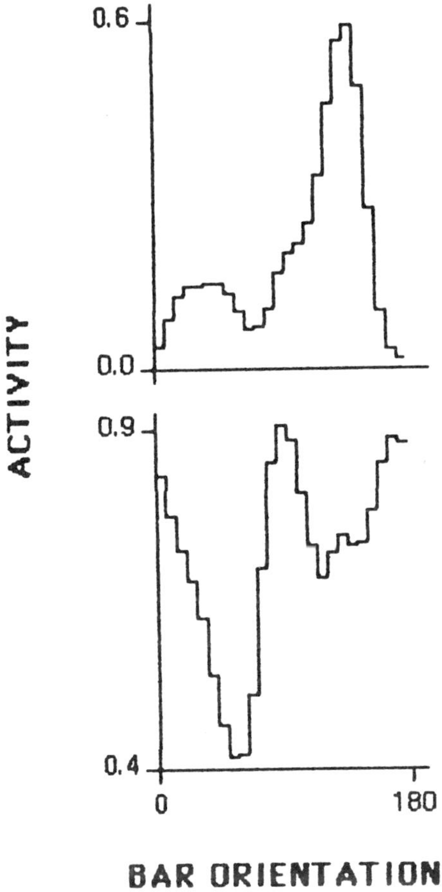

Figure 6. The response of a pair of hidden units to a rotating bar. The hidden units are those shown in Figure 3. After training was completed, learning was turned off and the network was stimulated by an input that consisted of a Gaussian bar passing through the center of the input array. The 1/e width of the bar was 0.2 of the array width—the same as the 1/e radius of the spot used for training. The bar was presented at 25 equally spaced orientations between 0° and 180°. The figure shows the activity of the hidden units as a function of bar orientation. Note that the responses of the two units are mirror reflections shifted by 90°. Orientation tuning is not perfect in that the units each have a smaller secondary response peak 90° from the main peak.

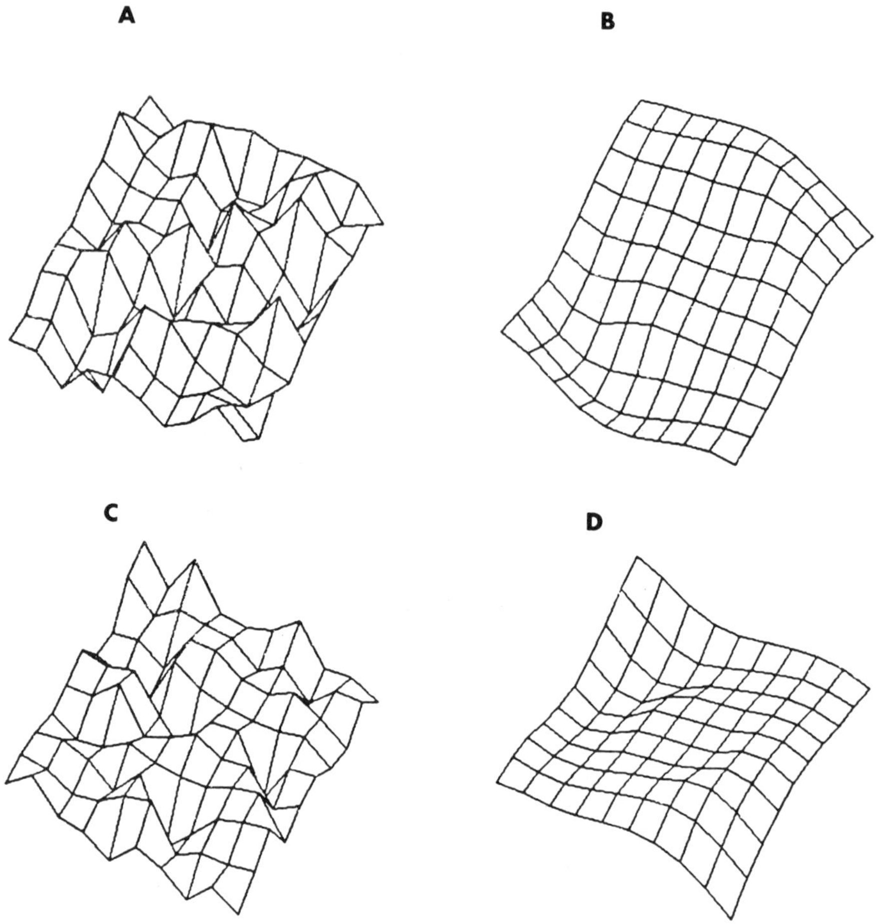

Figure 7. Hidden and output weights of the two units shown in Figure 3. (A) and (C) are the weights of hidden units 0 and 1, respectively. The plots represent the 100 weights on each hidden unit located at positions corresponding to the location of the input unit connected to the weight. (B) and (D) are graphs of the output weights plotted at the location of the output units. (B) and (D) are virtually identical curves rotated 90°. When (A) is rotated 90° with respect to (C), the two plots are similar but there are many small differences in detail.

dimensional plot of weight value versus position of the relevant unit. For the weights connecting the hidden units to the input, the relevant position is that of the input unit from which the weight gets activity. For the weights connecting the output units to the hidden units, the relevant position is the location of the output unit that receives the activity. The weights between the hidden and output layer form a smooth surface, shaped so as to enable the output units to use the information coded in the hidden units to generate a spot at the correct location on the output array. One might expect a similarly shaped, smooth pattern for the hidden unit weights if the hidden and output weight matrices were transposes. Instead, a very rough, oscillating pattern was observed. Many repeats, using different spot sizes and different numbers of hidden units, confirm that this rough pattern is quite general. When very small spots are used, the hidden unit weight pattern becomes quite smooth and closely resembles the output weight pattern. However, if this smooth hidden weight pattern is used as a starting point for training with a larger spot, a rough pattern, indicative of the larger spot, is eventually regenerated. The full significance of these rough, rapidly alternating patterns of the hidden unit weights is not yet known, but they may serve to encode spot size or more general spatial frequency information.

HIDDEN UNIT ENCODING OF STEREO IMAGES

The visual cortex contains many binocular neurons, some of which are involved in extracting depth information from stereo images on the retinas. Is it possible that identity mapping the two retinal images would give hidden units that were binocular and encode depth information? This is by no means obvious, since one solution to identity mapping two retinas would be to use one subset of hidden units exclusively for mapping one retina and the complementary subset for mapping the other retina. To see what solution identity mapping actually found, a simplified visual network was used. To avoid the complexities of three dimensions, linear retinas were used with depth as the second dimension. The stereo images for these linear retinas were generated as shown in Figure 8. On each training cycle, a location was picked at random, within a circle around the fixation point. This location was then projected to the retinas through the focal point of each eye. The activity for each unit of the input retinal array was computed as a Gaussian function of its distance from the location of this projection point. The depth of the chosen location, relative to the fixation point, is encoded in the disparity between its location on the two retinas. The flat, linear retinas used here are an approximation to a horizontal slice through the curved retinas of the eye. Each input retinal array has 20 units, and the network output layer is made up of two arrays identical to the input. Training and data display are the same as described previously for monocular square retinas.

The results obtained from computer runs with two, three, four, and nine hidden units are shown in Figure 9. The network with only two hidden units could not solve the problem. In this network the input stimuli were mapped to identically corresponding positions on the two output arrays; i.e., there was no disparity. For three and more hidden units, the disparity present in the input was recreated to some degree on the output. The accuracy with which disparity was reproduced was quite good with four units and virtually

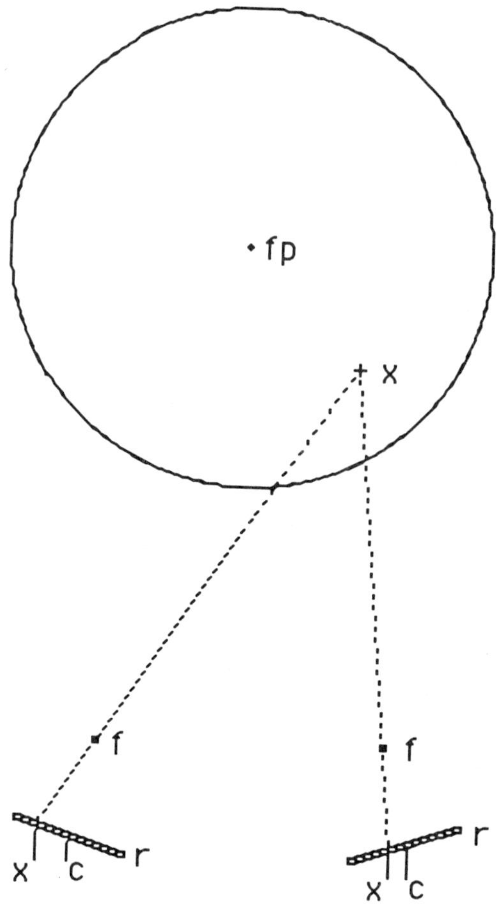

Figure 8. This diagram shows how a point located at *x*, in a two-dimensional world, is projected to two linear retinas, *r*. The diagram is not to scale. The actual scale used approximated human eyes fixated at a point, *fp*, about 20 cm straight ahead. The circle to which *x* is confined has a radius of 5 cm. The center of the retinas, where the point *fp* would project, is indicated by *c*. The vergence angle and location of *fp* were kept fixed. The activity of the units in *r* computed from

$$\textit{activity} = e^{-(dx/s)^2}$$

where dx = the distance from the center of the unit to x and s = 0.1 times the length of the retina.

perfect with nine units. Close examination of the data in Figure 9 reveals several interesting points. The contour lines are analogous to the lines of a coordinate system, with the output of each hidden unit representing the

Figure 9. The output of hidden units from networks taught the stereo identity mapping task. The data is displayed the same way as in previous figures of this kind. The significance of the patterns is discussed in the text.

value of a coordinate. In the case of two hidden units, the contour lines are angled in depth to compensate for the fact that, as a point gets further from the observer, more lateral movement is required to produce a fixed movement on the retina. However, the contour lines are exactly parallel, so the output of the hidden units contains no unambiguous depth information, which is why the output units could not learn to recreate the input disparity. For more than two hidden units the contour lines are no longer exactly parallel. This provides a coordinate system in depth that the output units can use in recreating the required disparity. In the case of nine hidden units, nonparallel pairs of the contour lines cover the whole area, which means that these units taken together have a distributed code for depth which can be used by the output units to recreate the disparity.

It is evident, upon inspection of Figure 9, that in no case did the network solve the problem by treating the two eyes separately. For example, in the case of four hidden units, a simple solution would be to dedicate two of the hidden units to each eye. If this had occurred, the result would have looked like a pair of copies of the two hidden unit case. What is actually observed is a different, distributed, binocular representation of the space for each hidden unit. It is not known why this distributed solution is found. It does not seem to be because it is more accurate, so it must arise because of some other, implicit constraint. Although we don't know why it happens, it could be of considerable significance, because it leads to a global representation in the hidden units. This means that objects are located in space relative to the whole observer rather than to the individual eyes. This type of global, distributed solution often occurs when identity mapping or related techniques are used to program networks.

USING HIDDEN UNIT REPRESENTATIONS FOR SPATIAL COMPUTATIONS

The computer simulation experiments described above demonstrate that the activity pattern of a small number of hidden units can be programmed to represent the location of a spot in an image. The question now arises whether these hidden unit representations can be used in spatial computations that require several input locations to produce an output. For example, consider the problem of predicting the future position of a moving object from its velocity. A simple version of this problem was used to see if a network could be programmed to use hidden unit representations. Velocity was represented as two spots separated by the distance a single spot would move in a fixed time interval. The network was trained to predict the location of the spot after the next time interval. This is analogous to predicting the next location of a spot in an apparent motion task. Assuming the coordinates of the three spots are given by (x_1, y_1), (x_2, y_2), and (x_3, y_3), the required computations are $x_3 = 2x_2 - x_1$ and $y_3 = 2y_2 - y_1$. Rather than using samples separated in time, a system with two separate input arrays was used so that both samples could be presented at the same time. A two-level system was used. The first level consists of two separate identity mapping networks, one for each input sample. The hidden units from these two networks supply the input to the second level. The second-level network has a few hidden units and an output array identical to the input arrays used in the first-level networks. This made it possible to teach the upper-level network by providing it with an image of a spot at the desired location. The first-level units were trained exactly as described previously. All three of these networks are independent with respect to training. In particular the upper-level network propagated no error to the lower-level networks. This means that the only way the upper-level network could solve the problem is to use the two different hidden unit representations of spot location generated by the lower-level networks. The whole system is diagrammed in Figure 10.

Each cycle of training consisted of randomly selecting the position of the two input spots and applying them to the input arrays of the left and right lower-level networks. Each lower-level network then used its own input spot as a learning target for identity mapping just as described previously. The upper-level network received its input from the hidden units of both lower-level networks. These inputs represent two spot locations coded in the activity of hidden units. The learning target for the upper network was the

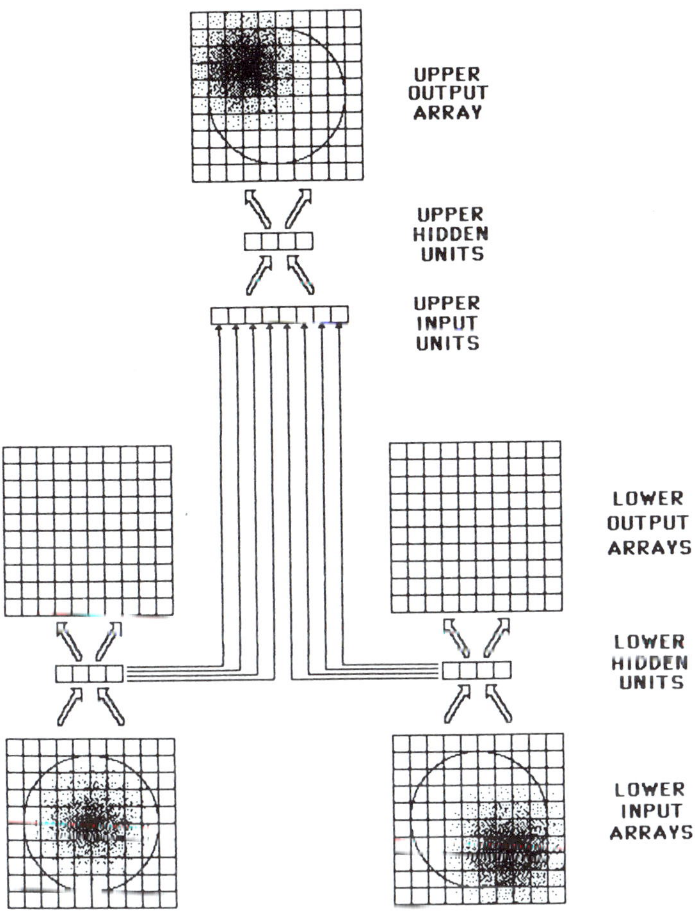

Figure 10. Two-level network used to learn "three-spots-in-a-row" and vector subtraction. The input arrays of the two lower-level networks show a typical pair of input spots for three-spots-in-a-row learning. The desired position of the third spot is depicted on the output array. Four hidden units are shown in this diagram, but various numbers of hidden units were used.

image of a spot located on the line defined by the two input spots. All spots were equidistant and in the order left network spot, right network spot, upper-level spot. Note this does not mean that the left network spot was always on the left side. Since all the image arrays were the same size, there is an additional constraint on the right network spot requiring it be placed so that the third spot will fall within the output array of the upper-level network. The net effect of all this was to train the output of the upper level to predict the next location of a point moving with the velocity indicated by the two input spots.

All three networks were trained together, which means that, initially, the upper-level network has meaningless inputs that only gradually take on meaningful values. Eventually, the upper-level network learns to predict the correct spot location with a root mean square center error less than the distance between units in the image arrays, when nine hidden units were used. In typical simulation runs of the two-layer system, a few hidden units in the upper level would not learn anything useful but would just stay at either zero or one all the time. This is an oddity of the parameters of the learning model, but it means that the accuracy observed is not the maximum obtainable from nine units.

To see how the network solved the problem, the activity of the hidden units in the upper level network was plotted separately with respect to the positions of the three spots. These plots are shown in Figure 11. The activity plots with respect to the position of the output spot show that each hidden unit represents one coordinate of the spot in a rectilinear reference frame. The orientations of these frames differ from unit to unit, as does the offset of the frame relative to spot space. When only two hidden units are used in the upper-level network, their coordinate systems are perpendicular. When more units are used, each takes on a different orientation. Plotting the activity of the upper-level hidden units with respect to spot position on each of the input arrays demonstrates that related coordinate systems have been associated with the input spaces. These coordinate systems have the same orientation as the upper-level network. They are rotated 180° with respect to each other, reflecting the need to subtract coordinate values to get the correct result.

The computation carried out by the network just described is of the same general kind involved in many aspects of spatial processing. It is analogous to vector addition or subtraction. Two-layer networks of the kind just used are not required for vector subtraction. A single-level network with one hidden layer was found to do a good job of vector subtraction, using a similar three spot protocol. Here, the computation, defined for points in a Cartesian frame, is $x_3 = x_2 - x_1$ and $y_3 = y_2 - y_1$. The single-level network solved the problem in the same way as the two-level network, but the result was more accurate, presumably because of the absence of error introduced in the first

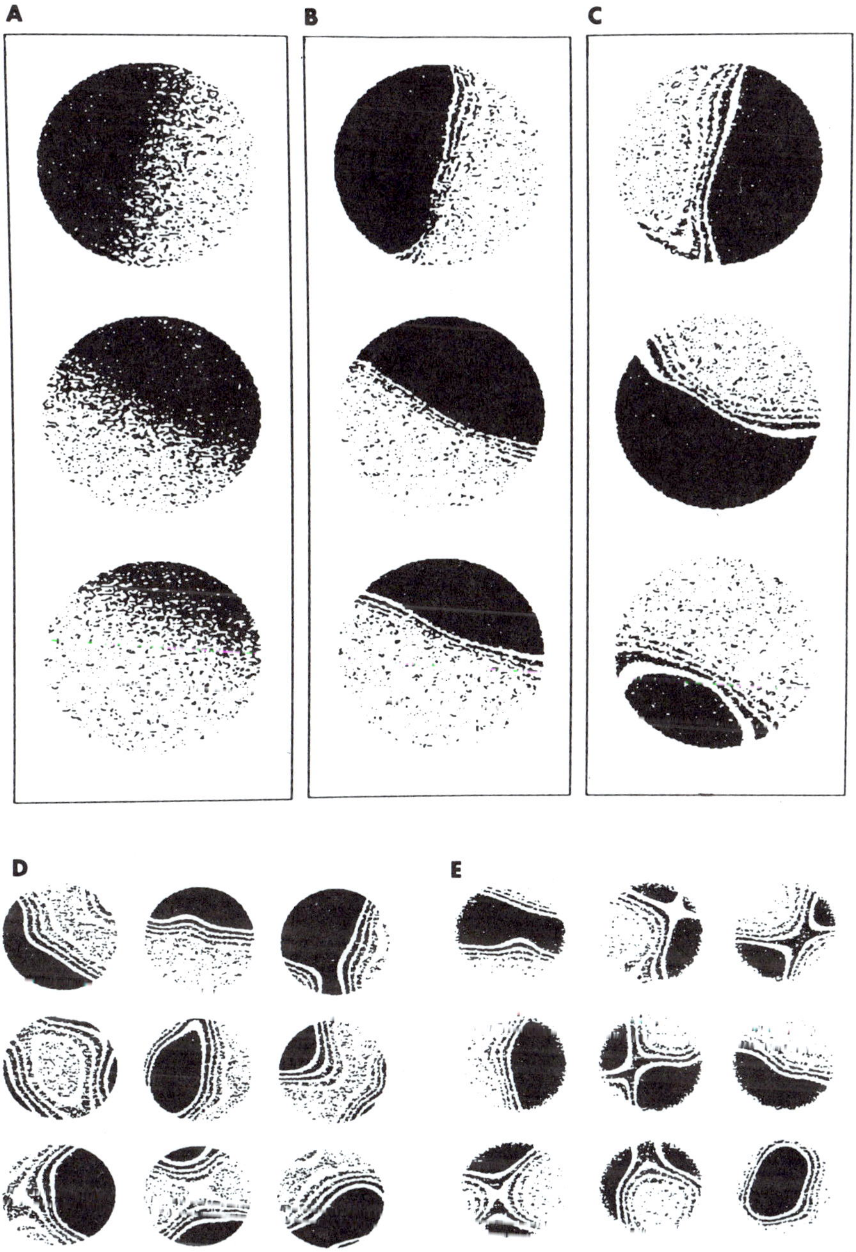

Figure 11. Results from a three-spots-in-a-row learning trial. (A) shows the response of three upper-level hidden units plotted with respect to the location of the third, i.e., the output spot. (B) shows the response patterns of the same three hidden units plotted with respect to the location of the left input spot while the right input spot is kept fixed at the center of its array. (C) is the same as (B) but with right and left interchanged. (D) and (E) show the patterns of activity for the nine hidden units in the left and right lower-level networks.

level of hidden unit encoding. The hidden units in this network represent a directed line segment connecting two points. If the two points were samples of the position of a moving object, then this would be a representation of the velocity of the object.

DISCUSSION

The concept that learning can be used to program systems to carry out specific computations is not new, but its practical implementation had to await the discovery of a truly powerful learning procedure. The work described here shows that back propagation has the power to program networks of neural-like units to do some of the kinds of computations involved in spatial behavior. One of the most important points to come out of this study is that, when the functions being learned are continuous, there is good generalization from a finite set of examples to the general function. Although the training set was finite, it was very large, generally several hundred thousand cycles. In the studies described here, no effort was made to tune the learning procedure to lower this number. The main goal was to demonstrate that it works at all.

Back propagation learning requires a target pattern so that error can be computed. In a strict sense this pattern is supplied by a "teacher"; however, for the problems studied here, the appropriate patterns are available in a natural way from the perceptron process. In the case of identity mapping this is trivially true, because the input and target patterns are the same. In the case where two samples of a moving spot's position was used to predict a third position, this third location is available after a time delay. This means that a system with appropriate time delays can learn this computation using three successive observations of the same moving stimulus. The fact that external teachers are not required for many back propagation learning tasks is of considerable biological significance. One might imagine that a good deal of evolutionary effort has gone into configuring nervous systems so that they will be programmed automatically by experience.

The power of the back propagation model, together with the ease with which many functions can be learned, raises the question of how relevant it is to real neurophysiological mechanisms. It is still too early to give a solid answer to this question. However, there are certain features of the learning procedure used here that are very unphysiological. Most prominent among these is the apparent need to propagate the error back through synaptic weights and along an axon in the wrong direction at high speed. This problem can be overcome by reconfiguring the network used for learning so that error is transmitted back through the network as a more normal neural-like signal conducted along a path of its own (Zipser & Rumelhart, in press).

Even if back propagation is not used by the nervous system, the networks that it programs may still be of interest, because they can reveal the way that computations are carried out and the accuracy to be expected. The reason this may be so is that the back propagation procedure, by minimizing error, tends to find optimal solutions. To produce biologically plausible results, the starting architecture of the system must resemble the part of the nervous system being modeled. In the work described here, the resemblance between the systems used and those known to be involved in the brain is very slight, so the results obtained are not directly applicable. Nonetheless, the results have enough biological flavor to indicate that more detailed modeling would yield more accurate predictions.

In the work described here, only simple Gaussian spots were used as stimuli. A variety of techniques have been used to see how the results would differ if more complex shapes were used as input patterns. For example, spots with random sizes as well as random positions were used. Learning tests were also run on patterns consisting of two or three spots, at all different, randomly picked locations. In all of these cases the ability of a set number of hidden units to reproduce the input image on the output declined with image complexity. However, the center of gravity of the complex images was preserved on the output with good accuracy. For example, the center error for images of three spots at random locations was .043, just about 3 times the center error observed with one spot images.

The results on stereo identity mapping are interesting, because they suggest that learning techniques may produce networks that can deal with stereo without the need for relaxation, as has been suggested elsewhere (Szeliski, 1986). The real problem is whether a way can be found to program networks to resolve disparity ambiguities that does not require an explicit, external teacher. I have done some preliminary studies with multipoint stimuli that indicate that the identity mapping paradigm used here will work using only modest numbers of hidden units. The hidden units are binocular but tend to specialize on limited regions of the retinas. It is not yet clear whether this approach can be extended to deal with difficult problems such as random dot stereograms.

REFERENCES

Andersen, R.A., Essick, G.K., & Siegel, R.M. (1985). The encoding of spatial location by posterior parietal neurons. *Science, 230*, 456–458.

Rumelhart, D.E., Hinton, G.E., & Williams, R.J. (1986). Learning internal representations by error propagation. In D.E. Rumelhart, J.L. McClelland, & the PDP Research Group, *Parallel distributed processing: Explorations in the microstructure of cognition. Vol. 1. Foundations*. Cambridge, MA: MIT Press/Bradford Books.

Szeliski, R. (1986). *Cooperative algorithms for solving random dot stereograms* (Tech. Rep. No. CMU-CS-86-133). Pittsburgh, PA: Carnegie-Mellon University, Department of Computer Science.

Zipser, D., & Rumelhart, D.E. (in press). Neurobiological significance of new learning models. In E. Schwartz (Ed.), *Computational neuroscience*. Cambridge, MA: MIT Press/Bradford Books.

CHAPTER 11

Neural Representation of Conceptual Knowledge*

JEROME A. FELDMAN
Computer Science Department
University of Rochester

How it is that the brain codes, stores and retrieves memories is among the most important and baffling questions in science. [Thompson, 1986]

INTRODUCTION

In addition to its compelling scientific interest, the neural substrate of memory is a question of considerable practical importance. There are obvious applications in neurology, but it goes well beyond that. Our understanding of how human minds incorporate and process information has a profound influence on many aspects of social intercourse including formal and informal education, psychotherapy, public information, and interpersonal communication [Roediger, 1980]. Research in the behavioral and brain sciences entails implicit assumptions about neural encoding. Cognitive science is particularly concerned with how conceptual information is treated by human brains.

Recent advances in the behavioral, biological, and computational sciences may yield a major improvement in our understanding of the neural coding of memory. Neurobiology is making remarkable strides in elucidating the structure of the nervous system and the details of its functioning. The

* Many of the ideas in this chapter arose in discussions with Geoff Hinton over the years and were crystallized in working with Mark Fanty on his dissertation proposal. Also providing valuable feedback on early drafts were Dana Ballard, Gary Dell, Horst Greilich, Brian Madden, and John Maunsell.

This work was supported in part by ONR grant No. N00014 81 K 0655, and in part by NSF Coordinated Experimental Research grant No. DCR-8320136.

behavioral sciences are developing deep structural models of language, vision, etc. and employing increasingly sophisticated experimental and simulation techniques to refine them. Computer science has produced powerful devices and profound theories of representation and computation which are supporting the study of how the complex structural theories of the behavioral scientists could be carried out by the information processing mechanism revealed by neurobiology. The neural encoding of conceptual knowledge is one central question in this enterprise.

This entire chapter is predicated on the direct neural encoding of conceptual structure—could it be otherwise? In a conventional digital computer, the wiring diagrams tell us virtually nothing about the information structure stored in the system; the computer is a general-purpose interpreter. Almost all its hardware is passive memory which is idle except when accessed by the interpreter. Many cognitive scientists still like to think of the brain in this way and, for them, the study of neural encodings makes no sense. One explores the high level structure of cognition in a way that is independent of any embodiment. A great deal can be learned in this way, and this chapter relies heavily on the results of such studies. The problem with this stance is that the speed of the system (brain) relative to that of its elements (neurons) places very severe constraints on the possible organization of knowledge. Human reactions, over a wide range of tasks, take a few hundred milliseconds, or about 100 times the switching time of individual neurons. Thus, for example, an interpreter of the standard kind would be too slow by many orders of magnitude. While there are in principle a wide range of possible realizations of intelligence, we know that our brains are constrained to use one that is quite direct. One can also view this chapter as an exploration of the relationships among three kinds of structure: the structure of the brain, the structural characterization of behavior, and the conceptual structure of knowledge. The questions addressed in this chapter are (a) What do we now know about the range of possible encodings of knowledge in the brain? and (b) What does this suggest for experimental and theoretical research? All of this falls within a research domain characterized by terms like those in Table 1.

We will pursue these questions following what has become the paradigmatic method of cognitive science—converging constraints. Neither

Table 1. Common Terms

Connectionist
Neural Model
Massively Parallel
Cell Assembly
Pattern of Activation
Parallel Distributed Processing

behavioral, biological, nor computational results alone greatly restrict theorizing, but, taken together, these results preclude many models that look plausible from a narrow perspective. The 100-step rule for simple behaviors is an elementary example of how converging constraints can lead to insights.

For concreteness, let us suppose that the problem is to describe how a human brain could represent and exploit the information in a standard encyclopedia. We assume that any representation will employ some notion of *concept* which corresponds roughly to a dictionary word-sense. The only other primitive needed is the *relation*. We assume that all the required knowledge can be captured as a collection of relations among concepts. The encyclopedia contains pictures, so some appearance models are required, but these can also be expressed in terms of concepts and relations. There is no attempt to define formally the notion of concept and relation, because the various representational schemes discussed will entail properties of the primitives. What we will do is show how different neural encoding models relate to constraints arising from various branches of science. To begin, we will focus only on the representation of concepts, ignoring relations as most of the literature continues to do.

Even leaving relations aside, the literature contains a wide range of notions of what a concept is and how it might be represented neurally. The simplest view restricts consideration to very concrete nouns such as *horse* or *chair*, ignoring more complex concepts such as *game*, *yesterday*, *active*, or *love*. Even in this quite restricted context, the representation of concepts has proven to be a deep problem for philosophers, linguists, psychologists, etc. [Smith & Medin, 1981]. Our point of departure is the computational models of artificial intelligence, which have made explicit many of these issues in conceptual knowledge [Charniak & McDermott, 1985]. A minimum requirement is that the representation support the answering of questions about concepts (and later about their relations). One particularly simple kind of question concerns the structure of the object itself, e.g., how many legs has some chair. Models which treat concepts as unstructured collections of attributes will need to provide additional mechanisms for answering even these simple questions. Many studies of neural concept representations ignore entirely the issue of how concepts are used, but no serious attempt to understand real cognition can afford to do so.

The range of possible representations of concepts in the brain is constrained by a combination of computational considerations and neurobiological findings. The 100-step rule for simple tasks already eliminates any conventional computer model. Other pertinent facts include the relatively small number of neurons (about 10^{11}, or one hundred billion), the large number of connections between neurons (about 10^4 per unit), and the low rate of information transfer. It may seem that 10^{11} is not a small number, but when one considers the 10^6 input fibres from each eye, a computer scientist immediately detects a major constraint. For example, dedicating one unit to test for a possible line between any pair of points in the retina would take more neurons ($(10^6)^2$) than there are. The information rate between individual neurons at a firing rate of 100 spikes per second is about five bits, or enough to encode one letter of the alphabet. (If complex messages are being conveyed, it is not by individual neurons.) Much of the computational power of the brain derives from its great connectivity, and a challenge to theory is to explain how this power is realized. Although this chapter concentrates on representation more than learning, there are also constraints on plasticity that are pertinent. The major finding is that the growth of new fibers in adults is much too slow and constrained to account for learning, and that there is no generation of new neurons. There is sufficient converging evidence [Lynch, 1986] to allow us to assume that long-term concept and relation memory comes about through changes in connection strength, but rapid connection change is problematical. There is also reason to believe that skill learning may involve significantly different mechanisms than concept learning [Thompson, 1986].

The best way to begin a serious discussion of neural encoding of concepts is to tie down two simple theories that embody the extreme ends of the range of possible answers. The most compact representation possible would be to have a unique unit dedicated to each concept. If we assume that a unit corresponds to one neuron, then this is the *grandmother cell* or *pontifical cell* theory. The other extreme would be to have each concept represented as a pattern of activity in all of the units in the system. This is well known as the *holographic* model of memory and is the most highly distributed theory that we will consider. In addition to the pure theory based on optical holograms, we will call holographic any model that has all the units in a system encoding each concept [Willshaw, 1981]; most of these are matrix formulations. Nothing much would change in either theory if a unit corresponded to a dendritic subtree instead of an entire neuron. The discussion will proceed by moving in from the extremes to examine the range of plausible encoding models and some of their properties. A great deal of excellent work has been done employing the end-case assumptions, but neither of them could actually be right.

The extreme opposing models of neural representation lead to radically different views of many aspects of cognitive science. Table 2 presents a number of contrasting terms that arise, respectively, from the punctate and fully distributed views of neural coding. Not all of these items will be meaningful to every reader, but everyone should recognize some striking contrasts in perspective. All of these contrasting notions will find their way into the discussion. One term that is possibly misleading in Table 2 is *distributed representation*. The problem is that people have been using this term to denote

Table 2. Contrasting Terms

punctate	diffuse
local	distributed representation
grandmother neuron	hologram, spin-glass
disjoint codes	homogeneous code
detector	filter
labelled line	pattern of activity
active memory	passive memory
reduction	emergence
hierarchy	complete connectivity
recruiting	adapting
general computation	correlation

everything from a fully holographic model to one where two units help code a concept; thus, the term has lost its usefulness. The various contrasting terms often accompany significant differences in research goals and strategies. One intriguing idea that we will pursue is that many of these apparent conflicts are basically alternative ways of looking at the same set of phenomena, vaguely reminiscent of atomic physics and thermodynamics.

We should first dispense with an abstract argument that equates the punctate and hologram models. It is true, in a sense, that an encoding having one active unit per concept is a pattern of activity in the mass. But this identification is too abstract to be meaningful. Another proposed way to identify holographic and punctate representations comes from linear algebra and the idea of alternate coordinate axes for a vector space. If, as in many models, the output of a unit is the (thresholded) linear combination of its inputs, one can view this unit as a (very large) vector, v, whose coordinates are the outputs of each predecessor unit. There is, in principle, another set of bases for the vector space for which this vector, v, is an axis and can therefore be represented by one nonzero coordinate. This argument fails for three reasons. Even for strictly linear input combinations, the output threshold destroys the applicability of linear algebra, and there is no biologically plausible way to eliminate nonlinearity. In addition, no single transform would work unless the set of concepts were independent and therefore small (cf. Section 2). More importantly, the computational properties of the two representations are radically different, as with the example just above.

Punctate Models

The next step is to show why neither the pure punctate nor the holographic model is consistent with the facts. We will start with the punctate model, because the story here is simpler.

The extreme end of the compact representation position is to assume each concept is represented by exactly one neuron. This view received considerable support from single unit recording research, which found that units in sensory areas responded best to a relatively narrow class of stimuli [Hubel & Weisel, 1979]. The punctate encoding is also called *labelled lines*, emphasizing the fact that, in this encoding, each axon will be conveying a specific message when it is active. The most influential expression of this position was Horace Barlow's "neuron doctrine," which is worth quoting in its outline form [Barlow, 1972].

> The following five brief statements are intended to define which aspect of the brain's activity is important for understanding its main function, to suggest the way that single neurons represent what is going on around us, and to say how this is related to our subjective experience. The statements are dogmatic and incautious because it is important that they should be clear and testable.
>
> *First dogma*
>
> A description of that activity of a single nerve cell which is transmitted to and influences other nerve cells, and of a nerve cell's response to such influences from other cells, is a complete enough description for functional understanding of the nervous system. There is nothing else 'looking at' or controlling this activity, which must therefore provide a basis for understanding how the brain controls behavior.
>
> *Second dogma*
>
> At progressively higher levels in sensory pathways information about the physical stimulus is carried by progressively fewer active neurons. The sensory system is organized to achieve as complete a representation as possible with the minimum number of active neurons. [Cf. Figure 7]
>
> *Third dogma*
>
> Trigger features of neurons are matched to the redundant features of sensory stimulation in order to achieve greater completeness and economy of representation. This selective responsiveness is determined by the sensory stimulation to which neurons have been exposed, as well as by genetic factors operating during development.
>
> *Fourth dogma*
>
> Just as physical stimuli directly cause receptors to initiate neural activity, so the active high-level neurons directly and simply cause the elements of our perception.

Fifth dogma

> The frequency of neural impulses codes subjective certainty: A high impulse frequency in a given neuron corresponds to a high degree of confidence that the cause of the percept is present in the external world.

The five dogmas are much less controversial in their original form than in most of the strawman characterizations derived from them. Obviously enough, much of their force is in defense of the direct encoding position shared by all connectionist models. The most relevant point in our discussion is that individual neurons are the appropriate object for study in determining how the brain does its work. This is commonly identified with the most compact, punctate, style of neural model and is an appropriate introduction to this end of the spectrum.

Figure 1 (after Shastri, 1985) presents a vastly oversimplified version of how a punctate system might encode and exploit conceptual knowledge. The memory network is a category-based hierarchy with each concept and property-name represented by a rectangular unit. The triangular nodes stand for intermediate units which become active when two of their three input lines are active. Suppose the system has a routine that retrieves its knowledge of food tastes as an aid to ordering wine, such as that cartooned in the lower half of the figure. If activation is spread simultaneously to the *main course* of the meal and to the desired property *has-taste*, exactly one triangular evidence node, b_1, will receive two active inputs and this will lead to the activation of the concept *salty*. This is the required answer, but for technical reasons an intervening clean-up network is needed where the answer is acutally put to use. One interesting feature of Shastri's model is that the same memory network is able to classify a salty, pink food as ham—the triangular evidence nodes work in both directions.

While oversimplified, Figure 1 does convey much of the flavor of punctate (and other compact) connectionist models and their appeal to some scientists and rejection by others. The main point is that everything is quite explicit; the concepts, properties, and even the rules of operation are simple and direct. This makes it relatively easy to express and test specific models either at the neural level or more abstractly, as in our example. No one believes that the brain uses exactly the structure of Figure 1, but any highly compact concept representation could behave in essentially the same way. The very explicitness of all this is what makes many scientists reject punctate models either for abstract study or as a theory of brain structure. While any particular structure or theory can be encoded (just like circuit design), nothing may be learned about the general properties of intelligence. Moreover, how could such a hard-wired system develop and adapt in living brains? We

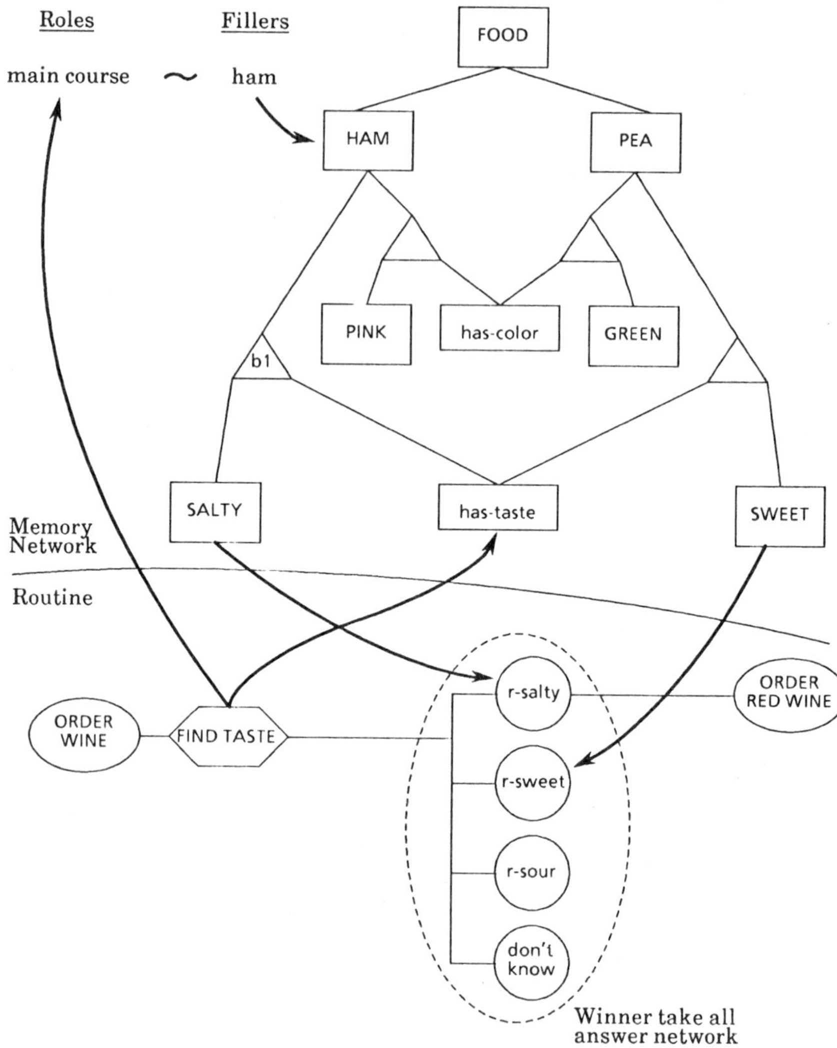

Figure 1. Connectionist Retrieval System

will discuss this last issue in Section 3; the others are questions of scientific taste and judgment. But the answer to how the brain represents knowledge is not a matter of taste, and we next explore some arguments that it cannot be in the punctate style of Figure 1.

The first point is that a large number of neurons ($\sim 10^5$) die each day, and these are distributed throughout the brain. If each concept were represented by exactly one neuron, one would expect to lose at least some concepts (at random) each day. This argument is often taken to be conclusive evidence against the compact model, but there is a slight variant of the punctate view that is proof against the death of individual units. Suppose that, instead of one unit per concept, the system dedicated three or five, distributed somewhat physically. All of the theories and experiments would look the same as in the one unit case, but the redundancy would solve the problem of neuron death. While the number of neurons dying is large, the fraction is quite small ($\sim 10^{-6}$), so the probability of losing two of the representatives of a concept in a lifetime is quite low ($\sim 10^{-7}$). There is also considerable redundancy at the system level, several separate ways to perform a given function. As we will see, there are other reasons for rejecting the punctate model, but the death of neurons is not critical.

Another, and more telling, argument against a purely punctate model is that there would not be enough units to represent everything. It is fine to envision a unit that is active when you recognize your grandmother's picture in the encyclopedia, but what about all your other memories of her? Is there a separate unit for each age, each outfit, each body position, etc.? There are two aspects to this argument, the sheer number of things to be represented and the relational structure among them. Merely representing all the possible concepts is not a problem if efficient encodings are used. For example, a group of 100 binary units can represent $2^{100}(\sim 10^{30})$ distinct patterns, which is far more than the number of concepts required. A million such groups is ample memory and is a small fraction of the neurons available ($\sim .1\%$). The most efficient coding information is not the main design criterion, but there are constraints which preclude some encodings as too wasteful. Supporters of the compact end of the spectrum have conceded this and have developed coding techniques which permit the encoding of significantly more information without a major change in computational architecture.

In parallel with these theoretical developments, experimentalists have been reframing their view of the activity of single units. For some time, the idea of a neuron as a "detector" of one kind of event has been declining, but no alternative term has yet evolved. The psychophysicist's idea of a *filter* is the diffuse equivalent (all units filter all signals) and equally misleading. Although no new word has been established, many experimentalists now (correctly, in my view) view sensory neurons as having responsivity of different fineness to a variety of stimulus dimensions. Single unit neurophysiology studies are finding effects of stimuli beyond the classical receptive field [Allman, Miezen, & McGuiness, 1985], as is inevitable in an interacting system of units. And, of course, any idea that all of intelligence could be understood simply by finding the neuron for each concept was always silly. It is clear that Barlow's dogmas could use some revision, and Appendix A suggests one variant which seems reasonable. For example, dogma five should reflect the fact that firing frequency encodes some stimulus feature information as well as confidence. Figure 2 shows the firing rate of a visual system

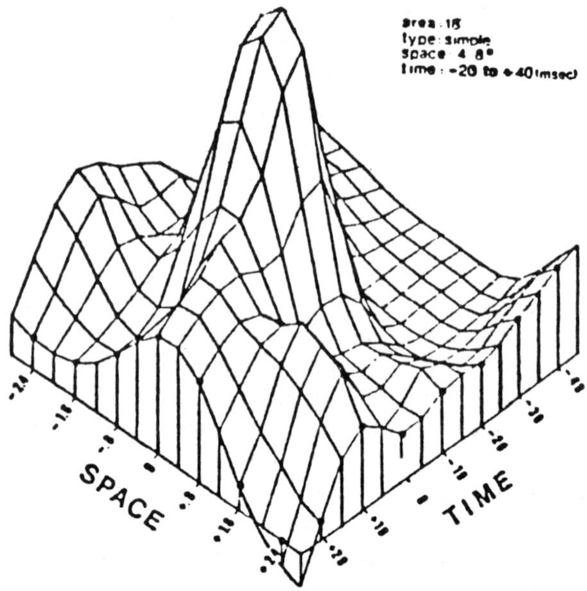

Figure 2. Response of a cat visual (area 18) cell to a binocular stimulus as a function of temporal and spatial offset. Spatial offset is from -2.4 to 2.4 degrees, temporal from -20 to +40 m.sec. (From Gardner, Douglas, & Cynader, 1985.)

neuron as a function of binocular temporal and spatial offset of a target, and is typical. When one includes the firing rate (instead of just on/off) as part of the information code, the range of distinct entities representable goes up significantly [Ballard, 1986b]. For example, the relative activity rate of three color channels is enough to specify a wide range of hue and intensity combinations. Summarizing the discussion to this point: While the purely punctate view is unsupportable, there is no numerical problem with a theory that has each concept represented by the activity of a few units. Such an encoding shares many of the properties of the punctate model and is consistant with single unit experiments in a wide range of brain structures.

Another argument used against compact models of neural representation arises from bulk activity experiments. A single small stimulus can give rise to activity in a significant fraction of the total population. There are several reasons why this fact does not preclude compact representations. For one thing, the simplicity of a stimulus looks different from various encoding schemes. A small dot, in a Fourier encoding, activates receptors for all spatial frequencies. In a parameter-space encoding [Ballard, 1986a], a single feature provides evidence for many higher-level features and these may be suppressed more slowly than when more information is present. Finally, there does appear to be some nonspecific gating [Barber, 1980] which activates an entire structure when any input appears. None of this is to say that the punctate position is correct, only that arguments from bulk activity do not preclude its viability.

While the cell-death and bulk activities arguments against punctate representation are easily accommodated, it is easy to see that there really could not be enough neurons to have one for each concept of interest. One clear example arises in early vision. It is well known that the visual system is sensitive to at least the following local stimulus properties: orientation, intensity, hue, depth, motion direction, and size. A system that could resolve 10 values for each of the six dimensions would require 10^6 units to represent all combinations of values. But there are about 10^6 separate points at the narrowest (retinal ganglion) level that must be represented, so the total requirement would be 10^{12}, which is too many. Similar combinatorial arguments can be made at higher level conceptual levels, as in all the memories of one's grandmother. At these higher levels, the structure of the knowledge itself provides considerable encoding economy (Section 3), but for early processing additional mechanisms are required. In fact, quite a lot is known about the response properties of cells in visual cortex, and there is a clean computational story that both helps organize the experimental data and contributes to our goal of constraining the possible models of neural encoding.

The basic idea is depicted in Figure 3 for the case of two stimulus dimensions. Suppose (as turns out to be the case) that units respond nonuniformly to various stimulus dimensions. For example, the vertical rectangle in the lower left depicts the responsiveness of a cell that is five times more sensitive to size than to orientation and the horizontal rectangle the opposite. The nice point is that the joint activity of two such cells can code the stimulus space as finely as the finest dimension of either (cf. the crossed rectangles), while requiring significantly fewer units. This computational mechanism goes by the name *coarse–fine coding* and appears to describe a good deal of neural computation. In general, given K stimulus dimensions, each with a desired resolution of N values, the punctate encoding requires N^K units. A course–fine encoding with the coarse dimensions D requires a total of

$$T = K \cdot N \left(\frac{N}{D} \right)^{K-1}$$

units. This is because there will be K separate tilings (covers) of the K-dimensional space, but each will be covered coarsely and this requires (N/D) units in all dimensions but one, which has N units. This formula still grows exponentially, but is significantly smaller for the cases of interest. For example, our early vision example had K = 6 and N = 10. With D = 5, this yields 1920 units per point instead of the 1,000,000 for the pure punctate encoding. Since I believe these ideas to be central, a number of related issues will be discussed next.

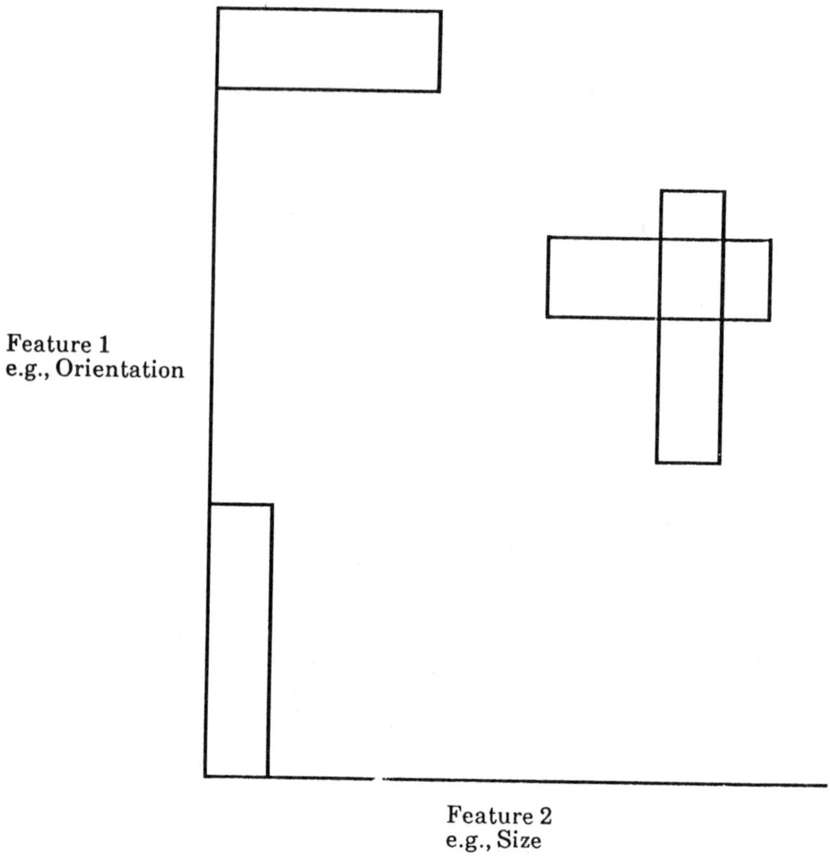

Figure 3. Coarse-Fine Coding

The critical point in the construction is the overlap of receptive fields, not their asymmetric shape. Essentially the same arguments can be made for symmetric overlapping fields, and this is known as *coarse coding* [Hinton, 1981b]. Computational ideas of this kind have been known for some time to provide a nice account of hyperacuity, the ability of people to resolve details finer than the spacing of their receptors. In both coarse and coarse–fine coding there is a price to be paid for saving all these units. If two stimuli that overlap co-occur, the system will be unable to resolve them. This situation is depicted in Figure 4. Suppose one stimulus is encoded by the two rectangles labelled X and the other by the two rectangles labelled Y. The intersections labelled X and Y encode the desired information, but the "ghosts" labelled G would be equally active. One of these expresses the conjunction of Y-size and X-orientation and the other the converse. This is an instance of *cross-talk* in neural encodings. Cross-talk is the fundamental problem of shared encodings and appears to be a critical limiting factor on distributed models,

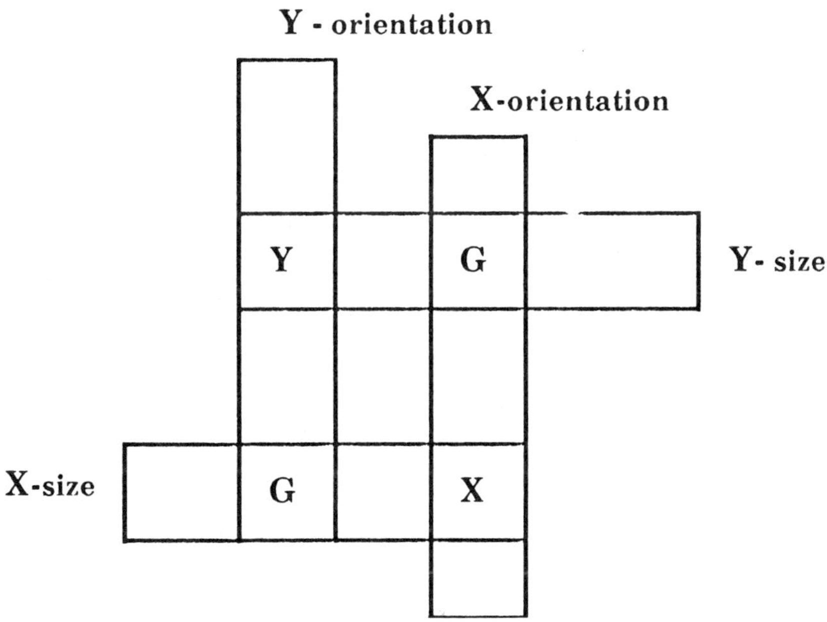

Figure 4. Ghosts in the Machine

as we will see. For the simple case of coarse coding, some analysis of the trade-offs has been carried out [Sullins, 1985].

A final point on coarse coding is that the multiple-dimension, fine-grained information is left implicit in the representation. The joint activity of several units encodes the desired information, but how can subsequent computation make use of it? This is another critical issue in shared encodings and a major reason why no holographic model has gone more than one level. In the case where the information is carried by the activity of a small number of units (K ~ 10), there is a simple and biologically plausible solution. Suppose that the inputs to a unit (neuron) were not all treated uniformly, but were grouped into *sites*, each of which computed separately a combined input value. If each site computed the logical AND of all its inputs, this would be an ideal receiver for the joint output of coarse coding. A cartoon of how this might go is shown in Figure 5, which shows a fragment of an abstract multiplication table for integers between 1 and 9. Only one inhibitory link is shown. Units representing possible answers (e.g., 12) have separate conjunctive receiving sites and take on the activation of the maximum (or logical OR) of the sites. There are two reasons why this general idea provides a major saving over punctate encodings. One reason is that one computational role of units has been brought down to sites, significantly increasing the feasible numbers (by perhaps 1000). The other is that not all combinations of outputs need to be explicitly represented. Although it is well beyond the current story, one can

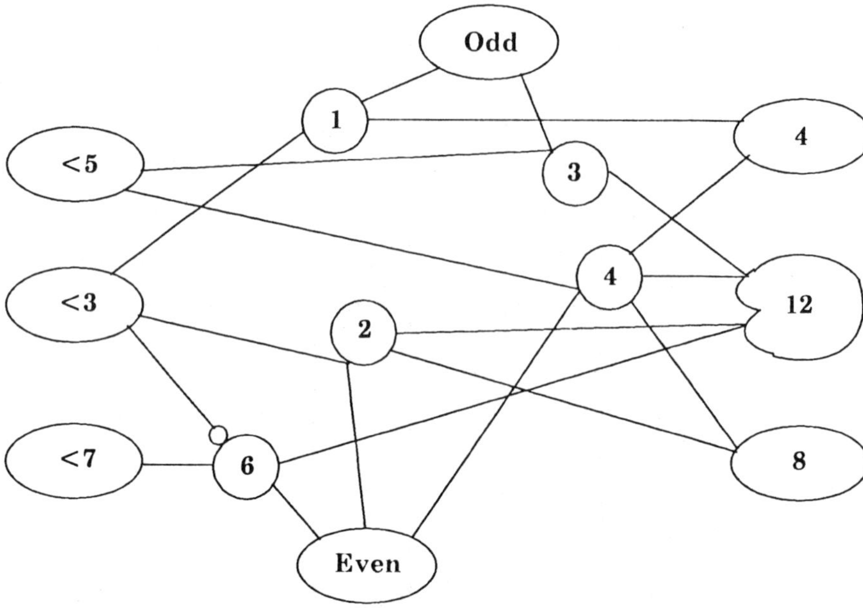

Figure 5. Multiplication Table for Integers

envision how the coarse–fine coding units could combine to represent the feature space with different resolutions.

It is interesting to consider if we could eliminate the circular nodes carrying the punctate number representations and deal directly with the distributed representation based on number range and odd–even parity. The obvious solution is to link directly the appropriate combinations to the receiving sites. This works fairly well, but does encounter some problems. Suppose 1 and 4 were simultaneously active. Our distributed representation of 2 is <3 and even, both of which are active in this case, and this could lead to activation of the answer 8. This problem can be fixed, e.g., by having separate sites for the first and second arguments, but it does point out the delicacy of doing computation with distributed representations.

One final discussion will close off this path of consideration. The coarse–fine coding examples used overlapping encodings, but were based on the minimum possible number of cells to cover some feature space. Suppose instead we allowed for redundancy in the coverage, say three separate tesselations of the space. In terms of Figure 3, a second covering could be similar bars at 45° and 315° to the axes [Ballard, 1986b]. We could still use separate receiving sites for each desired stimulus, but the computation would be not just a logical AND. In fact, a thresholded sum of activity might be quite plausible as the sites' way of computing the likelihood of its combination being present. This would combine the error tolerance and information reduction ideas in a simple and plausible way. Edelman has come to essentially the same kind of model through a very different route, starting from his expertise in immunology [Edelman, 1981]. Since this is how I believe the brain really works, I will next attempt to show how the holographic approach narrows down (sic) to the same solution.

Holographic Models

Holographic models have been fervently supported by biologists, psychologists, and theoreticians. There is no comparable fervor for compact models. In fact, several researchers preach fully distributed models while employing punctate ones, often in the same work. To cite one well-known example, the elegant travelling salesman model of Hopfield and Tank [1986] is purely punctate and unmappable to Hopfield's holographic memory proposals [1982]. One major contributing factor in the popularity of this view was the early work of Karl Lashley, who found that, for a variety of tasks, the deficit exhibited by lesioned rats was best explained by the total amount of cortex removed—the "law of mass action." Lashley summarized his view of memory representation in the classic 1950 paper "In Search of the Engram." The following quotation continues to motivate much current work [Lashley, 1950]:

> It is not possible to demonstrate the isolated localization of a memory trace anywhere within the nervous system. Limited regions may be essential for learning or retention of a particular activity, but within such regions the parts are functionally equivalent. The engram is represented throughout the area. . . . Briefly, the characteristics of the nervous network are such that when it is subject to any pattern of excitation, it may develop a pattern of activity, reduplicated throughout an entire functional area, by spread of excitations, much as the surface of a liquid develops an interference pattern of spreading waves when it is disturbed at several points. . . . Consideration of the numerical relations of sensory and other cells in the brain makes it certain, I believe, that all of the cells of the brain must be in almost constant activity, either firing or actively inhibited. There is no great excess of cells which can be reserved as the seat of special memories. The complexity of the functions involved in reproductive memory implies that every instance of recall requires the activity of literally millions of neurons. The same neurons which retain the memory traces of one experience must also participate in countless other activities.
>
> Recall involves the synergic action or some sort of resonance among a very large number of neurons. . . . From the numerical relations involved, I believe that even the reservation of individual synapses for special associative reaction is impossible.

Since Lashley's time, the intricate specialized structure of mammalian cortex has been elucidated, and no one currently holds the view that all of cortex is holographic. (Reading Lashley's article makes it clear that he would reach very different conclusions on current experimental evidence.) Since all of the primary sensory and motor areas have been found to have specialized structure, the holographic hypothesis is currently restricted to "higher" brain areas whose functional organization is not yet understood. Another historical source of motivation for diffuse models was the wonderful 1949 book, *The Organization of Behavior*, by Donald Hebb. Hebb introduced the notion of *cell assemblies*, but was (appropriately for the time) vague about how they actually encode knowledge. Hebb definitely envisioned a dynamic pattern of activity, but there are two interpretations of this. The most literal would be to assume concepts are purely dynamic and are not tied to any particular tissue; this idea has been pursued a bit [von der Malsberg, 1985; Bienenstock, 1985], but without much success, and the experimental data is not encouraging. A more general notion of dynamic activity of cell assemblies is inherent in all current connectionist theories; the compact–diffuse question is about what fraction of a system is involved. A closely related question is structure; it does not seem useful to think of a computer as a *chip assembly*. We will return to this after examining theoretical hologram models.

Holographic models are theoretically attractive because of two properties: fault-tolerance and generalization. The brain clearly has these properties, and it is easy to see informally why holographic models do also. If all of the units of a large system are involved in coding one concept, the failure of some of them can be tolerated. Furthermore, two concepts which share much of their activity patterns will tend to behave similarly. More technically, the theoretical hologram models have all been based on one form or another of mathematical correlation. An excellent compendium of holographic-style models can be found in Hinton and Anderson [1981], and I will often cite this rather than the original papers.

The purest holographic model is, unsurprisingly, the optical hologram itself. A nice presentation can be found in Willshaw [1981], where it is also shown that the purely linear hologram is undesirable on both computational and biological grounds. The models typically studied are large rectangular matrices representing all the possible connections between m input units and n output units, which need not be distinct (cf. Figure 6). The most common case, and one of the easiest to consider, is where each unit compares the sum of its weighted inputs to its threshold, and emits 1 if the sum of inputs is greater and 0 otherwise. A concept is represented as a binary vector over all the input lines. Each element of the vector may be uninterpreted or can be thought of as the presence or absence of a microfeature characterizing the concept. (A punctate view of this would be to imagine each output unit as being responsible for computing one bit of the output, using all of the inputs according to its weights on them.) If the input and output are identified, the matrix becomes a pattern-completion machine. The basic idea is simple and derives from the mathematical notion of correlation. The correlation of two binary vectors u and v is simply:

$$\sum_{j=1}^{n} u_j v_j$$

and this is obviously maximized when u = v. In a well-distributed set of vectors, an input distorted by modest noise or omission will correlate best with the appropriate complete vector. This is the basic source of error resistance and generalization in holographic-style models. What makes them interesting is that a connection matrix with the appropriate weights can be learned with a simple, local procedure. Most of the learning discussion is in Section 3, but this case is so central that it will be done here.

For the auto-associative, pattern completion case we want weights w_{ij} such that

$$X_j - \sum w_{ij} X_i \Big|_0^1$$

assuming the threshold is included as another weight. The natural local weight-change rule is to increase the weight between two units when they both are active (output = 1). This is essentially Hebb's rule and can be interpreted as increasing the weight of units whose firing is correlated. Symbolically:

$$\Delta w_{ij} = (X_i \cdot X_j) \cdot \delta \qquad \text{where } \delta \text{ is a constant.}$$

If all the weights are initially set to zero and the autoassociation matrix is trained on some P sequences, then

$$w_{ij} = \delta \cdot \sum_{k=1}^{P} X_i^{(k)} \cdot X_j^{(k)}.$$

The resulting matrix can be shown to be optimal in some cases (orthonormal vectors) and is a good estimator in many others [Kohonen, 1984]. Thus a simple and biologically plausible local learning rule yields a matrix which reliably finds the best match to an input pattern under some distortion conditions (but not others). The binary, linear autoassociator is the simplest of wide variety of essentially similar models. Many variations have been tried on all aspects of the model [Hinton & Anderson, 1981; Kohonen, 1984; Barto & Anandan, 1984]. The more refined models use a variant of Hebb's rule which allows feedback from some external result to affect the weight change process [Kohonen, 1984; Barto, Sutton, & Brouwer, 1981; Anderson & Murphy, 1986]. The shortcomings of the methodology do not lie in the learning rules; they demonstrate that essentially anything that can be represented as a linear threshold matrix can be learned by correlation—the problem is that the representation itself is much too weak. There is a well-known relation between correlation matrices and classical pattern recognition, and the same basic inability to deal with structure, occlusion, and invariance applies in both cases.

There are a number of computational considerations which rule out any existing holographic encoding and make it extremely unlikely that one will be discovered. Before examining these in detail, we must eliminate a trivial path of escape sometimes sought by defenders of the holographic models. It is a truism of computation theory that any universal mechanism can simulate any other if computational costs are ignored. Since a large enough (associative or other) memory can be made universal, one can find a way to encode any computation as a holographic memory plus a small interpreter. But ignoring computational costs is precisely what we cannot do in the current enterprise—computational costs are one of the principle constraints on the viability of neural representation schemes. Much of the holographic work has been directed to purely passive memory networks, and we will examine these, but it is important not to forget that no passive memory will satisfy the basic time and competence constraints for conceptual knowledge. A sequential machine with a fast associative memory is still orders of magnitude off the required performance.

When computational issues are taken seriously, the holographic model is fatally deficient, even as a passive associative memory. The basic problems with any holographic representational scheme are cross-talk, communication, invariance, and the inability to capture structure. Essentially the same problems have prevented the development of holographic computer memories or recognition systems, despite considerable effort. Consider the problem of representing a concept like *grandmother* as a pattern of activity in all the units in some memory network. We can assume that different parts of the network represent various input modalities (e.g., vision) and they all lead to the same overall pattern. But notice what would happen if two (or more) concepts were presented at the same time, e.g., grandmother at the White House. The encodings for the two concepts would normally overlap and the system would get garbled. This is a massive instance of the cross-talk problem of Figure 4. Of course, we can reduce the probability of cross-talk by having fewer units active for each pattern. Suppose that there are punctate output detectors for each pattern in the holographic memory (nothing

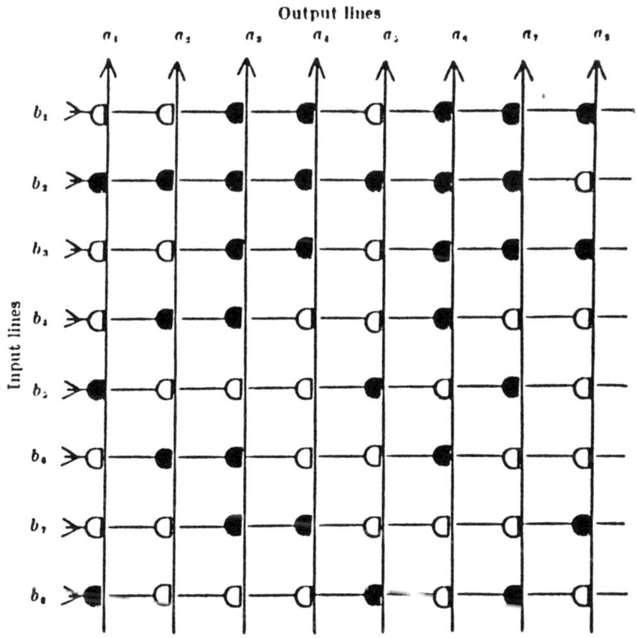

Figure 6. An Associative Net. The nodes that have been activated in the storage process are colored black. (From Willshaw, 1981.)

becomes easier if there are not). Willshaw [1981] addressed the question of how one could arrange the coding and detector thresholds so that only the desired detector would respond to each pattern. If one assumes that the cross-talk is randomly distributed (the best case), the system will be reliable for individual retrievals only if the number of units active for each pattern is proportional to the logarithm of the number of units in the diffuse memory. This means that a network of 1,000,000 units should use an encoding with about 20 active units per concept—this is suspiciously close to the number that would arise from a redundant, coarse-coded compact approach.

If a fully distributed encoding is used for concepts, there is an unsolved problem of how conceptual information could move from one subsystem, e.g., vision, to another, such as speech. The obvious encoding would be with a *bus* or group of links as wide as the holographic representation, which is unrealistic. In fact, there are many fewer long distance than local connections in the cortex. If only some of the units were linked, they would constitute a more compact representation of the concept, which violates the holographic assumption. Even so, unless these encodings were largely disjoint, communication over the bus would have to be sequential to avoid cross-talk. If cross-talk were present, people would make mistakes like seeing a horse and a chair and mistakenly saying something irrelevant like *apple sauce*. If microfeatures were the basis of the representation, then something like *rocking horse* should result. The idea that visual information is conveyed to other parts of the brain one concept at a time clearly violates timing constraints by orders of magnitude. The notion that *attention* can restrict the system to one concept at a time is not tenable. Consider how a cowboy movie is mapped from vision to concepts. What is the concept for each frame: the posse, the horse and rider and saddle and hat, etc.? How do they fit together to keep a gestalt of the scene and scenario? What is happening with all the other visual information? How does attention know what to transmit unless the structure of the scene and story line are already known at the vision end? And, of course, we would be back at a sequential machine model. There appears to be no alternative to assuming that, at least for communication, representation of concepts must be largely disjoint and thus compact.

The same communication problem would arise within the concept memory itself if one tried to build a knowledge structure like Figure 1 with a diffuse representation. If a concept like *salty* were represented only by a large pattern, the links for this entire pattern would have to go to all the places that related to saltiness—and be treated correctly at each of these. If a concept encoded by N units needed to be linked to M other concepts, a total of M*N links would be needed. One punctate intermediate unit can reduce the requirement to N + M connections. The more distributed the representation, the more serious this problem becomes. Again, any serious reduction in this wiring requirement would constitute a compact representation. And, as in the intermodal case, unless these representations were largely disjoint, concept processing would have to be sequential to avoid cross-talk. This eliminates the spreading activation and massively parallel processing which was the motivation for the whole idea.

Even so, no one has suggested how to represent any but the simplest concepts in the holographic style. For example, the problem of all the different historical views of one's grandmother is as difficult for the hologram as for the punctate model. In fact, it is far from obvious how to make the same distributed pattern active for alternative views of a chair, even without occlusion. One could, I suppose, assume that there are essential invariant features of one's grandmother that are derived (magically) from all the different pictures, stories, etc. involving her. Even so, how would one express the structure of concepts, like the fact the grandma has two legs, the left of which is slightly shorter. All concepts in any holographic structure that I know of are totally without internal structure. There is an idea of associating the components of a representation with microfeatures (and this will be discussed later) but these are still unstructured. Nor do any of the holographic proposals provide a way of answering even simple questions like the color of grandmother's hair (let alone at different ages). There are a number of other problems with holographic models, but this should suffice to show that this end of the compact–diffuse spectrum is no more viable than the punctate end.

The biological evidence against anything vaguely like a holographic model is equally compelling. This is only fair, since the holographic hypothesis denies any relevance to neuroanatomy and physiology. Since intricate specificity and detailed visuotopic, tonotopic, etc. maps have been discovered everywhere in the brain that has been examined, the only hope for the hologram is for some higher association areas. Even there, the anatomical structure has been found to be much like the sensory areas [Goldman-Rakic & Schwartz, 1982] and nothing like the connection of each input to each output required by matrix models. Simple counting arguments show that at most 10,000 neurons could be in a matrix-like network and the local connectivity precludes this possibility as well. Neurosurgery relies upon precision stimulation and recording to localize lesion sites. There may be a point to studying fully distributed encodings, but direct mapping to the brain is not among them. Even Kohonen [1980, p. 132], who is a major figure in holographic-style modelling says:

> There are good experimentally verified reasons to assume that for a particular sensory experience or other occurrence, the pattern of activity over the complete memory field consists of only a few activated local areas . . .

Another argument for highly distributed representations derives from the large number of input fibres ($\sim 10^4$) to cortical neurons. If all these fibers participate actively then, ipso facto, the representation is diffuse. There are three reasons why this argument fails. While there has been no definitive study on the number of presynaptic events required for neural firing, estimates gleaned from papers and conversations run from one event to a few dozen. No one has suggested that several thousand synapses must fire at once for an action potential. Also, we can see from Figures 5 and 7 that many of the connections could represent alternative ways of activating the same concept (e.g., from different points in space). Another way of looking at this is that the thousand-fold connectivity is capturing an "OR" of activation conditions rather than an "AND." Finally, as we show in Section 3, learning in a connectionist system requires the potential for many more connections than are ever made functional. The most striking physiological demonstration of this principle is in the neural reorganization studies of Merzenich, et al. [1984].

All of this may seem to be flogging a dead horse model, but purely holographic theories continue to be seriously proposed. There has been a very recent flurry of interest in spin-glasses as holographic memory models in the theoretical physics community [Hopfield, 1982; Toulouse, Dehaene, &

Changeux, 1985]. Any physical system will, in isolation, reach some stable state, and each of these states could be looked upon as encoding a different concept; this is obviously a massively parallel system. The spin-glass story is beyond the scope of this article, but the key idea is that spin-glasses are idealized materials which can take on many stable states in ways that are mathematically interesting. The hope is that analysis of these will provide insights into the behavior of the brain and/or the design of parallel hardware. This might well happen, but the current spin-glass models share the inherent deficiencies of all holographic representations and have no direct applicability to the neural representation of concepts.

There are two basic ways of adding the required structure to hologram-like models. One can either add structure to the collection of units (cf. Figure 7), or one can try to construct structures out of components, each of which is an unstructured cell assembly. Both approaches lead to systems quite like the redundant, coarse-coded approach. Consider the latter idea: Nothing as concrete as Figure 1 has been attempted, but one envisions about 100 (Palm) units in an assembly jointly encoding some unspecified number of concepts. But we know that the cross-talk and coarse-coding constraints limit the range of possible number of concept nodes to a few hundred. Of course, the units in the assembly cannot represent the detailed structure of the concept, because they are shared by unrelated concepts. The structure is represented by connections (also never specified) between cell assemblies which presumably play a role equivalent to the evidence links in Figure 1. This move preserves the sovereignty of the cell assembly, but only in a titular fashion—the work is all done by the connections among assemblies.

The alternative move, adding structure to the assembled masses, comes in many variants and is treated in the next section.

THE MIDDLE GROUND AND BEYOND

> It is difficult to interpret such findings, but I think that they point to the conclusions that the associative connections or memory traces of the conditioned reflex do not extend across the cortex as well-defined arcs or paths. Such arcs are either diffused through all parts of the cortex, pass by relay through lower centres, or do not exist. [Lashley, 1950, p. 461]

> It is clear that various forms and aspects of learning and memory involve particular systems, networks and circuits in the brain and it now appears possible to identify these circuits, localize the sites of memory storage and analyze the cellular and molecular mechanisms of memory. [Thompson, 1986]

We have seen that neither the purely punctate nor the fully holographic model are at all plausible as theories of how the brain represents conceptual knowledge. A variety of arguments have all converged on the idea that concepts are represented by overlapping activity among a modest number (3–100) of units, and that the structure within and across these groups cannot be uniform or arbitrary. This is all fairly close to the compact end of the original spectrum, but the story is more complex than that. There are important lessons to be learned from the work on highly distributed models and many difficult problems which have eluded all theories. To simplify the discussion, we will adopt the standard convention and treat all compact representations as punctate. The first section showed both that the purely punctate theory cannot be biologically valid and also that the redundant coarse-coded variant is plausible and computationally very close to punctate version. It is simply much easier to understand punctate systems and all current models, however distributed, use punctate encodings of some elements for clarity. It is understood that direct neurobiological applications will have to take the detailed encoding more seriously.

From my perspective, there is only one conceptual difference between the current compact and diffuse models of neural knowledge representation and this difference can be seen as one of research strategy. Consider the punctate model of the appearance of horses shown in Figure 7. This is taken from a paper of mine that attempts to show how vision yields such hierarchical conceptual descriptions [Feldman, 1985]. At the lowest level are feature pairs, which we assume are derived by early vision networks. (The fact that there are feature pairs rather than individual features or n-tuples of them arises from technical considerations not relevant here (cf. Feldman, 1985; McClelland & Rumelhart, 1986). The remaining structure is organized as a hierarchy (cf. Table 1 and dogma 2) where activation propagates towards the top pontifical cell which is active when a horse is recognized. Even in this unrealistically punctate version, the recognition of a horse is a pattern of activation in many of these units. Since the connections are two-way, mentioning the name of a horse will cause (some) activation in all the nodes comprising visual and other descriptions of horses. Looking at this another way, we see that it makes no sense to talk about the activation of a single concept in such structure—activation automatically spreads to encompass a subnetwork. Notice also that this structure will recognize a horse even when some features are missing or garbled if the other features plus context are sufficiently strong. This captures the error tolerance and some of the generalization ability that were most attractive about correlation matrices. Computer experiments of moderate complexity along these lines have been successfully carried out [Sabbah, 1985; Shastri, 1985]. The current version of holographic correlation theory uses microfeatures (and feature pairs) essentially identical to the bottom row of Figure 7. The difference is that a correlation model would not have the hierarchical structure but would represent *horse* exclusively by the pattern of activity of the feature-pair units. But consider how one might expand the correlation model to represent horse heads,

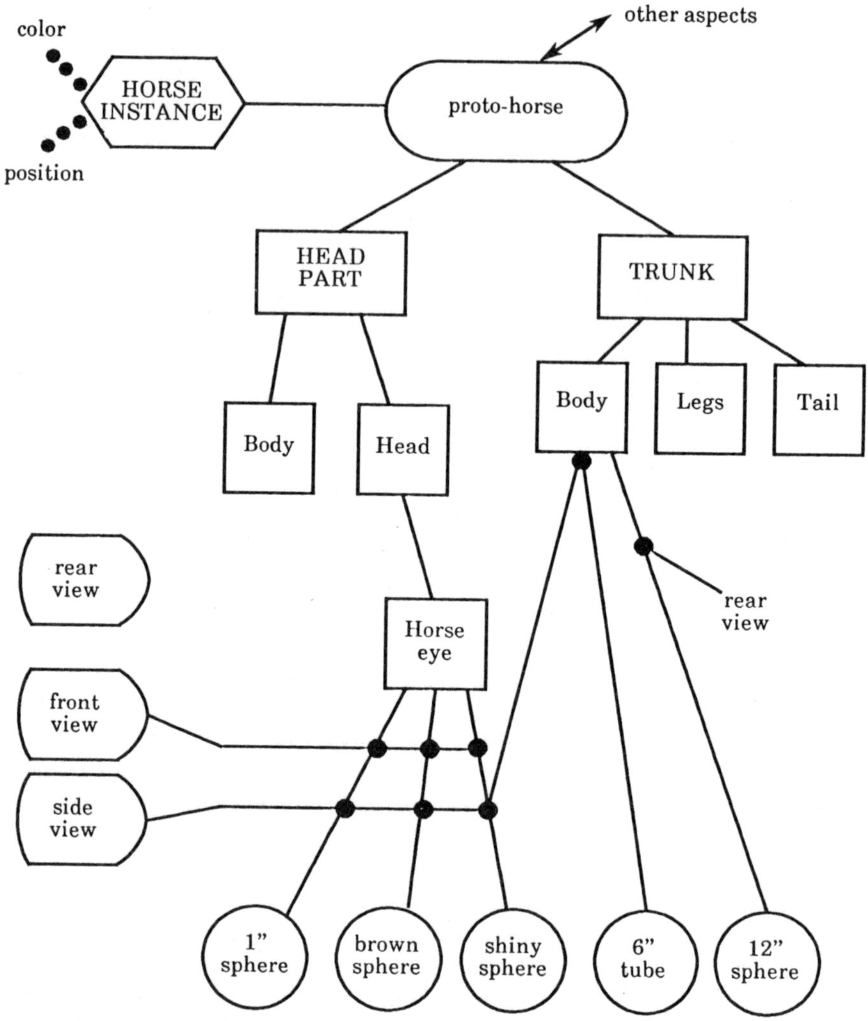

Figure 7. Punctate Model of a Horse. (Taken from Feldman, 1985.)

eyes, etc., as well. No one has done this, but it is hard to imagine any solution that did not separate out groups of features and combine them to form higher groups. Now, we have seen that combining diffuse encodings is a problem whose only known solution is to focus the representation. So why are all these smart people so excited about highly distributed representations?

There are two important problems that have been effectively studied using diffuse encodings—generalization and learning. Some insight into the situation can be gained by examining the reading model of Figure 8, taken from McClelland and Rumelhart [1981]. This is one of the earliest and most

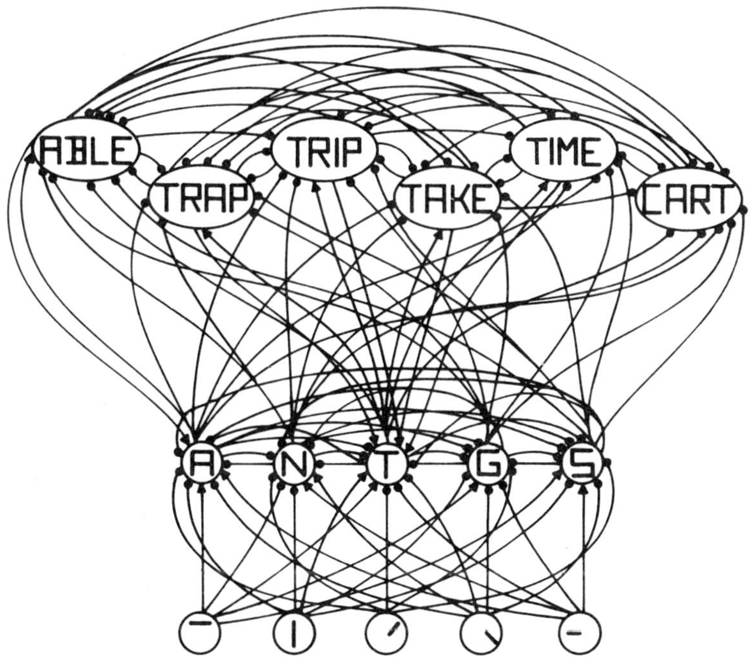

Figure 8. Reading Model (after McClelland & Rumelhart, 1981). A few of the neighbors of the node for the letter *T* in the first position in a word, and their interconnections.

successful applications of connectionist modelling to psychology. The model is basically punctate, incorporating units for specific shapes, letters, and words in a highly structured excitation–inhibition network. The model was used to successfully model a wide range of experimental data, including the word-superiority effect previously considered paradoxical. Subjects can more reliably recognize a letter, e.g., *a* in the context of a word, e.g., *fast*, than in isolation, and the model shows clearly how feedback from the word level can aid decisions at the letter level. But the model did more than this, and therein lies our tale. It turns out that the model (and the subjects) can also recognize letters better in the context of pronounceable nonwords than in isolation. There is no explicit representation of these nonwords, but the units for the similarly spelled words combine to provide enough boost to the target letter. Anderson and Hinton [1981, p. 28] describe this in the following way:

> Thus, the pronounceable non words are represented by distributed patterns of activity at the word level.

There is much to be learned from an examination of this statement. It is, of course, literally not true. There are many aspects of the nonwords (their length, pronunciation, etc.) which are not represented at all and one would not expect someone to recognize repeated occurrences of a nonword from this model. What is true is that, *for this task*, the network behaves as if it had a representation of the pronounceable nonwords. In fact, it is not just pronounceable nonwords that show the enhancement effect. The concepts with the distributed representation would have to be something like "collections of common pieces of words." It seems to me to be much better to understand this "emergence" as a general property of evidential networks, than to postulate extra concepts having a distributed representation. In general, the research that has been described as studying diffuse representations can be understood as really concerned with system properties of neural networks. Structured networks of punctate elements (like an electronic circuit) have *emergent properties*, as anyone who has tried to understand or repair one can testify.

What actually happens in the demonstrations of distributed representations is the following. A problem is picked in which certain properties (e.g., surface syntactic categories) are the ones required for a particular class of general answers (e.g., case roles). The distributed representation is chosen to be the required properties, and thus the system shows (or learns) the appropriate generalization. Since the combination rule is always linear threshold, care must be taken with the choice of primitives and some precoding may be necessary. Other generalizations over the same domain (e.g., past tense formation) are based on different properties and are thus studied in separate experiments using entirely different distributed representations. A great deal of elegant and important work has been done in this style, but the basic relation between properties and generalizations seems to have been misunderstood.

The generalizations based on property vectors do capture relations that would not arise naturally from hierarchical representations such as that in Figure 7. In the horse example, one might want to capture the generalization that animals with hair are mammals. We could link *horse* to *mammal*, but this is of no direct use in classifying novel hairy animals. The point is that, if all uses of a concept were forced to go through some focus, the system would be unable to yield many crucial generalizations. There are systematic relationships among properties that are best captured directly. The relations between syntactic and case roles, between pronunciation and morphology, and between hairiness and mammalhood are systematic and most compactly encoded by circumventing particular instances of words or mammals. There is an interesting duality here. If one knew enough hairy mammals, it is likely that activation of *hairy* would lead (indirectly) to the activation of *mammal*. One could view this as all the individual mammals constituting a distributed representation of the hairy-mammal relation, but this is a bad way to view network properties which are in general quite complex evidential relations [Shastri, 1985].

Learning

All of the preceding discussion has explored the ways in which conceptual knowledge might be represented in neural networks, ignoring the critical problem of learning. The question of the relative role of learning in intelligent behavior is as old and as basic as they come. It is now clear that humans come with an enormous amount of prewired structure, develop a great deal more in environmentally driven ways, and (except for administrators) continue to learn throughout life. The problem this presents for connectionist learning studies is a severe methodological one.

The basic difficulty is that we know human learning is based on an elaborate existing structure, but we know very little about the exact nature of this structure, particularly in the area of conceptual knowledge. Any study of learning either assumes some existing structure (at the risk of trivializing the learning aspects) or assumes no existing structure and is restricted to quite simple problems. The correlation matrix models of memory have presented an attractive research vehicle. The total initial connectivity and linear threshold rule are a minimal a priori structure, and the correlation method of updating is an easily analyzed learning rule. The results of such studies often take on a structure-from-chaos aspect which many people find attractive.

Obviously enough, these correlation matrices employ a diffuse encoding, and the appeal of the two ideas is strongly connected. But the ideas of correlation, feedback, and weight change are not restricted to diffuse representations, linear threshold rules, or complete connectivity matrices. In fact, complete connectivity is not biologically plausible, and the other assumptions also have problems. A great deal of recent work [Belew, 1986; Rumelhart & McClelland, 1986; Parker, 1985] involves using correlation type ideas on networks with much richer structure, while preserving domain independence.

It has been known for some time that any effective learning rule would have to include an input from the ultimate result of the computation; Hebb's correlation rule has no way to punish a connection that leads to a disaster. In a network that directly links inputs and outputs, like a matrix model or a 1-layer perceptron, it is easy to punish the offending links. For a system with more structure and indirect links, it has not been obvious how to assess who deserves the credit and who deserves the blame. Recent developments in learning theory, while not totally solving this problem, have enabled the study of learning in much richer structures [Rumelhart & McClelland, 1986].

The basic idea is to *back-propagate* error signals from the output (which is directly corrected) to successively lower levels of input. A simple version would be to have the weight change at layer n, $\Delta w_{ij}^{(n)}$, given by

$$\Delta w_{ji}^{(n)} = \epsilon \delta_i^{(n+1)} X_j^{(n)}$$

where $\delta_i(n+1)$ is the error attributed to unit i in layer $n+1$, $X_j^{(n)}$ is the output of units in layer n (now continuous valued) and ϵ is a learning constant. One can view this as changing w_{ji} in proportion to how the output of unit j *correlates* with the error attributed to unit i. The back-propagation algorithm has been used in a variety of studies and has been quite successful [McClelland & Rumelhart, 1986]. There are still problems with its range of application, convergence rate, and biological plausibility, but it does provide the best way known to study learning in structured networks. For all the reasons discussed above, this is a critical problem for our enterprise.

Learning in compact connectionist systems involves some additional considerations. A major problem in this formulation is "recruiting" the compact representation for new concepts [Wickelgren, 1979; Feldman, 1982; Shastri, 1985; Fanty, 1986]. It is all very well to show the advantages of representational schemes, as in Figures 1, 7, and 8, but how could they come about? This question is far from settled, but there are some encouraging preliminary results. The central question is how a system that grows essentially no new connections could find (recruit) compact groups of units to capture new concepts and relations. One relevant result concerns the probability of finding compact groups that link nodes in a random graph [Feldman, 1981]. It turns out that, for biologically reasonable parameters, the probability of a compact cluster ($\sim$ 20 units) is quite high. (Incidentally, the probability of finding a single unit is essentially zero, providing another argument against punctate theories.) The brain is not random, except perhaps very locally, but that actually is good news for recruiting concept representations. Current work on learning in more structured networks [Fanty, 1988; Hinton, in press] is examining this more closely.

Another issue is when the system needs to recruit descriptions of new concepts. A connectionist system can recognize a new concept, e.g., *pink Bentley*, without having an explicit representation for it—the joint activity of the components will suffice. But if specific names, properties, or relations of the new concept are required (pink Bentleys are native to Los Angeles), then a specific compact representation is required and is presumably recruited as described above.

Structure and Relations

Questions of structure have played a major role throughout the chapter. The structure of the nervous system, of domains of knowledge, and of concepts themselves have been central concerns. Despite the initial disclaimer, some consideration of relations has also appeared. Structure, in my opinion, is the dominant issue in neural knowledge representation. Most of the shortcomings of holographic style models can be traced to their unstructured nature. This has been recognized for some time, and people have made attempts to correct the problem. One idea is to identify the components of the giant vector with microfeatures of concepts. The cleverest choice is to make the microfeatures follow a conceptual hierarchy, so that more specific concepts (e.g., *horse*) share features with more general ones (*mammal, animal*) [Hinton, 1984; Anderson, 1983]. This implicitly captures one kind of structural relation among concepts, but leaves all the others untouched. And, if properties are all microfeatures of the representation, how could one incorporate new information like the fact that early horses were a few inches tall or that a particular horse had broken a leg? This move, like the generalization demos, can be seen as another way of exploiting relationships that exist among properties (cf. Shastri, 1985) and has nothing to do with holography.

When it comes to more general relations, there are two ways to try. One, which was mentioned earlier, assumes that relations among concepts are represented by links (axons) connecting the representations of the concepts involved. This inherently requires compact representations for the concepts, although many authors who talk about the idea in general terms would be repelled by the thought. I believe that resource considerations make this the only viable option, but no one has figured out how to make it work except in simple cases. One thing we do know is that hierarchical compact representations like Figure 7 can capture internal structural relations defining concepts, and that this goes far towards solving the questions of how to treat all the different views of a horse or your grandmother. Briefly, the conceptual knowledge about, e.g., horses, is a largely shared structure deriving much of its content from its relationship to other concepts. This is a kind of distributed representation, but not at all what is usually meant by the term. There is also strong evidence that we do not retain all the detailed memories of our grandmothers, but recreate them with a significant tendency towards regularization [Neisser, 1982].

If one is committed to a diffuse, unstructured representation for concepts, there is no direct way to realize relationships among them. Ignoring computational costs, one can design a holographic memory to store relational information in symbolic form, e.g., as triples like (Brother Billy Jimmy) [Hinton,

1981a; Kohonen, 1980]. But this is a move of desperation in neural modelling, since all of the problems of sequential symbolic computing reemerge. A sequential machine with a fast relation store will show none of the performance or context sensitivity that motivates connectionist models.

In summary, connectionist studies of pure learning minimize the preexisting structure and tend to study diffuse models. This has turned out to be very valuable and will continue to be. Like the studies of overall convergence, (e.g., Wilson & Cowan, 1972; Hopfield, 1982; Cohen & Grossberg, 1983), these are best done assuming no particular structure of the network. But the structure is there, and only compact representations can capture it.

CONCLUSIONS

Let us return to the analogous contrast between atomic physics and thermodynamics. Atomic physics (and chemistry) is concerned with precise structures and their interactions. Because of the complexity involved, cell biology would be an even better analog for the detailed structural concerns of compact models. Modern thermodynamics (statistical mechanics) derives its power from abstracting, from enormous systems of units, a very small number of state variables that characterize certain questions of interest. While the abstraction is only strictly true for systems without structure, the results of the theory have much wider application. One cannot do chemistry without knowing thermodynamics, but thermodynamics is a formalism that contains no way to say anything about chemistry on our planet. Similarly, bulk models of neural activity are likely to continue to yield valuable insights into neural functions, but it is fatal to ignore the detailed structure present. The challenge is to get global results into a form where they can be used to advantage in working out how conceptual knowledge is represented and exploited by the brain.

Structured networks of evidence-combining simple units have a number of attractive computational properties. Error resistance, context dependence, and the ability to assimilate conflicting information are natural properties of such systems. A reasonable degree of generalization follows from these properties. Simple weight-change rules can enable these systems to improve their performance significantly. The collective behavior of such systems can produce powerful computations not easily anticipated, as is true of any complex circuit. While certain concepts are represented explicitly, the system is not restricted to dealing only with those. A properly structured network can behave, in certain situations, as if it had concepts and relations implicitly represented. An explicit, compact handle on concepts is required when they have disjunction or internal structure or when they participate in relations or are communicated among subsystems. In addition, the nature of any emergent system properties depends heavily on which concepts are explicitly represented and on the detailed structure of the representation.

This suggests the following research priorities for the question of neural representation of knowledge. The detailed anatomy and physiology of nervous systems remains a top priority. The computational properties of connectionist models must be better understood both in specific circuits and as mass systems. Plasticity and learning, of course, remain in their central place, but the central problem is change of existing structure [Hinton, 1987]. The most difficult problems, however, appear to lie at the higher conceptual levels. All of the concepts and relations treated in this chapter are extremely simplistic. The technical level of research on concept and knowledge representation among linguists, philosophers, and symbolic AI types is enormously more sophisticated (e.g., Brachman & Levesque, 1985). If it is true that our brains do these things directly with neural nets, connectionist formulations should yield better characterizations of higher-level thought than symbolic logic, perhaps through the mediation of new scientific lan-

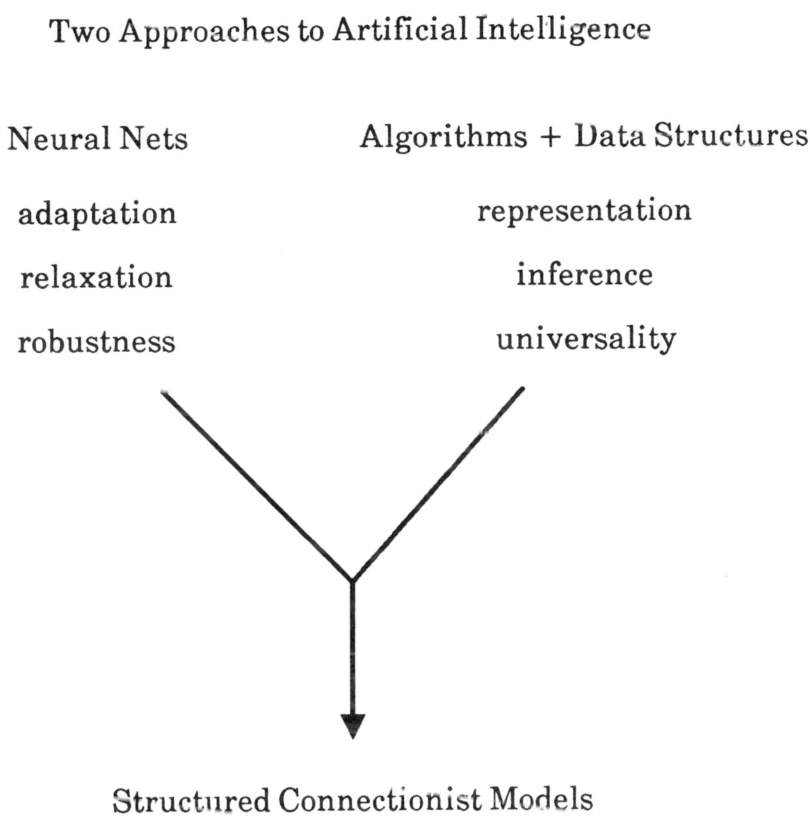

Figure 9. Two approaches to artificial intelligence.

guages. Expectations that this will happen without detailed consideration of the structure of the tasks and of the underlying hardware should be based on a time-frame of evolutionary scale. More likely, we will need to exploit many structural insights from artificial intelligence, neurobiology, and cognitive science and try to synthesize connectionist models that combine the strengths of each approach. Figure 9 depicts this idea graphically!

REFERENCES

Abeles, M. (1982). *Local cortical circuits.* Berlin: Springer-Verlag.

Allman, J., Miezen F., & McGuiness, E. (1985). Stimulus specific responses from beyond the classical receptive field: Neurophysiological mechanisms for local-global comparisons in visual neurons. *Annual Review of Neuroscience, 8,* 407–430.

Amari, S., & Arbib, M.A. (1982). In S. Levin (Ed.), *Competition and cooperation in neural networks* (Lecture notes in biomathematics, 45). Berlin: Springer-Verlag.

Andersen, R.A. (1986, March). Value, variable, and coarse coding by posterior periatal neurons. *The Behavioral and Brain Sciences* (Open Peer Commentary), 9(1), 90–91.

Anderson, J.A. (1983). Cognitive and psychological computation with neural models. *IEEE Transactions: Systems, Man and Cybernetics, 13,* 799–815.

Anderson, J.A., & Hinton, G.E. (1981). Models of information processing in the brain. In G.E. Hinton & J.A. Anderson (Eds.), *Parallel models of associative memory.* Hillsdale, NJ: Erlbaum.

Anderson, J.A., & Mozer, M.C. (1981). Categorization and selective neurons. In G. E. Hinton & J.A. Anderson (Eds.), *Parallel models of associative memory.* Hillsdale, NJ: Erlbaum.

Anderson, J.A., & Murphy, G.L. (1986). Comments on concepts. *Proceedings of the NSF Connectionism Workshop.* Carnegie Mellon University, Psychology Dept.

Ballard, D.H. (1986a, March) Cortical connections and parallel processing: Structure and function. *The Behavioral and Brain Sciences,* 9(1), 67–120.

Ballard, D.H. (1986b) *Interpolation coding: A representation for numbers in neural models* (Tech. Rep. No. 175). Rochester, NY: Department of Computer Science, University of Rochester.

Barber, C. (1980). *Evoked potentials.* Baltimore, MD: University Park Press.

Barlow, H.B. (1972). Single units and sensation: A neuron doctrine for perceptual psychology? *Perception, 1,* 371–392.

Barnden, J.A. (1984). On short term information-processing in connectionist theories. *Cognition and Brain Theory,* 7(1), 25–5.

Barto, A.G., & Anandan, P. (1984). *Pattern recognizing stochastic learning automata* (Tech. Rep. No. 84–30). Amherst, MA: Computer and Information Science, University of Massachusetts at Amherst.

Barto, A.G., Sutton, R.S., & Brouwer, P.S. (1981). Associative search network: A reinforcement learning associative memory. *Biological Cybernetics, 40,* 201–211.

Beurle, R.L. (1956). Properties of a mass of cells capable of regenerating pulses. *Philosophical Transactions of the Royal Society, Series B. 240*, 55–94.

Bienenstock, E. (1985). Dynamics of central nervous system. In J.P. Aubin & K. Sigmund (Eds.), *Proceedings, workshop on dynamics of macrosystems*, Laxenburg, Austria: Springer-Verlag.

Brachman, R., & Levesque, H. (1985). *Readings in Knowledge Representation*, Palo Alto, CA: Morgan-Kaufman.

Charniak, E., & McDermott, D. (1985). *Introduction to Artificial Intelligence*. Reading, MA: Addison-Wesley.

Cohen, M.A., & Grossberg, S. (1983). Absolute stability of global pattern formation and parallel memory storage by competitive neural networks. *IEEE Transactions: Systems, Man and Cybernetics, 13*, 815–825.

Desimone, R., Schein, S.J., Moran J., & Ungerleider, L.G. (1985). Contour, color, and shape analysis beyond the striate cortex. *Vision Research, 25*, 441–452.

Edelman, G.M., (1981). Group selection as the basis for higher brain function. In F. O. Schmitt, F.G. Worden, G. Adelman, & S.G. Dennis (Eds.), *The organization of cerebral cortex*. Cambridge, MA: MIT Press.

Edelman, G.M., & Mountcastle, V.B. (1978). *The mindful brain: Cortical organization and the group-selective theory of higher brain function*. Cambridge, MA: MIT Press.

Fanty, M.A. (1988). *Learning in structured connectionist networks*. Doctoral dissertation and Tech. Rep. 252. Rochester, NY: Computer Science Dept., Univ. Rochester.

Feldman, J.A. (1981). A connectionist model of visual memory. In G. E. Hinton & J. A. Anderson (Eds.), *Parallel models of associative memory*, Hillsdale, NJ: Erlbaum.

Feldman, J.A. (1982). Dynamic connections in neural networks. *Biological Cybernetics, 46*, 27–39.

Feldman, J.A. (1985). Four frames suffice: A provisional model of vision and space. *The Behavioral and Brain Sciences, 8*, 265–289.

Feldman, J.A., & Ballard, D.H. (1982). Connectionist models and their properties. *Cognitive Science, 6*, 205–254.

Feldman, J.A., & Shastri, L. (1984). Evidential inference in activation networks. *Proceedings of the Cognitive Science Conference*, Boulder, Colorado

Gardner, J.C., Douglas R.M., & Cynader, M.S. (1985). A time-based stereoscopic depth mechanism in the visual cortex. *Brain Research, 328*, 154–157.

Geman, S. (1981). Notes on a self-organizing machine. In G.E. Hinton & J.A. Anderson (Eds.), *Parallel models of associative memory*, Hillsdale, NJ: Erlbaum.

Goldman-Rakic, P.S., & Schwartz, M.L. (1982). Interdigitation of contralateral and ipsilateral columnar projections to frontal association cortex in primates. *Science, 216*, 755–757.

Gross, C.G., Rocha-Miranda, C.E., & Bender, D.B. (1972). Visual properties of neurons in inferotemporal cortex of the macaque. *Journal of Neurophysiology, 35*, 96–111.

Hebb, D.O (1949). *The organization of behavior*. New York: John Wiley.

Hillis, W.D. (1985). *The connection machine*. Cambridge, MA: MIT Press.

Hinton, G.E. (1981a). Implementing semantic networks in parallel hardware. In

G. E. Hinton & J.A. Anderson (Eds.), *Parallel models of associative memory.* Hillsdale, NJ: Erlbaum.

Hinton, G.E. (1981b). Shape representation in parallel systems. *Proceedings of the 7th International Joint Congress on Artificial Intelligence* (pp. 1088–1096). Vancouver, Canada.

Hinton, G.E. (1984). *Distributed representations* (CMU-CS-84-157). Department of Computer Science, Carnegie Mellon University, Pittsburgh, Pennsylvania.

Hinton, G.E. (in press). Connectionist learning procedures. *AI Journal.*

Hinton, G.E., & Anderson, J.A. (1981). *Parallel models of associative memory.* Hillsdale, NJ: Erlbaum.

Hopfield, J.J. (1982). Neural networks and physical systems with emergent collective computational abilities. *Proceedings, National Academy of Sciences of the United States of America, 79,* 2554–2558.

Hopfield, J.J., & Tank, D.W. (1986). Computing with neural circuits: A model. *Science,* 233, 4764, 625–632.

Hubel, D.H. & Wiesel, T.N. (1979). Brain mechanisms of vision. *Scientific American, 241*(3), 150–162.

Just, M.A., & Carpenter, P.A. (1985). Cognitive coordinate systems: Accounts of mental rotation and individual differences in spatial ability. *Psychological Review, 92,* 137–72.

Knapp, A.G., & Anderson, J.A. (1984). Theory of categorization based on distributed memory storage. *Journal of Experimental Psychology: Learning, Memory and Cognition, 10,* 616–637.

Kohonen, T. (1977). *Associative memory: A system-theoretical approach.* Berlin: Springer-Verlag.

Kohonen, T. (1980). *Content-addressable memories.* Berlin: Springer-Verlag.

Kohonen, T. (1984). *Self-organization and associative memory* (Springer Series in Information Sciences, 8). Berlin: Springer Verlag.

Kohonen, T., Oja, E., & Lehtio, P. (1981). Storage and processing of information in distributed associative memory systems. In G.E. Hinton & J.A. Anderson (Eds.), *Parallel models of associative memory.* Hillsdale, NJ: Erlbaum.

Lashley, K. (1950). In search of the engram. In *Physiological mechanisms in animal behavior* (pp. 454–483). Symposia of the Society for Experimental Biology, No. 4. New York: Academic Press.

Lashley, K.S. (1929). *Brain mechanisms and intelligence.* Chicago, IL: University of Chicago Press.

Lippmann, R.P. (1987, April). An introduction to computing with neural nets. *IEEE Acoustics, Sound and Signal Processing,* 4–22.

Lynch, G. (1986). *The neurobiology of learning and memory.* Cambridge, MA: MIT Press.

McClelland, J.L., & Rumelhart, D.E. (1981). An interactive activation model of the effect of context on language learning (Part I). *Psychological Review, 88,* 375–401.

McClelland, J.L., & Rumelhart, D.E. (Eds.). (1986). *Parallel distributed processing: Explorations in the microstructure of cognition. Vol. 2: Applications.* Cambridge, MA: Bradford Books/MIT Press.

Merzenich, M.M., Nelson, R.J., Stryker, M.P., Cynader, M.S., Schoppmann, A., & Zook, J.M. (1984). Somatosensory cortical map changes following digit amputation in adult monkeys. *The Journal of Comparative Neurology, 224*, 591–605.

Minsky, M. (1980). K-Lines: A theory of memory. *Cognitive Science, 4*(2), 117–133.

Minsky, M. & Papert, S. (1972). *Perceptrons* (2nd ed.). Cambridge, MA: MIT Press.

Mumford, D. (1986, March). Two tests for the value unit model: Multicell recordings and pointers. *The Behavioral and Brain Sciences*, Open Peer Commentary, 9(1), 102–103.

Neisser, U. (Ed.). (1982). *Memory observed.* San Francisco, CA: W.H. Freeman.

Newell, A. (1983). Intellectual issues in the history of artificial intelligence. In F. Machlup & U. Mansfield (Eds.), *The study of information: Interdisciplinary messages.* New York: John Wiley.

Palm, G. (1980). On associative memory. *Biological Cybernetics, 36*, 19–31.

Palm, G. (1985, June). Associative networks and their information storage capacity. *Cognitive Systems, 1*(2), 107–185.

Palm, G. (1986). What are the units of neural representation? submitted for publication.

Palm, G., & Bonhoeffer, T. (1984). Parallel processing for associative and neuronal networks. *Biological Cybernetics, 51*, 201–204.

Parker, D.B. (1985). *Learning-logic.* (Tech. Rep. No. 47). Cambridge, MA: Center for Computational Research in Economics and Management Science, MIT.

Phillips, C.G., Zeki, S., & Barlow, H.B. (1984). Localization of function in the cerebral cortex. *Brain, 107*, 328–361.

Pribram, K.H., Nuwer, M., & Baron, R. (1974). The holographic hypothesis of memory structure in brain function and perception. In R.C. Atkinson, D.H. Krantz, R.C. Luce, & P. Suppes (Eds.), *Contemporary developments in mathematical psychology.* San Francisco, CA: W.H. Freeman.

Roediger, H.L. III (1980). Memory metaphors in cognitive psychology. *Memory and Cognition, 8*(1), 231–246.

Rumelhart, D.E., & McClelland, J.L. (1981). An interactive activation model of the effect of context in language learning (Part 2). *Psychological Review, 89*, 60–94.

Rumelhart, D.E., & McClelland, J.L. (Eds.). (1986). *Parallel distributed processing: Explorations in the microstructure of cognition. Vol. 1: Foundations.* Cambridge, MA: Bradford Books/MIT Press

Sabbah, D. (1985). Computing with connections in visual recognition of Origami objects. *Cognitive Science, 9*, 25–50.

Schank, R.C., Collins, G.C., & Hunter, L.E. (1986). Transcending inductive category formation in learning. *The Behavioral and Brain Sciences*, 9(4), 639–651.

Sejnowski, T J. (1986). Open questions about computation in cerebral cortex. In J.L. McClelland & D.E. Rumelhart (Eds.), *Parallel distributed processing: Explorations in the microstructure of cognition. Volume 2: Applications*, Cambridge, MA. Bradford Books/MIT Press.

Shastri, L. (1985). *Evidential reasoning in semantic networks: A formal theory and its parallel implementation* (Doctoral dissertation and Tech. Rep. No. 166). Rochester, NY: Computer Science Department, University of Rochester.

Shaw, G.L., Silverman, D.J., & Pearson, J.C. (1985). Model of cortical organization embodying a basis for a theory of information processing and memory recall. *Proceedings, National Academy of Sciences of the United States of America*, 82, 2364–2368.

Smith, E.E., & Medin, D.L. (1981). *Categories and concepts.* Cambridge, MA: Harvard University Press.

Sullins, J. (1985). *Value cell encoding strategies* (Tech. Rep. No. 165). Rochester, NY: Computer Science Department, University of Rochester.

Sutton, R.S., & Barto, A.G. (1981). Toward a modern theory of adaptive networks: Expectation and prediction. *Psychological Review, 88*(2), 135–170.

Thompson, R.F. (1986). The neurobiology of learning and memory. *Science, 233,* 941–947.

Toulouse, G., Dehaene, S., & Changeux, J-P. (1985). *A spin glass model of learning by selection.* Unpublished manuscript.

van Heerden, P.J. (1963). A new optical method of storing and retrieving information. *Applied Optics, 2,* 387–392.

von der Heydt, R., Peterhans, E., & Baumgartner, G. (1984). Illusory contours and cortical neuron responses. *Science, 224,* 1260–1262.

von der Malsburg, C. (1985). Nervous structures with dynamical links. *Berichte der Bunsen-Gesellschaft fur Physikalische Chemie., 89*(1), 703–710.

Weinberger, N.M., & Diamond, D.M. (1986). Dynamic modulation of the auditory system by associative learning. In G. Edelman, J. Cowan & E. Gall (Eds.), *Auditory function* (Neurosciences Institute Publication). New York: John Wiley.

Wickelgren, W.A. (1979). Chunking and consolidation: A theoretical synthesis of semantic networks, configuring in conditioning, S-R versus cognitive learning, normal forgetting the amnesic syndrome, and the hippocampal arousal system. *Psychological Review, 86,* 44–60.

Wilensky, R. (1986). *Some problems and proposals for knowledge representation.* (Rep. No. UCB/CSD 86/294, Computer Science Division). Berkeley, CA: University of California.

Willshaw, D. (1981). Holography, associative memory, and inductive generalization. In G.E. Hinton & J.A. Anderson (Eds.), *Parallel models of associative memory.* Hillsdale, NJ: Erlbaum.

Wilson, H.R., & Cowan, J.D. (1972). Excitatory and inhibitory interactions in localized populations of model neurons. *Biophysical Journal, 12,* 1–24.

APPENDIX

Interim Neuron Doctrine

The following five brief statements are intended to define which aspect of the brain's activity is important for understanding its main function, to suggest the way that single neurons represent what is going on around us, and to say how this is related to our subjective experience. The statements are dogmatic and incautious because it is important that they should be clear and testable.

First dogma

A description of that activity of a single nerve cell which is transmitted to and influences other nerve cells, and of a nerve cell's response to such influences from other cells, is a complete enough description for functional understanding of the nervous system. There is nothing else 'looking at' or controlling this activity, which must therefore provide a basis for understanding how the brain controls behaviour. ***Since significant behaviors involve many individual nerve cells, functional understanding of the nervous system will require scientific languages for characterizing the behavior of networks of neurons.***

Second dogma

Efficient coding of information is a central problem of the sensory system. At progressively higher levels in the sensory pathways, information about the physical stimulus is more abstract and is represented by progressively fewer active neurons.

Third dogma

Trigger features of neurons are matched to the redundant features of sensory stimulation in order to achieve greater completeness and economy of representation. This selective responsiveness is determined by the sensory stimulation to which neurons have been exposed, as well as by genetic factors operating during development.

Fourth dogma

Just as physical stimuli directly cause receptors to initiate neural activity, so the active ***networks of intermediate and*** high-level neurons directly and simply cause the elements of our perception.

Fifth dogma

Frequency coding is the primary basis of neural communication. Sensory neurons respond with high frequency discharge to external stimuli which fit into a narrow range of possibilities; the higher the discharge, the more narrow the range of possible causes.

Zeroth dogma

Intelligent behavior and its neural realization are incredibly complex. A functional understanding of this will require organizational principles from the behavioral and computational sciences as well as biology.

Author Index

Subject Index